REPRODUCTIVE TOXICOLOGY AND INFERTILITY

REPRODUCTIVE TOXICOLOGY AND INFERTILITY

Editors

Anthony R. Scialli, M.D.

Department of Obstetrics and Gynecology
Georgetown University Medical Center
and
Reproductive Toxicology Center
Columbia Hospital for Women
Washington, D.C.

Michael J. Zinaman, M.D.

Department of Obstetrics and Gynecology
Georgetown University Medical Center
Washington, D.C.

McGraw-Hill, Inc.
Health Professions Division

New York St. Louis San Francisco Auckland
Bogotá Caracas Lisbon London Madrid
Mexico Milan Montreal New Delhi Paris
San Juan Singapore Sydney Tokyo Toronto

Reproductive Toxicology and Infertility

1234567890 DOCDOC 98765432

ISBN 0-07-105438-3

This book was set in ITC Galliard by Huron Valley Graphics, Inc. The editors were Jane Pennington and Steven Melvin; The production supervisor was Clare Stanley. The cover was designed by José Fonfrias.

R.R. Donnelley and Sons, Inc. was printer and binder.

Library of Congress Cataloging-in-Publication Data

Reproductive toxicology and infertility / editors, Anthony R. Scialli, Michael J. Zinaman.

p. cm.

Includes bibliographical references and index.

ISBN 0-07-106438-3 :

1. Reproductive toxicology. 2. Infertility. I. Scialli, Anthony R., date . II. Zinaman, Michael J.

[DNLM: 1. Environmental Exposure—adverse effects. 2. Infertility. 3. Reproduction—drug effects. WQ 205 R4298]

RA1224.2.R475 1993

616.6'92—dc20

DNLM/DLC

for Library of Congress 92-49769

CIP

To Sergio

CONTENTS

CONTRIBUTORS

Ted L. Anderson, M.D.
Department of Obstetrics and Gynecology
Vanderbilt Medical Center
Nashville, Tennessee
(Chapter 7)

Gary J. Chellman, Ph.D.
Syntex Research
Institute of Toxicological Sciences
Palo Alto, California
(Chapter 4)

Erma Z. Drobnis, Ph.D.
Assistant Adjunct Professor, Department of Obstetrics and Gynecology
Division of Reproductive Biology and Medicine
University of California at Davis
Davis, California
(Chapter 5)

Leon Earl Gray, Jr., Ph.D.
Chief, Developmental Reproductive Biology Section
Health Effects Research Laboratory
United States Environmental Protection Agency
Research Triangle Park, North Carolina
(Chapter 8)

Gary Klinefelter, Ph.D.
Health Effects Research Laboratory
Environmental Protection Agency
Research Triangle Park, North Carolina
(Chapter 8)

Grace K. Lemasters, Ph.D.
Department of Environmental Health
University of Cincinnati
Cincinnati, Ohio
(Chapter 11)

Belinda A. Marascalco, M.D.
Division of Reproductive Endocrinology and Infertility
Department of Obstetrics and Gynecology
Georgetown University Medical Center
Washington, D.C.
(Chapter 2)

Preston C. Sacks, M.D.
Center for Fertility and Reproduction
Columbia Hospital for Women Medical Center
Washington, D.C.
(Chapter 6)

Steven M. Schrader, Ph.D.
Chief, Functional Toxicology Section
Experimental Toxicology Branch
Division of Biomedical and Behavioral Science
National Institute for Occupational Safety and Health
Cincinnati, Ohio
(Chapter 9)

Anthony R. Scialli, M.D.
Department of Obstetrics and Gynecology
Georgetown University Medical Center
and
Reproductive Toxicology Center
Columbia Hospital for Women Medical Center
Washington, D.C.
(Chapter 10)

Sherry G. Selevan, Ph.D.
Reproductive and Developmental Toxicology Branch
United States Environmental Protection Agency
Washington, D.C.
(Chapter 11)

James A. Simon, M.D.
Chief, Division of Reproductive Endocrinology and Infertility
Department of Obstetrics and Gynecology
Georgetown University Medical Center
Washington, D.C.
(Chapter 7)

David Vantman, M.D.
Assistant Professor, Department of Obstetrics and Gynecology
University of Chile
Santiago, Chile
(Chapter 1)

Peter K. Working, Ph.D.
Liposome Technology, Inc.
Menlo Park, California
(Chapter 4)

Koji Yoshinaga, Ph.D.
Adjunct Professor, Department of Anatomy
Georgetown University School of Medicine
Washington, D.C.
(Chapter 3)

PREFACE

This multiauthored book is directed at clinicians needing an understanding of ways in which drugs and other potential toxicants may alter male and female reproductive competence. The audience includes physicians and scientists with an interest in clinical infertility problems and the book, therefore, has a major clinical orientation. At the same time, work derived from animal and in vitro models that is important in understanding reproductive toxicology is introduced and placed in the appropriate clinical context.

The first section of the book provides background information necessary for an understanding of the reproductive process in mammals. Male and female reproductive systems are separately reviewed. Because loss of the conceptus at or prior to implantation presents as infertility, this important portion of early embryology is also discussed.

The second section builds on this background with descriptions of the ways in which normal reproduction may become deranged. Clinical disease states are considered as examples of the kinds of disorders encountered, but particular emphasis is given to those disorders known to be, or theoretically likely to be, associated with toxicant exposure.

The third and final section describes methods by which fertility can be monitored and by which reproductive toxicity might be identified. Animal methods are described but most of the section will consider clinical and epidemiologic methods that are currently in use or under consideration.

INTRODUCTION

The fitness of a species depends on the ability of its members to reproduce. It is surprising, then, that the human species, which dominates the earth, has so much reproductive failure among its members. Perhaps as many as 15 percent of couples have difficulty achieving a clinical pregnancy, and, once conceived, at least 20 percent of these clinical pregnancies will end without the birth of a child. Preclinical losses (occult pregnancies) may represent an even greater proportion of pregnancy failures.

Fertility doctors have long known that a certain amount of reproductive failure is built into the system; the per-cycle fecundity of couples is only 25 percent under the best of circumstances. Nonconceptive menstrual cycles are a part of normal reproductive physiology. Physicians cannot change that, but they can attempt to identify and treat the conditions that increase the rate of nonconceptive cycles. It is, for this purpose, instructive to study physiology, that is, to learn why things go right in order to be better prepared to fix them when they go wrong.

Many of the scientists who tell us the most about normal reproductive function are toxicologists rather than physiologists. The role of reproductive toxicology as an aid to clinicians is not immediately evident and needs some explanation. A historical perspective is helpful in this regard.

Scholars in antiquity had all manner of explanations for the generation of the young of animals. These amuse us now but were seriously considered in their day. It was during the nineteenth century that reproductive physiology advanced as a modern science, with contributors such as Enrico Sertoli and Franz von Leydig. Among the investigators of that era was Gregor Mendel, whose work in genetics was rediscovered at the beginning of the twentieth century. Partly a result of a genetic-based determinism, and partly because the placenta was considered an effective barrier to bad influences, reproductive difficulties were considered failures of the fitness of the individuals afflicted with them rather than as dysfunction of specific organ systems.

In the 1930s, Fred Hale, a worker at the Texas Agricultural Experimental Station, ran a series of experiments in pigs that revolutionized our thinking about reproductive problems. Hale administered a vitamin A–deficient diet to pregnant sows in an attempt to disrupt normal reproductive physiology. The resultant piglets had anomalies of orbit development, showing the success of this manipulation. Hale's experiment followed decades of work with chick and avian eggs. It had long been known in the poultry industry that abnormal development and other reproductive disasters could be induced by varying the temperature and other conditions of incubation, because chicks have no placenta and limited homeostatic mechanisms. Thus, the Hale experiment made an impression on those who viewed mammalian reproduction as privileged. There followed a series of experiments in other laboratories, using a number of manipulations to disrupt normal developmental events.

In 1941, an Australian ophthalmologist, Norman Gregg, described congenital cataracts in a child whose mother had contracted German measles during pregnancy. Twenty years later, German physicians called attention to limb reduction defects in children with antenatal exposure to the sedative thalidomide. These two episodes made the world aware that chemical and other agents can produce developmental abnormalities and have stimulated scientific and regulatory interest in the prevention of such abnormalities. A decade after thalidomide, the wives of men on a company softball team noted that they were having trouble conceiving. The husbands had in common, besides an interest in softball, exposure to 1,2-dibromo-3-chloroproprane (DBCP), a pesticide used to kill worms. Follow-up studies on this group showed DBCP exposure to be associated with oligospermia and male infertility.

These episodes made clinically relevant the relationship between toxicologists and physiologists because human "experiments" such as these clearly showed a toxicant-induced disruption in reproductive physiology. There arose from these episodes a renewed interest in identifying mechanisms by which normal physiologic events go awry. Herein lies the purpose of this book. *Mechanisms* has become the catchword of the modern era; if we can understand mechanisms of disease, we can develop therapies or even strategies for prevention.

For this reason we have organized a reproductive physiology book with a toxicology accent, because we believe that toxicology studies are likely to provide the basis for understanding reproductive disorders and perhaps for identifying disorders associated with environmental exposures. In fact, we devote our final section to the issue of monitoring for environmentally induced reproductive failure. At present, monitoring is inconsistently performed in workplaces and other environments that are considered to impose risks to reproduction. It is not clear how much reproductive failure will be shown to be due to such environmental exposures; however, it is evident that we will need to monitor for such toxicity if we hope to identify it when it occurs. One of the limitations in this effort is the inherent variability in normal measures of reproductive competence, such as menstrual cyclicity and semen analysis parameters. It is gratifying that current research is being directed to the characterization of the normal range and interindividual reproducibility of reproductive endpoints. This is not a text on the diagnosis and treatment of infertility, nor is it a catalog of genetic causes of gonadal failure, but nonetheless we expect it will be useful to physicians and geneticists who take care of infertile couples. It will also be helpful for investigators in the biological sciences to put into clinical perspective the basic physiology research that is described here. Finally, we envision that students of the reproductive sciences will benefit from the approach we have adopted. Students are a wonderful audience; they have needs and requirements of a text that put authors and editors to the test. We believe students will benefit from the presentation of physiology as an applied science, rather than as a strictly bench science. It is our hope that all our readers, physicians, scientists, and students, will gain new insights into mechanisms of reproductive success and failure that are becoming so useful in clinical work.

Production of this book has been made more enjoyable by the participation of our contributors, who are among the best and the busiest in the field. We tip our hats to these scientists and clinicians who have done so much to bring this important area of research up to full speed.

ANTHONY R. SCIALLI, M.D.

MICHAEL J. ZINAMAN, M.D.

PART

I

THE BASIS OF FERTILITY

CHAPTER 1

Male Reproduction: An Overview

David Vantman

The human male reproductive system can be divided into the endocrine and the canalicular compartment. The endocrine compartment, which includes the brain, the hypothalamus, the anterior pituitary gland, and the testes is responsible for the development and maintenance of the secondary sexual characteristics as well as stimulation of spermatozoa development. The canalicular compartment, which includes the seminiferous tubules and adnexal glands under the regulation of the endocrine system, produces and stimulates the maturation of the spermatozoa. The regulatory mechanisms of interaction between both compartments is extremely sensitive to environmental changes; accordingly, many substances are capable of altering this equilibrium.

The knowledge of the physiologic events that regulate both compartments will allow us to understand the possible mechanisms of toxicity of different substances and therefore the appropriate treatment that may be developed.

PHYSIOLOGY OF TESTICULAR FUNCTION

The testis is regulated by two gonadotropins, luteinizing hormone (LH) and follicle-stimulating hormone (FSH), which are synthesized and released in the pituitary under the regulation of the hypothalamic peptide luteinizing hormone–releasing hormone (LHRH or GnRH).

The hypothalamus is anatomically linked to the pituitary both by a portal vascular system and by neural pathways (Green and Harris, 1949). The portal vascular system provides a mechanism for the delivery of releasing hormones from the brain to the pituitary and thus provides the major pathway by which the brain controls anterior pituitary function. Reverse flow through this hypophyseal–portal circulation may also allow pituitary hormones to reach the brain by a direct pathway rather than through the general circulation (Oliver et al, 1977). LHRH, a decapeptide, is synthesized in the hypothalamus and travels to the anterior pituitary gland via the hypophyseal–portal circulation. It stimulates acutely the release of both FSH and LH by a calcium-dependent mechanism that is independent of cyclic AMP (cAMP) or cGMP (Conn et al, 1979; Naor and Catt, 1980). LHRH also has a more long term effect on the regulation of gonadotropin synthesis.

In men, LH and probably FSH are released in episodic bursts, with pulses occurring approximately every 2 h. For LH this episodic release has been clearly demonstrated because the half-life of the hormone in the circulation is 1 h. By contrast, episodic release of FSH is less apparent because its half-life in the circulation is approximately 3 h. LH pulses result from the episodic release of LHRH from the hypothalamus. The rate and amplitude of LHRH pulsatile stimulation alters the plasma level of both LH and FSH and in turn influences the state of testicular function.

LH and FSH are glycoproteins composed of two polypeptide chains designated α and β. The distinct immunologic and functional characteristics of these hormones are determined by their unique subunits (Vaitukaitis et al, 1976). Both hormones are secreted by the same basophilic cell types in the pituitary. LH interacts with specific high-affinity receptors on the Leydig cells, and this leads to increased testosterone synthesis (Dufau and Catt, 1978). LH stimulation of the steroidogenic process involves activation of the intracellular enzymes adenyl cyclase and protein kinase C (Podesta et al, 1978).

Although the entire role of FSH is not defined, completion of spermatogenesis requires the presence of high intratesticular levels of testosterone. FSH action requires binding to surface receptors on the Sertoli cell. This binding results in an increase in cAMP concentration in Sertoli cells, activation of the cAMP-dependent protein kinase and an increase in the rates of RNA and protein synthesis, including those of androgen-binding protein (ABP) and the aromatase enzyme complex that converts testosterone to estradiol (Dorrington et al, 1972; Fakunding and Means, 1977; Means et al, 1976; Dorrington and Armstrong, 1975). ABP is thought to help maintain high intratesticular levels of testosterone.

FSH does enhance secretion of testosterone. This synergism of FSH and LH action has not been well explained but probably involves interaction between the Sertoli and Leydig cells.

REGULATION OF SECRETION OF LHRH AND GONADOTROPINS

The rate of secretion of LH is controlled by the action of sex steroids at the hypothalamus and pituitary. Both testosterone and estradiol can inhibit LH secretion. Testosterone can be aromatized to estradiol in the hypothalamus and the pituitary, but the two hormones are thought to act independently.

Testosterone decreases LH levels primarily by reducing pulse frequency. By contrast, estradiol decreases LH levels by reducing LH pulse amplitude without changes in pulse frequency. Dihydrotestosterone also exerts negative feedback on LH secretion similar to that of testosterone. Testosterone also appears to have a negative feedback action on LH secretion at the pituitary level resulting in diminished LH response to acute LHRH stimulation in the presence of normal levels of plasma estradiol (Caminos-Torres et al, 1977).

Serum FSH concentration increases selectively in proportion to the loss of germinal elements in the testis (Van Thiel et al, 1972; Baker et al, 1976). A specific nonsteroidal FSH inhibitor (inhibin) from the seminiferous tubule has been thought to regulate FSH secretion. However, studies in the castrated rat indicate that FSH and LH concentrations can both be maintained chronically within the normal range by testosterone alone in the absence of other testicular factors. In addition, it appears that estradiol produces a selective decrease in LH levels. When testosterone is present in subphysiologic concentrations but estradiol is maintained at physiologic levels, this change in the T/E ratio leads to selective elevation of FSH levels with normal levels of LH, thus providing an alternative explanation to the inhibin hypothesis (Sherins, 1987).

ANDROGEN PHYSIOLOGY AND BIOCHEMISTRY

Five enzymes (or enzyme complexes) are involved in the conversion of cholesterol to testosterone. In this process the side chain of cholesterol is cleaved in two steps to reduce the size of the molecule from 27 to 19 carbons, and the A ring of the steroid is oxidized to the Δ4-3-keto configuration. The predominant pathway in the human testis appears to be the Δ5 pathway.

The rate-limiting reaction in testosterone synthesis is the mitochondrial conversion of cholesterol to pregnenolone by the cholesterol side chain cleavage enzyme (CSSC), which is regulated by LH. Testosterone levels are pulsatile in response to the episodic release of LH, and peak levels occur approximately 30 min after the episodic LH burst.

Although testosterone is the major secretory product, dihydrotestosterone, androsterone, androstenedione, progesterone, and 17-hydroxyprogesterone are also secreted by the testis. Testosterone serves as a prohormone for estradiol and dihydrotestosterone production. Normally 75 percent of the estradiol produced in the male is derived from the conversion of testosterone to estradiol (aromatization) in the peripheral tissues (Siiteri and MacDonald, 1973), and only 25 percent of the estradiol is secreted directly by the Leydig cells. Dihydrotestosterone is formed by the action of 5α-reductase in the peripheral tissues (Wilson, 1975).

Testosterone circulates normally bound to the β-globulin testosterone binding globulin (TeBG) (60 percent), and to albumin (35 percent); approximately 5 percent remains unbound. It is this free fraction that is available to the target tissues. TeBG is produced in the liver, and its concentration is responsive to the prevailing levels of testosterone and estradiol. Testosterone reduces TeBG concentration, while estradiol markedly increases TeBG concentration.

Testosterone and its metabolites are excreted primarily in the urine (90 percent) in the form of 17-ketosteroids and as a series of polar compounds, including diol and triol conjugates (Brooks, 1975).

ANDROGEN ACTION

The actions of androgens are summarized as follows:

1. Regulation of gonadotropin secretion by the hypothalamic–pituitary system
2. Initiation and maintenance of spermatogenesis
3. Formation of the male phenotype during sexual differentiation
4. Promotion of sexual maturation at puberty (Griffin and Wilson, 1985)

SPERMATOGENESIS

Spermatogenesis is the progressive maturation of germ cells from spermatogonia to mature spermatozoa. Observations of the seminiferous epithelium taken from testis biopsies reveal that specific cell types are repeatedly seen only in association with certain other cells. These observations of grouping of cells are known as stages in the cycle of the seminiferous epithelium. There are six stages of cell associations in men (Heller and Clermont, 1963). Each stage represents a particular degree of maturation. Stages I through VI constitute one cycle. It has been estimated that in men it takes 4.6 cycles or 74 days for a mature sperm to develop from an A_p spermatogonium (Heller and Clermont, 1963).

The ultrastructural features of the transformation of the human spermatid into a spermatozoon consist of a highly coordinated reorganization of the nucleus and cytoplasm and development of the flagellum. The chromatin becomes progressively more dense, and the nucleus comes to occupy an eccentric position adjacent to the cranial pole of the spermatid, separated from it by an acrosomal cap. The latter begins as a cytoplasmic vacuole containing flocculent material. The cilia structure that serves as the core of the sperm tail develops from a centriole near the Golgi apparatus and ultimately consists of nine outer fibers and two inner fibers. The mitochondria form a helix around the cilia from the neck to the annulus of the tail. The terminal region of the tail consists of the axial filament surrounded by the cell membrane.

Energy for sperm motility is derived from the hydrolysis of ATP generated in the mitochondrial sheath. The axial structure of the tail contains a central pair of microtubules surrounded by nine doublet tubules and nine dense fibers; the doublets are attached to the central tubules by a series of radial spokes, to each other by dynein arms, and to the axonemal membrane by Y links. Motility is believed to involve a sliding action of the microtubules. The dynein arms contain dynein, which has powerful ATPase activity (Satir, 1979). Sliding is thought to be generated by interaction of the dynein arms and to be restricted by the radial spokes (Linck, 1979; Gibbons, 1979).

Ejaculated sperm meet most of their energy requirements by utilizing fructose from the seminal vesicles, although mammalian sperm can also anaerobically metabolize glucose and mannose to lactic acid. The sperm mitochondria have the enzymes of both the Embden–Meyerhof glycolytic pathway and the Krebs cycle.

CONTROL OF SPERMATOGENESIS

Spermatogenesis requires an intact hypothalamic–pituitary axis with the normal pattern of secretion of the two gonadotropic hormones (LH and FSH). LH stimulates the secretion of testosterone from the Leydig cells in the testis. Testosterone diffuses from the interstitial space into the spermatogenic tubule, where it affects spermatogenesis either directly or indirectly through its effects on Sertoli cell function.

Following hypophysectomy in men, spermatogenesis is arrested at the spermatogonial stage and no spermatocytes are formed; spermatogenesis can be restored or initiated by treatment with human menopausal gonadotropin plus human chorionic gonadotropin (hCG), and, once restored, spermatogenesis can be maintained by hCG alone (Setchell, 1982). This phenomenon, together with the finding that in otherwise normal subjects with suppressed FSH activity spermatogenesis can be restored by hCG alone (Bremner et al, 1981), suggests that FSH may be required for initiation but not for maintenance of spermatogenesis.

THE ANATOMY OF THE REPRODUCTIVE TRACT

Spermatozoa are released into the tubular lumen of the seminiferous tubules along with fluid secreted by the Sertoli cells. Sperm then pass through the efferent ductules, which emerge from the upper pole of each testis. These ductules drain the secretion of the testis and at this point are usually known as the rete testis.

After their emergence from the testes, the efferent ductules join together to form the first major duct system, namely the epididymal duct. The walls of the efferent ductules are made up of myoid cells that contract spontaneously; however, this contractile tissue differentiates into smooth muscle in the epididymis, and its activity comes increasingly under the control of the autonomic nervous system as the duct descends. The epididymal duct has an active epithelium that is highly ciliated. The density of ciliation is progressively reduced in the more distal parts of the duct.

At its termination, each epididymal duct thickens considerably to form the vas deferens. Each duct is lined by columnar epithelium that also is ciliated. On entering the abdominal cavity, each vas deferens joins from above its corresponding seminal vesicle. At this point each vas deferens dilates to form the ampulla of the vas.

The seminal vesicles each consist of a tightly convoluted muscular blind-ended tubule. These tubular structures that form each seminal vesicle narrow to form the duct of the seminal vesicle. Each of these ducts joins with its corresponding vas deferens to form the ejaculatory duct. The ejaculatory ducts penetrate the prostate gland between the median and lateral lobes opening into the posterior urethra.

The prostate is a glandular structure that is enclosed within a muscular coat. It is made up of three lobes, a median and two lateral lobes, that surround the posterior urethra. The secretions of the prostate, which are expelled during emission, leave the gland by a series of small ducts, all of which enter the prostatic urethra around the veromontanum.

The bulbourethral glands are two small lobulated glands that secrete small amounts of mucus and open into the membranous urethra. The urethral glands also secrete small amounts of mucus, and these small glands open into the distal end of the penile urethra.

ROLE OF THE EPIDIDYMIS

Caudal epididymal sperm are more fertile than caput sperm in guinea pigs (Young, 1931). This observation has been confirmed in several species (Orgebin-Crist et al, 1975). Most investigators feel that the changing environment and physiology of the epididymis are critical to the maturation of the spermatozoa.

During transit through the epididymis, the spermatozoa and their milieu undergo many dramatic changes. In addition to the changes in morphology, chemistry, motility, fertilizing potential, and metabolism, spermatozoa also undergo alterations in permeability, antigenicity, and surface membrane charge during epididymal passage (Bedford, 1975).

The epididymis depends on androgens for maintenance of its histologic integrity (Maneely, 1959). The onset of sperm motility and fertility potential occurs during epididymal transit. The secretion of carnitine, glycerol phosphorylcholine (GPC), and sialic acid into the fluid are all androgen-dependent processes.

The luminal fluid changes significantly in the epididymis. A large portion of the fluid secreted from the seminiferous tubules flows into the rete testis and is absorbed in the efferent ductules and proximal epididymis. Inositol, total lipid, and total protein concentrations decrease as the luminal fluid moves through the epididymis (Jesee and Howards, 1976; Setchell and Waites, 1975). In contrast, potassium concentration rises. GPC, sialic acid, carnitine, hypotaurine, and acetylglucosamide are secreted in large quantities into the fluid, but their functions are only poorly understood.

Morphologic changes in epididymal spermatozoa include changes that occur in the acrosomal membranes as the spermatozoa undergo epididymal transit. Biochemical modifications in sperm occurring during epididymal maturation include an increase in cAMP content, a loss of nonnuclear protein, shifts in free amino acid concentrations, a loss of palmitic acid, decreases in the amounts of many phospholipids, increases in the amounts of unsaturated fatty acids, changes in electrophoretic properties, and stabilization of the desoxyribonucleoprotein complex.

TRANSPORT OF SPERMATOZOA

Mature spermatozoa are released from Sertoli cells into the lumen of the seminiferous tubule, wherein they traverse approximately 6 m of ducts before they leave the urethral meatus. This ductal system can be subdivided into several components. From the seminiferous tubule, the spermatozoa first travel into the rete testis and then leave the rete testis via the efferent ductules, finally passing into the epididymis. Sperm movement into the epididymis may be attributed to several factors: positive fluid pressure from the rete testis, fluid currents established by the beating of cilia along the walls of the efferent ducts, peristaltic contractions of the afferent ducts, contractions of the seminiferous tubules (Niemi and Kormano, 1965), and the contractions of the tunica albuginea.

Prior to or during ejaculation, the spermatozoa move from the cauda epididymis into the vas deferens. The distal portion of the vas deferens is termed the ampulla. The ampulla terminates in the prostatic urethra after it has been joined by the duct of the seminal vesicle, forming the ejaculatory duct.

The epididymal duct (Gaddum, 1969) and the efferent duct (MacMillan and Aukland, 1960) contract spontaneously. These contractions are thought to aid in sperm transport even during periods of sexual abstinence, whereas the contractions of the vas deferens and distal epididymis propel the spermatozoa during ejaculation.

SEMEN ANALYSIS

The standard clinical semen evaluation parameters are ejaculate volume, sperm concentration, sperm motility, and sperm morphology (Eliasson, 1975).

The normal volume of the ejaculate varies markedly and is highly dependent on the abstinence interval. The human ejaculate can be divided into several fractions. The first preejaculatory fraction comes from the mucous urethral (Littré's) and bulbourethral (Cowper's) glands. The fraction that follows is a low-viscosity, opalescent liquid that contains no spermatozoa and is derived from the prostate. The third, or principal, fraction is a mixture of prostatic fluid, gel from the seminal vesicles, and spermatozoa from the vas deferens and distal epididymis. The terminal fraction contains gel and spermatozoa from the seminal vesicles. The prostate and seminal vesicles provide more than 95 percent of the total volume of the ejaculate. The normal ejaculate ranges from 2 to 5 ml.

Sperm Concentration

The concentration of spermatozoa in the semen has traditionally been considered to be the most important clinical predictor of male fertility. According to the World Health Organization (WHO), the lower concentration accepted as normal is 20×10^6 cells/ml (WHO 1987).

Sperm Motility

The flagellar activity of the sperm cell is needed for normal transport through the female reproductive tract and to reach the site of fertilization. Assessment of sperm motility can therefore provide important information on sperm function. Sperm motility is usually rated in two ways: the percentage of motile cells and the quality of sperm movement. Sperm movement has been rated traditionally on a scale of 0 to 4+ (Eliasson, 1975). A normal value for sperm motility in the semen is at least 50 to 60 percent motile cells, and normal quality is usually 2+ or greater. This analysis can be subjective. In recent years, investigators have been using computer-assisted sperm analysis (CASA) to evaluate sperm movement characteristics (Vantman et al, 1988).

Sperm Morphology

Human sperm cytology serves as a sensitive index of the normalcy of the germinal epithelium (MacLeod, 1964). The clinical assessment is usually expressed as a percentage of "normal" cells, and a normal value is generally above 50 percent. MacLeod showed that the morphologic picture of a given individual is remarkably constant, and any variation in morphologic characteristics of the semen reflects an insult to the testis, such as a viral infection, heat, or radiation.

Capacitation

Ejaculated sperm recovered from semen shortly after collection are incapable of fertilizing oocytes. Spermatozoa must undergo a sequence of maturation events called capacitation before fertilization takes place. However, the biochemical and biophysical events that constitute capacitation are unknown. Three events or capabilities are normally associated with capacitated sperm, including the ability to undergo the acrosome reaction, the ability to bind to the zona pellucida of an unfertilized egg, and the acquisition of an altered form of motility called hyperactivation (Yanagimachi, 1970).

The spermatozoa undergo capacitation in the female reproductive tract, probably in the oviduct. A decapacitation factor present in seminal plasma may prevent the premature capacitation of sperm (Chang, 1957). Capacitated rabbit spermatozoa become decapacitated and lose their fertilizing ability when they are brought in contact with either the seminal or the epididymal plasma of a male rabbit. When the seminal plasma is removed, this inhibition is reversed and the sperm can be recapacitated and again rendered fertile. The decapacitation factor of the rabbit seminal plasma has been identified as a thermostable glycoprotein of about 350,000 daltons (Reyes et al, 1975; Eng and Oliphant, 1978). The role of decapacitation factor in the human remains to be clarified, but it has been postulated that the human female tract is able to destroy the decapacitation factor.

Capacitation probably triggers the events leading to the acrosome reaction. The process consists of the loss of the spermatozoon's outer acrosomal membrane and the plasma membrane associated with it. This loss results in the release of several enzymes needed for the fertilization process. Spermatozoa must penetrate the cumulus, corona radiata, and zona pellucida of the oocyte to be able to fuse with the oocyte plasmalemma. The attachment and movement of spermatozoa through the zona pellucida may be assisted by the various enzymes released during the acrosome reaction (Edwards, 1980). Acrosin, originally identified as a trypsinlike enzyme, has been assigned a key role in penetration of the zona pellucida (Hurtree and Srivastana, 1965; Zaneveld and Williams, 1970). Hyaluronidase may play a significant role in the cell dispersal of the cumulus oophorus and subsequent penetration of the zona pellucida by the spermatozoa.

ASSAYS TO EVALUATE SPERM FUNCTION

Traditional semen analysis has been used for many years to predict male fertility potential; however, the utility of semen analysis has been controversial. In the past few years it has been shown that men with low sperm count can impregnate their wives (Burris et al, 1988). Therefore, in order to predict male fertility potential, many aspects of sperm function have to be evaluated. Accordingly, we will discuss some of the assays employed to evaluate male fertility potential.

Sperm–Cervical Mucus Interaction

Some of the semen function related to transport in women may be assessed by observations of sperm–cervical mucus interaction (Moghissi, 1976). This assessment takes into account the interaction of male and female reproductive functions, and the results have been shown to correlate with fertility (Hanson et al, 1982). The precise functions of the sperm cells measured by in vitro tests of sperm–cervical mucus interaction are not fully appreciated, but they involve the surface of the spermatozoon as well as the motility of the sperm cell.

Human Spermatozoa–Zona-Free Hamster in Vitro Penetration Test

It is known that the zona pellucida is the principal block to interspecies fertilization between human spermatozoa and oocytes of the golden hamster (Yanagimachi et al, 1976). Consequently, after removal of the hamster zona pellucida by enzyme treatment, human sperm are able to fuse with the hamster vitellus. Both capacitation and the acrosome reaction are needed for this interspecies fertilization, and the hamster egg test is therefore a suitable assay for these sperm functions (Overstreet, 1982).

There are several clinical reports of a close association between male fertility and the success of sperm fusion with zona-free hamster egg in vitro (Rogers et al, 1979; Karp et al, 1981); however, false-positive and false-negative results have been described. Many cases of male infertility will not be detected by these assays if the disorder is related to abnormalities of sperm transport or the postfertilization events of embryonic development.

Sperm Binding to the Human Zona Pellucida

It has been proposed that sperm binding to the human zona pellucida and penetration of the perivitelline space of ova from cadaveric human ovaries can be used as an assay to evaluate male fertility potential (Overstreet and Hembree, 1976). Lately, the hemizona assay has been developed; it consists of the incubation of capacitated sperm with oocytes that failed to fertilize in an in vitro fertilization program. One-half of the oocyte is used with donor sperm and the second half with the patient's spermatozoa (Burkman et al,

1988; Vantman et al, 1989). The goal is to determine the number of sperm binding to the zona pellucida. The utility of both assays in determining male fertility potential is under investigation.

Computer-Assisted Sperm Analysis

In recent years different companies have developed software that used with the appropriate hardware, is capable of objectively studying sperm movement characteristics. This equipment is currently at the stage of being a research tool; however, some data suggest that these measurements will be useful in distinguishing patients with motility disorders (Vantman et al, 1989). The major utility of this technology will be to determine the regulatory factors involved in the control of sperm movement.

REFERENCES

Baker HWG, Bremner WJ, Burger HC et al: Testicular control of follicle-stimulating hormone secretion. *Recent Prog Horm Res* 32:429–476, 1976.

Bedford JJ: Maturation, transport and fate of spermatozoa in the epididymis, in Hamilton DW, Greep RO (eds): *Handbook of Physiology* (sec 7: *Endocrinology:* vol 5: *Male Reproductive System*). Baltimore, Williams & Wilkins, 1975, pp 303–317.

Bremner WJ, Matsumoto AM, Sussman AM et al: Follicle-stimulating hormone and human spermatogenesis. *J Clin Invest* 68:1044–1052, 1981.

Brooks RV: Androgens. *Clin Endocrinol Metab* 4:503–520, 1975.

Burkman LJ, Coddington CC, Franken DR, Kruger TF, Rozenwaks Z, Hodgen GD: The hemizona assay (HZA): Development of a diagnostic test for the binding of human spermatozoa to the human hemizona pellucida to predict fertilization potential. *Fertil Steril* 49:688–697, 1988.

Burris AS, Clarck RV, Vantman D, Sherins RJ: Low sperm concentration does not preclude fertility in men with isolated hypogonadotropic hypogonadism following gonadotropin therapy. *Fertil Steril* 50: 343, 1988.

Caminos-Torres R, Ma L, Snyder PJ: Testosterone-induced inhibition of the LH and FSH responses to gonadotropin-releasing hormones occurs slowly. *J Endocrinol Metab* 44:1142–1153, 1977.

Chang MC: A detrimental effect of rabbit seminal plasma on the fertilizing capacity of sperm. *Nature* 179:258–259, 1957.

Conn PM, Morrell DV, Dufau ML et al: Gonadotropin-releasing hormone action in cultured pituicytes: Independence of luteinizing hormone release and adenosine 3′,5′-monophosphate production. *Endocrinology* 104:448–453, 1979.

Dorrington JH, Armstrong DT: Follicle-stimulating hormone stimulates estradiol-17β synthesis in cultured Sertoli cells. *Proc Natl Acad Sci USA* 72:2677–2681, 1975.

Dorrington JH, Vernon RG, Fritz IB: The effect of gonadotropins on 3′,5′ cyclic-AMP levels of seminiferous tubule. *Biochem Biophys Res Commun* 46:1523–1528, 1972.

Dufau ML, Catt KJ: Gonadotropin receptors and regulation of steroidogenesis in the testis and ovary. *Vitam Horm* 36:461–592, 1978.

Edwards RG: *Fertilization in the Human Female.* New York, Academic Press, 1980, pp 586–592.

Eliasson R: Analysis of semen, in Berhman SJ, Kistner RW (eds): *Progress in Infertility.* Boston, Little, Brown, 1975, pp 691–714.

Eng LA, Oliphant G: Rabbit sperm reversible decapacitation by membrane stabilization with a highly purified glycoprotein from seminal plasma. *Biol Reprod* 19:1083–1094, 1978.

Fakunding JL, Means AR: Characterization and follicle-stimulating hormone activation of Sertoli cell cyclic AMP–dependent protein kinases. *Endocrinology* 101:1358–1368, 1977.

Gaddum P: Observations of the fetal mouse epididymis in culture. 168:188, 1969.

Gibbons BH: Studies on the mechanism of flagellar movement, in Fawcett DW, Bedfor JM (eds): *The Spermatozoon.* Baltimore, Urban & Schwarzenberg, 1979, pp 91–97.

Green JD, Harris GW: Observation of the hypophysioportal vessels of the living rat. *J Physiol (Lond)* 108:359–361, 1949.

Griffin JE, Wilson JD: Disorders of the testes and male reproductive tract, in Wilson JD, Foster DW (eds): *Williams Textbook of Endocrinology.* Philadelphia, WB Saunders, 1985, pp 259–311.

Hanson FW, Overstreet JW, Katz DF: A study of the relationship of motile sperm numbers in cervical mucus 48 hours after artificial insemination with subsequent fertility. *Am J Obstet Gynecol* 143:85–90, 1982.

Heller CG, Clermont Y: Spermatogenesis in men: An estimate of the duration. *Science* 140:184, 1963.

Hurtree EF, Srivastana PN: The clinical composition of the acrosomes of ram spermatozoa. *J Reprod Fertil* 9:47–60, 1965.

Jesee S, Howards SS: Survey of the sperm, sodium, and potassium concentrations in the epididymis of the hamster. *Biol Reprod* 15:626, 1976.

Karp LE, Williamson RA, Moore DE, Sky KK, Plymate SR, Smith WD: Sperm penetration assay: Useful test in evaluation of male fertility. *Obstet Gynecol* 57:620–623, 1981.

Linck RW: Advances in the ultrastructural analysis of the sperm flagellar axoneme, in Fawcett DW, Bedfor JM (eds): *The Spermatozoon.* Baltimore, Urban & Schwarzenberg, 1979, pp 99–115.

MacLeod J: Human seminal cytology as a sensitive indicator of the germinal epithelium. *Int J Fertil Cell* 9:281, 1964.

MacMillan EW, Aukland T: The transport or radiopaque medium through the initial segment of the rat epididymis. *J Reprod Fertil* 1:139, 1960.

Maneely RB: Epididymal structure and function: A historical and critical review. *Acta Zool* 40:1, 1959.

Manual De Laborayorio de la oms para el examen del semen humano y de la interaccion entre el semen y el moco cervical. Editorial Medica Panamericano (eds): *Valores Normales de las Variables del Semen.* p 41, 1987.

Means AR, Fakunding JL, Huckins C et al: Follicle-stimulating hormone, the Sertoli cell, and spermatogenesis. *Recent Prog Horm Res* 32:477–527, 1976.

Moghissi KS: Postcoital test: Physiologic basis, technique, and interpretation. *Fertil Steril* 27:117–129, 1976.

Naor Z, Catt KJ: Independent actions of gonadotropin releasing hormone upon cyclic GMP production and luteinizing hormone release. *J Biol Chem* 255:342–344, 1980.

Niemi M, Kormano M: Contractility of the seminiferous tubule of the postnatal rat testis. *Ann Med Exp Biol Fenn* 43:46, 1965.

Oliver C, Mical RS, Porter JC: Hypothalamic–pituitary vasculature: Evidence for retrograde blood flow in the pituitary stalk. *Endocrinology* 101:598–604, 1977.

Orgebin-Crist MC, Danzo BJ, Davies J: Endocrine control of the development and maintenance of sperm fertilizing ability in the epididymis, in Hamilton DW, Greep RO, (eds): *Handbook of Physiology* (sec 7: *Endocrinology;* vol 5: *Male Reproductive System*). Baltimore, Williams & Wilkins, 1975, p 319.

Overstreet JW: Evaluation of sperm function by tests of sperm ovum interaction in vitro, in Spira A, Jouannet P, (eds): *Human Fertility Factors.* Paris, INSERM, 1982, pp 263–280.

Overstreet JW, Hembree WC: Penetration of the zona pellucida of nonliving human oocytes by spermatozoa in vitro. *Fertil Steril* 27:815, 1976.

Podesta EJ, Dufau ML, Solano AR et al: Hormonal activation of protein kinase in isolated Leydig cells: Electrophoretic analysis of cyclic AMP receptors. *J Biol Chem* 253:8994–9001, 1978.

Reyes A, Oliphant G, Brackeet BG: Partial purification and identification of a reversible decapacitation factor from rabbit seminal plasma. *Fertil Steril* 26:148–157, 1975.

Rogers BJ, Van Campen HV, Ueno M, Lambert H, Bronson R, Hale R: Analysis of human spermatozoal fertilizing ability using zona-free ova. *Fertil Steril* 32:664–670, 1979.

Satir R: Basis of flagellar motility in spermatozoa: Current status, in Fawcett DW, Bedfor JM (eds): *The Spermatozoon.* Baltimore, Urban & Schwarzenberg, 1979, pp 81–90.

Setchell BP: Regulation of spermatogenesis and possible sites for contraceptive action, in Jeffcoate SL, Saudler M (eds): *Progress Towards a Male Contraceptive.* New York, Wiley, 1982, pp 1–18.

Setchell BP, Waites GMH: The blood–testis barrier, in Hamilton DW, Greep RO (eds): *Handbook of Physiology* (sec 7: *Endocrinology;* vol 5: *Male Reproductive System*). Baltimore, Williams & Wilkins, 1975, pp 143–172.

Sherins RJ: Disorders of male reproduction, in *A Review of Endocrinology: Diagnosis and Treatment.* Syllabus, 1987, p 582.

Siiteri PK, MacDonald PC: Role of extraglandular estrogen in human endocrinology, in Greep RO, Astwood EB (eds). (sec 7: *Endocrinology;* vol 2: *Female Reproductive System;* part 1). Washington, DC, American Physiological Society, 1973, pp 615–629.

Vaitukaitis JL, Ross GT, Braunstein GD et al: Gonadotropins and their subunits: Basic and clinical studies. *Recent Prog Horm Res* 32:289–331, 1976.

Van Thiel DH, Sherins RJ, Myers GH Jr et al: Evidence for a specific seminiferous tubular factor affecting follicle-stimulating hormone secretion in man. *J Clin Invest* 51:1009–1019, 1972.

Vantman D, Koukoulis G, Dennison L, Sherins, RJ: Computer assisted semen analysis: Evaluation of method and assessment of the influence of sperm concentration on linear velocity determination. *Fertil Steril* 49:510–515, 1988.

Vantman D, Banks SM, Koukoulis G, Dennison L, Sherins RJ: Assessment of sperm motion characteristics from fertile and infertile men using fully automated computer assisted semen analyzer. *Fertil Steril* 51:156, 1989.

Wilson JD: Metabolism of testicular androgens, in Greep RO, Astwood EB (eds): (sec 7: *Endocrinology;* vol. 5: *Male Reproductive System*). Washington, DC, American Physiological Society, 1975, pp 491–508.

Yanagimachi R: The movement of golden hamster spermatozoa before and after capacitation. *J Reprod Fertil* 23:193–196, 1970.

Yanagimachi R, Yanagimachi H, Rogers BJ: The use of the zona free animal ova as a test-system for the assessment of the fertilizing capacity of human spermatozoa. *Biol Reprod* 15:471–476, 1976.

Young, WC: A study of the function of the epididymis. *J Exp Biol* 8:151, 1931.

Zaneveld LJD, Williams WL: A sperm enzyme that disperses the corona radiata and its inhibition by decapacitation factor. *Biol Reprod* 2:363–368, 1970.

CHAPTER 2

Female Reproduction: An Overview

Belinda A. Marascalco

The ability for the human to reproduce can be altered on many levels by external influences. These influences may be as simple as changes in courtship behavior or as profound and complex as changes in the molecular workings of the cell. A basic understanding of the female reproductive process is necessary to place these alterations in proper perspective. This process can be summarized as gamete production and transport, fertilization, embryo transport, and implantation. This is a complex sequence of events, and despite significant research gains using lower mammals and, recently, nonhuman primates, insight into these events remains sketchy: How is it that the dominant follicle is selected each month? What propels the sperm to the right destination? What initiates completion of meiosis in the maturing oocyte? How is the oocyte retrieved and transported?

What follows will be a brief tour of human female reproduction; the intent is to arm the reader with basic knowledge to understand the ensuing chapters, not to dwell on controversy. The approach is anatomic, with the development and function of each part of the female reproductive tract considered separately.

OVARY

Development and Morphology

Within 4 weeks of conception, primordial germ cells migrate from the yolk sac epithelium to the genital ridges, a thickening of cells that form medial to the müllerian and wolffian bodies (Witschi, 1948). It is here that ovarian differentiation will take place beginning at about 7 weeks as the mitotically active germ cells replicate to a number of 6 to 7 million oogonia and then begin the process of meiosis (Baker, 1963). Meiosis is a special type of cell division in which the chromosomal content of each daughter cell is halved.

This is accomplished by initial replication of the 46 chromosomes and then two reductive divisions, meiosis 1 and meiosis 2. The oogonia that begin meiosis (now called oocytes) remain arrested in meiosis 1 until the periovulatory luteinizing hormone (LH) surge (Baker, 1982; Fowler and Edwards, 1973). Once meiosis is entered, the oocytes are no longer able to replicate, and the numbers of oocytes present at 20 weeks' gestation will continue to decrease until near exhaustion at menopause. This loss occurs by either atresia or release at ovulation. The adult ovary consists of an inner medulla and an outer cortex, which contains the oocyte–follicle complexes. These complexes will at puberty number 300,000 to 400,000 and by the time menopause has been reached will number only a few thousand. Conception ultimately occurs, provided a mature oocyte is present at a time when the female reproductive tract is able to respond to the needs of the traveling oocyte and ultimately the implanting embryo. This depends on the proper hormonal environment. The process of folliculogenesis links or integrates the maturation and release of an oocyte with peptide and steroid production by the granulosa–theca cells. This in turn depends on the presence of the gonadotropins, follicle-stimulating hormone (FSH), and LH and thus, a functioning hypothalamic–pituitary–gonadal axis.

Folliculogenesis

The primordial follicle consists of an oocyte with its nucleus arrested in the first meiotic division and surrounded by a basement membrane and a single layer of granulosa cells. Most of the follicular complexes reach this state late in fetal life or within the first few months of postnatal life. Folliculogenesis starts at this point, is initially gonadotropin-independent, occurs in the presence of low estradiol levels, and most often ends in atresia (Erickson, 1978; Goebelsmann, 1986a). This early process is known as the preantral, or primary oocyte, stage and involves enlargement of the oocyte, formation of the zona pellucida, and addition of another layer of granulosa cells. Once puberty is entered, these pools of primary follicles are stimulated by rising gonadotropin levels to become secondary follicles and are recruited for possible selection as the dominant follicle. These secondary follicles are characterized by the antrum, a central, fluid-filled cavity that forms around the granulosa-ensheathed oocyte. The fluid, rich in sex steroids, peptides (growth factors), mucopolysaccharides, gonadotropins, and many other compounds (renin, prostaglandins, oxytocin) reflects local production by the theca–granulosa cells, as well as transudation and even concentration from the vascular compartment (McNatty et al, 1979; Zimmermann et al, 1989).

Theca cells (surrounding the granulosa layer) differentiate and hypertrophy, and it is their increasing vascularization that allows connection of the follicular complex and the beginning of gonadotropin-dependent growth and selection of a dominant follicle. In the primate 28-day menstrual cycle, selection of this follicle is believed to occur by days 5 to 7 as evidenced by the increased production of estradiol by the ovary that will contain the dominant follicle (diZerega and Hodgen, 1981). The production of estrogen depends on cooperation between the theca and granulosa cell compartments, which are selectively responsive to LH and FSH, respectively. LH stimulates the production of androstenedione and testosterone from cholesterol by the thecal cell. FSH stimulates the granulosa cells to aromatize the thecal-produced androgens to estrone and estradiol (E_2) (Ryan and Smith, 1965). The mechanism of selection is hypothesized to depend on rapidly increasing amounts of E_2 in the ovarian microenvironment, which stimulates granulosa cell mitosis and increases the number of FSH receptors. The end result is an increase in FSH-induced aromatization of thecal androgens to estradiol. The follicle that manages to capture the lead with this E_2–FSH augmentation effect is able to continue with development to the preovulatory stage despite declining blood levels of FSH (Fritz and Speroff, 1982). The rest of the recruited pool, increasingly estrogen- and FSH-receptor-poor, will become atretic. Another mechanism that facilitates dominance includes selective vascularization of the thecal cells of the dominant follicle, allowing the dominant follicle better access to FSH, the levels of which are declining (diZerega and Hodgen, 1981).

All the aforementioned mechanisms help to explain maintenance of the dominant follicle but really do not adequately explain the difference in behavior exhibited among follicles and between ovaries, despite the same exposure to the gonadotropins. For the answer to this, many have invoked the existence of intraovarian regulators that would allow "gonadotropin action to be fine-tuned in situ" (Adashi, 1989). These possible regulators include growth factors, neuropeptides, and cytokines. Current theory is focusing on their ability to either amplify or attenuate the FSH and LH signals via autocrine or paracrine pathways, and, additionally, their ability to coordinate cell populations (somatic and germ) within the follicular unit (Adashi, 1991). Some of the factors linked with folliculogenesis include the insulinlike growth factors I and II (IGF-I, IGF-II), transforming growth factors alpha and beta (TGF-α, TGF-β), inhibin, epidermal growth factor (EGF), fibroblast growth factor (FGF), müllerian inhibitory substance (MIS), and angiogenic factors (Adashi, 1989). Knowledge of the IGFs as intraovarian regulators has progressed to the point of clinical application and therefore will be briefly reviewed.

In animals, specifically rodents, IGF-I is produced by granulosa cells of developing follicles, and receptors for IGF-I have been located on granulosa and theca–interstitial cells (Oliver et al, 1989; Davoren et al, 1986; Adashi et al, 1988). The primary role of IGF-I appears to be enhancement of gonadotropin hormonal action. For example, IGF-I enhances FSH-induced estrogen, cAMP, and proteoglycan production, as well as LH receptor production in rat granulosa cells (Davoren et al, 1985; Adashi et al, 1984; Adashi et al, 1985; Adashi et al, 1986a; Adashi et al, 1986b). In humans, IGF-I receptors have been identified on granulosa cells of graafian follicles, and IGF-I has been found in human follicular fluid. As in rodents, it acts synergistically with FSH in the production of estradiol by human granulosa cells (Gates et al, 1987; Eden et al, 1988; Erickson et al, 1989). Although studies have not shown IGF-I production by human granulosa cells, granulosa cells do synthesize IGF-binding proteins that may modulate IGF action in target tissues (Voutilainen and Miller, 1987; Sukkari et al, 1989; Jalkanen et al, 1989; Giudice et al, 1990; Geisthovel et al, 1989; Giudice et al, 1991; Ui et al, 1989). Clinically, growth hormone administration in women undergoing ovulation induction with human menopausal gonadotropins (hMG) decreases the amount of hMG needed and has been found to improve ovarian response in protocols for superovulation induction in gonadotropin-resistant patients (Homburg et al, 1988; Blumenfeld and Lunenfeld, 1989; Volpe et al, 1989). The mechanism is presumably via stimulation of IGF-I by growth hormone, either locally in the ovary or elsewhere in the body (liver).

Ovulation and the Corpus Luteum

Once the dominant follicle has been selected and proper growth has occurred, then rising estradiol levels, followed by a small rise in progesterone levels, elicit an LH surge. Thirty-four to 40 h following the onset of the surge, release of the oocyte will follow (Pauerstein et al, 1978; Garcia et al, 1981; Hoff et al, 1983). The midcycle LH surge initiates resumption of meiosis by the oocyte (cAMP-mediated), transforms granulosa cells to produce progesterone (in addition to estrogen), and stimulates prostaglandin and collagenase secretion (Yoshimura and Wallach, 1987; Goebelsmann, 1986b; LeMaire et al, 1975; Lumsden et al, 1986). Prostaglandins and collagenase assist ovum release by stimulating theca externa smooth muscle contraction and follicular wall breakdown, respectively. Although resumption of meiosis by the oocyte is clearly tagged to and dependent on the LH surge, the process remains poorly understood. Because removal of oocytes from their intrafollicular environment allows spontaneous resumption of meiosis, an oocyte maturation inhibitor (OMI) has been hypothesized (Pincus and Enzmann, 1935; Stone et al, 1978; Tsafriri and Channing, 1975; Edwards, 1965). This OMI requires intact cumuli around the oocyte to be effective (Tsafriri and Pomerantz, 1986). There is also evidence in bovine and rodent species that müllerian inhibitory substance, which is secreted by granulosa cells and found in follicular fluid, may act as a meiotic inhibitor (Vigier et al, 1984; Necklaws et al, 1986; Takahashi et al, 1986).

A small midcycle FSH surge occurs concomitantly with the LH surge. This surge stimulates the conversion of another proteolytic enzyme, plasminogen, to plasmin and facilitates the expansion of the cumulus (Beers and Strickland, 1976). Once extrusion of the oocyte and surrounding granulosa cells has occurred, the remaining granulosa cells become vascularized and luteinized. The collapsed and convoluted follicular wall takes on the characteristic yellow color of the corpus luteum (CL). The CL has a finite life span of 12 to 16 days unless conception and implantation take place, in which case, human chorionic gonadotropin (hCG), produced by the fetal–placental unit and working via LH receptors, will prolong the CL until approximately the ninth week of pregnancy (Lenton et al, 1984; Yoshimi et al, 1969). The CL secretes peptides (inhibin), androgens, estrogens, and progestins, but the latter, specifically progesterone, is most important to successful implantation and maintenance of pregnancy (Csapo et al, 1972; McLachlan et al, 1990). How progesterone confers uterine receptivity is complex. Many peptides important in growth and differentiation are induced in the endometrium during the luteal phase and in the decidua, and their successful secretion in response to progesterone may be pivotal to a successful nidation (Arceci et al, 1989; Tamada et al, 1990). Progesterone levels peak approximately 8 days after the LH surge, and the production of progesterone requires continued LH stimulation (Vande Wiele et al, 1970; Casper and Yen, 1979). Inadequate progesterone secretion or a short-lived CL have been implicated in early pregnancy loss and infertility (Strott et al, 1970; Sherman and Korenman, 1974; Wentz, 1979; Moszkowski et al, 1962).

Luteolysis

Decline of the CL is marked by lowering E_2 and progesterone levels, and the causative factor(s) remain unclear. The decline may be secondary to a relative gonadotropin deficiency or to local intraovarian estrogen, prostaglandins, or an as-yet-unnamed peptide (Schoonmaker et al, 1981; O'Gray et al, 1972). If fertilization occurs, the preimplantation embryo is capable of secreting hCG, and if implantation occurs, the hCG becomes detectable in the maternal serum 8 or 9 days postovulation (Catt et al, 1975; Mishell et al, 1974).

FALLOPIAN TUBE

Development and Anatomy

In the embryonic female the fallopian tubes develop from the müllerian duct system (paramesonephric duct system), which appears about 5 weeks after fertilization (Hunter, 1930). The development of this system is dependent on the absence of testicular tissue, which, if present, would secrete MIS. MIS in the fetus is responsible for müllerian duct regression (Jost, 1970). The fallopian tube is composed of an outer serosal layer, a middle muscle and connective tissue layer, and an inner stroma and epithelial layer. It may be divided into four parts, the interstitium, isthmus, ampulla, and infundibulum. The interstitial segment is 1 cm long and runs through the wall of the uterus. The isthmus is 2 to 3 cm long and exhibits the best-developed musculature and densest adrenergic innervation of the entire tube (Hodgson and Eddy, 1975). The ampulla is 5 to 8 cm long and provides the location for fertilization. The infundibulum is characterized by undulating, fingerlike processes called fimbriae, which are responsible for ovum pickup (Colborn, 1986). The lumen of the tube varies from 0.3 mm in its most proximal portion to 3 mm at the infundibular opening. The epithelium is made up of ciliated and secretory cells. The ciliated cells are present throughout the tract but predominate in the fimbrial epithelial folds (Ferenczy et al, 1972; Patek et al, 1972). In primates, the secretory cells are distributed throughout the ampulla and in the isthmus (Patek, 1974). Both cell types change structurally and functionally during the menstrual cycle in response to the changing hormonal milieu (Verhage et al, 1979; Bonilla-Musoles et al, 1983). Under the influence of increasing estrogen levels in the late follicular phase, secretory activity and the number of ciliated cells peak (Mastroianni et al, 1961; Verhage et al, 1979). Under the influence of progesterone in the luteal phase, the frequency of beat of the ampullary cilia peaks (Critoph and Dennis, 1977).

Function

This seemingly simple organ must accomplish many tasks, some of them quite divergent. Almost simultaneously, the tube must assist egg transport from the infundibulum to the ampullary–isthmic (AI) junction and sperm movement from the other direction. It must hold the egg at the AI junction while gating sperm access, and once fertilization has occurred, allow for safe passage of the embryo to the uterus. Additionally, the tubal secretions must sustain the final maturation of the oocyte, facilitate capacitation of the sperm prior to fertilization, and nurture the embryo after fertilization. How the tube executes these multitude of tasks is not well understood, but two explanations may be helpful: (1) form follows function, for example, morphologic heterogeneity allows functional heterogeneity, and (2) the tube, gametes, and embryo act in concert with each other, with events being mediated via a paracrine dialogue.

In the human, once the egg or secondary oocyte appears at the surface of the ovary, it must find its way into the tubal lumen. Using direct observations culled at the time of surgery, considering gross and microscopic anatomy, and extrapolating from mammalian data, direct fimbrial–ovarian contact probably occurs such that the cilia-rich fimbrial folds engulf the granulosa-shrouded oocyte and help propel it toward the AI junction (Westman, 1926; Blandau, 1969; Caffier, 1936; Doyle, 1951; Doyle, 1956). Tubal transit time from capture to arrival in the uterus will take approximately 80 h, with 72 of those spent in the ampulla (Croxatto et al, 1978). Movement of the ovum to the site of fertilization is postulated to be based on ciliary and smooth muscle activity, both of which are greatest at the time of and right after ovulation (Borell et al, 1957; Blandau, 1973). The spread of myoelectric activity along the tube varies with the time of the cycle. On cycle day 12, activity spreads from both ends of the oviduct toward the AI junction, and on cycle days 14 and 15 it spreads from the ampulla to the isthmus (Talo and Pulkkinen, 1982).

Once sperm are deposited in the vagina, the time of transport to the site of fertilization may be anywhere from minutes to hours (Overstreet and Katz, 1977; Thibault, 1973; Settlage et al, 1975). The mechanism of transport is both active and passive, with sperm movement serving to keep the sperm suspended in uterine and tubal fluids, and uterine myometrial contractions helping to propel them forward (Hunter, 1988). Periovulatory waves of contractions moving principally toward the ampullary–isthmic junction have been documented in the tubal isthmus in laboratory animals (Blandau and Gaddum-Rosse, 1974; Battalia and Yanagimachi, 1979). Compelling data in laboratory animals have led some investigators to hypothesize that the isthmus serves as a reservoir from which sperm are released once the appropriate signals are received from ovum and/or the hormonal environment (Harper, 1973a; Harper, 1973b; Overstreet and Cooper, 1975; Hunter, 1981; Hunter, 1987). Support for this so-called ovum signal was recently presented in a study in which a strong correlation was found between the ability of human follicular fluid from a particular follicle to cause sperm "chemotaxis" and the ability of the egg obtained from the same follicle to fertilize (Ralt et al, 1991).

For successful fertilization to take place, the ovum (under the influence of the LH surge) must have resumed meiosis and be in metaphase of meiosis 2. It is at this point that the first polar body has been extruded, there are 23 chromosomes (each with two chromatids), and proper nuclear and cytoplasmic maturation have taken place. Completion of meiosis 2 and ultimate achievement of 23 single chromosomes within the oocyte nucleus do not occur until fusion with the sperm. Prior to attaching to and penetrating the zona pellucida, the sperm must undergo a series of events. The first of these is called capacitation; it occurs in the female reproductive tract, involves alterations in the cell membrane, and is clinically identified by hyperactive motility. The second, which is part of the same process, is called the acrosome reaction and includes the fusion of the acrosomal membrane with the plasma membrane, formation of vesicles, and release of proteolytic enzymes that aid in penetration (Chang, 1951; Austin, 1952; Yanagimachi and Noda, 1970). Once one sperm has penetrated the zona pellucida and fused with the oocyte cell membrane, exocytosis of cortical granules from the ooplasm causes the zona to become impervious to other sperm (Wassarman, 1987; Sathananthan et al, 1982). The mechanism for retention of the ovum at the AI junction has not been clearly defined. The dense adrenergic innervation, well-developed musculature, and relative decrease in luminal diameter of the isthmus in comparison with the ampulla allow the possibility of a functional sphincter. The cholinergic and adrenergic innervation of the oviduct are modulated by both steroids and locally produced neural and humoral peptide factors (Sachy et al, 1989; Donnez et al, 1985; Kannisto et al, 1986). Substance P and gastrin-releasing peptide have been detected in the nerve fibers of tubal muscle, and substance P has been shown to induce a contractile response at the AI junction (Kannisto et al, 1986; Forman et al, 1985). Relaxation of tubal musculature occurs in the luteal phase under the direct and indirect effects of progesterone (Coutinho et al, 1975). These indirect effects include an increase in the ratio of prostaglandin E to prostaglandin F, an increase in beta adrenoreceptors, and an increase in immunoreactivity of vasoactive intestinal peptide in the isthmus (Wilhemsson and Lindblom, 1980; Lindblom et al, 1983; Helm et al, 1987). All these may stimulate relaxation of the AI junction and assist movement of the preembryo to the uterine cavity.

Once fertilization takes place, nurture of the cleaving zygote becomes of paramount importance. It is in this context that the secretory function of the oviduct has gained recent attention. Oviductal fluid, which may be actively secreted or transudated from the vascular bed, differs from plasma in energy substrate types, electrolyte concentrations, macromolecular content, and pH. This composition varies with the phase of the cycle and the location in the tube (Leese, 1988; Borland et al, 1980; David et al, 1973). The substances within this fluid that are most likely to confer embryotrophic action are the peptide and protein macromolecules (Sayegh and Mastroianni, 1991). These proteins may be oviductal-specific such as the glycoproteins isolated from human oviductal fluid or non-oviductal-specific such as the placental proteins (PPs). These PPs were initially isolated from the endometrium but also have been identified in human oviducts (Waugh and Lippes, 1989; Sjoberg et al, 1986; Tiitinen et al, 1986). The function of both groups of proteins is unknown. Supporting the idea that oviductal cells are embryotrophic are the improved quality and number of human embryos developing to the blastocyst stage when cocultured on human tubal ampullary cells versus standard culture medium (Bongso et al, 1989). Interestingly, this positive effect is not specific for oviductal cells. It is also seen in other culture systems such as Vero cell monolayers and fetal bovine fibroblasts (Menezo et al, 1990; Wiemer et al, 1989). What these systems may have in common is the secretion of factors important to growth and differentiation. Factors such as EGF and TGF-β have been shown to be genetically expressed in the preimplantation mouse embryo (Rappolee et al, 1988). EGF receptor function has been demonstrated in older embryos, but whether or not EGF and TGF-β receptors are expressed in the preimplantation embryo remains to be clarified (Nexo et al, 1980). Proof that gametes or embryos locally affect tubal epithelial synthesis is more difficult to establish. Gupta and colleagues showed that the rabbit ampulla had a higher lactate concentration when fertilized eggs were present in this portion of the tube (Gupta et al, 1970).

The tube can be envisioned as a series of microenvironments that are generated and maintained by the changing estradiol and progesterone concentrations. These microenvironments are critical to gamete transport, fertilization, embryo development, and embryo transport. Under normal conditions the embryo arrives in the uterine cavity approximately 2 to 3 days after fertilization at the morula stage of development.

UTERUS

Development and Anatomy

As was described for the fallopian tubes, the uterus originates from the müllerian (paramesonephric) duct system, which begins developing from invaginations of dorsal coelomic epithelium at 5 weeks of fetal age. Initiation and maintenance of this growth depends on the absence of MIS, which is produced by testicular tissue and is responsible for müllerian duct regression (Jost, 1970). At their point of origin the ducts are solid buds that grow and elongate in a medial and caudal direction while canalizing from the point of origin. At about 8 weeks the ducts meet and fuse, and gradually the intervening septum is reabsorbed, forming a uterovaginal cavity that will give rise to the epithelium and glands of the uterus and cervix. Surrounding mesenchyme will give rise to uterine musculature (Jones, 1985).

The uterus is a muscular, pear-shaped organ covered by the visceral peritoneum and composed of three outer muscle layers and an inner stromal and glandular layer known as endometrium. The endometrium varies in thickness from 1 to 6 mm depending on the time of the menstrual cycle. During menstrual bleeding the most superficial layer of the endometrium, the zona functionalis, is shed. Both the endometrial and muscle layers of the uterus experience marked physiologic changes in response to the changing hormonal environment of the menstrual cycle and pregnancy. This plasticity allows different activities by the same structure, for example, the relatively quiescent uterine muscle of pregnancy versus the contractile organ of labor, the endometrium that facilitates sperm transport and capacitation versus the endometrium that nourishes the blastocyst and facilitates implantation. Historically, these changes have been studied in relation to estradiol and progesterone, focusing on their structure, receptors, and biological effect. Indeed, by working through cytoplasmic receptors that regulate the transcription of specific genes, they are the primary effectors in the continual cycle of growth, proliferation, differentiation, desquamation, and regeneration. Their absolute amounts and relative concentrations are critical to the proper formation of secretory endometrium (Good and Moyer, 1968). However, it is increasingly clear that many growth factors, cytokines, and binding proteins may be important mediators of many of these biological processes (Tabibzadeh, 1991; McLachlan et al, 1991). We will briefly review the changes seen in the uterus during the menstrual cycle (Noyes et al, 1950).

During the follicular phase, under the influence of estrogen, there is proliferation of both epithelium and stroma, with mitotic figures noted in both areas. In addition, glycogenesis occurs, and glycogen storage is noted by day 10 as the glands become long and tortuous. As proliferation of the glandular epithelium continues, the cells become heaped up on each other in a pseudostratification pattern. The first sign of ovulation (progesterone effect) is the appearance of subnuclear vacuoles at 48 h postovulation. These represent the accumulation of glycogen under the influence of both E_2 and progesterone. The secreted glycogen will be broken down to provide energy substrates for the floating blastocyst. Under persistent progesterone stimulation, the stroma becomes more vascularized and edematous and reaches peak secretory activity in the midluteal phase, in time for the arriving embryo. Six days postovulation, scanning electron microscopy reveals the development of bulbous protrusions known as *pinopods* in the microvilli-containing cells of the luminal epithelium (Martel et al, 1989). In addition to increasing the surface area, these pinopods may help to appose the walls of the uterine lumen by endocytosis of uterine fluid (Enders and Nelson, 1973). Implantation of the embryo has been described as a four-part process. This includes apposition of the embryo next to the endothelium, attachment via cell adhesion molecules and glycoproteins such as laminin and fibronectin, intrusion between the epithelial cells, and, finally, penetration of the basement membrane via proteases elaborated by the trophoblast (Kliman et al, 1989; Martin and Arias-Stella, 1982; Queenan et al, 1987). A number of growth factors (EGF, TGF-β, IGF-I), cytokines (interleukins, colony-stimulating factors), and growth factor–binding proteins (IGF-binding proteins) are secreted by epithelium, stroma, or decidualized cells and may facilitate implantation and embryo differentiation (Tamada et al, 1990; Tabibzadeh, 1991; Rosenfeld et al, 1990). Their individual actions (taken as a whole) include mitogenesis, differentiation, neovascularization, and immunoregulation, all important aspects of the implantation process. Data on their receptor production in uterine and trophoblastic tissues implies reception as well as production (Hofmann and Anderson, 1990; Ghahary and Murphy, 1989). This supports the importance of their roles but does not clarify whether their participation in the implantation process is obligatory.

If implantation does not take place, then at the end of the secretory phase the glands will become exhausted and collapse. The premenstrual phase is marked by a polymorphonuclear and mononuclear leukocyte infiltrate. Menstruation is triggered by falling progesterone levels and follows an ischemic phase marked by vasoconstriction of arterioles and coiled arteries (Markee, 1940).

HYPOTHALAMIC–PITUITARY AXIS

Folliculogenesis, steroid production, and proper ovarian function are dependent on episodic secretion of LH and FSH from the anterior pituitary. Gonadotropin secretion is dependent on release of gonadotropin-releasing hormone (GnRH) from the medial basal hypothalamus. This decapeptide reaches the anterior pituitary gland via blood flow through hypothalamic–pituitary portal vessels. The synthesis and release of GnRH (from cell bodies located in the CNS) is modified by neurotransmitters that may either stimulate release (catecholamines) or inhibit release (β-endorphins) (Plant, 1986). In general, the relationship between GnRH and the pituitary hormones LH and FSH is sequential, that is, a pulse of GnRH elicits a pulse of LH and FSH (Santen and Bardin, 1973; Clarke and Cummins, 1982). The pulsatile nature of the GnRH secretion is critical to pituitary response. Continuous infusion of GnRH causes a reduction in gonadotropins after an initial stimulation (Knobil, 1980). The release of LH and FSH can be further altered by the frequency of the GnRH pulses and by the steroid and peptide concentrations bathing the gonadotrophs. These alterations lead to changes in gonadotropin pulse frequency and amplitude during the menstrual cycle. The follicular phase is characterized by more-frequent but lower-amplitude pulses, and the luteal phase by lower-frequency, higher-amplitude pulses (Brody et al, 1982; Filicori et al, 1986). Interestingly, in women with hypothalamic amenorrhea, the normal menstrual cycle can be recapitulated by GnRH pulses set at a constant frequency (Crowley and McArthur, 1980). This has led to the concept that the complex changes of the menstrual cycle can be explained by the pituitary's response to ovarian commands (Knobil, 1980). In this scheme, the pulsatile release of GnRH plays a passive but crucial role. It should be noted, however, that recreating a normal menstrual cycle with a fixed-frequency GnRH analog may not reflect normal physiology.

The key moments in the menstrual cycle can be broken down into stimulation and inhibition of gonadotropin secretion. These changes can be explained by estrogen and progesterone acting either directly on the pituitary, or indirectly via effect on hypothalamic centers. Ovarian-derived peptides, such as inhibin and activin, clearly inhibit and stimulate (respectively) the secretion of FSH, but their importance in normal physiology has yet to be established (Ying, 1988). Inhibition of gonadotropin secretion is seen in the luteal phase via negative feedback by both estradiol and progesterone (Nippoldt et al, 1989; Soules et al, 1984). The inhibition of LH pulse frequency and therefore GnRH pulse frequency appears to be mediated by endogenous opioids (Van Vugt et al, 1984). In ovariectomized monkeys, estradiol and progesterone replacement increases β-endorphin levels in portal blood, and in humans, oral administration of an opioid antagonist during the luteal phase increases LH pulses (Gindoff et al, 1988). Negative-feedback mechanisms are also seen in the late follicular phase when FSH levels seem to be selectively inhibited by rising estradiol (and inhibin) levels. These changes may aid in the selection and maintenance of the dominant follicle.

Positive-feedback response of the hypothalamic–pituitary unit is illustrated by the midcycle LH surge, which appears to be provoked by rising estradiol levels (March et al, 1979; Lobo, 1991). Progesterone values begin to rise 10 to 12 h prior to the peak of the LH surge, and in the presence of preovulatory estradiol levels, serve to enhance, hasten, and time the LH release (Collins et al, 1984; March et al, 1979; Batista et al, 1989). GnRH governs this LH surge in primates, but its role in periovulatory gonadotropin dynamics in humans is not as clear (Neill et al, 1977; Collins et al, 1984).

The basic components of female reproduction have been described. Success depends on proper function and coordination of four organ systems, the reproductive tract, ovary, pituitary, and hypothalamus. Despite our lack of understanding in some areas of this elegant process, there is enough basic knowledge to guide us in our study of pathophysiologic and toxicologic processes.

REFERENCES

Adashi EY (ed): Putative intraovarian regulators, in Speroff L (ed): *Seminars in Reproductive Endocrinology* (vol 7, no 1). New York, Thieme Medical Publishers, 1979, preface.

Adashi EY: The ovarian life cycle, in Yen SSC, Jaffe RB (eds): *Reproductive Endocrinology* (3d ed). Philadelphia, WB Saunders, 1991, pp 209–214.

Adashi EY, Resnick CE, Svoboda ME, Van Wyk JJ: Somatomedin-C enhances induction of luteinizing hormone receptor by follicle-stimulating hormone in cultured rat granulosa cells. *Endocrinology* 116:2369, 1984.

Adashi EY, Resnick CE, Brodie AMH et al: Somatomedin-C mediated potentiation of follicle-stimulating hormone induced aromatase activity of cultured rat granulosa cells. *Endocrinology* 117:2313, 1985.

Adashi EY, Resnick CE, Svoboda ME et al: Somatomedin-C as an amplifier of follicle stimulating hormone action: Enhanced accumulation of adenosine 3,5-monophosphate. *Endocrinology* 118:149, 1986a.

Adashi EY, Resnick CE, Svoboda ME et al: Independent and synergistic actions of somatomedin-C in the stimulation of proteoglycan biosynthesis by cultured rat granulosa cells. *Endocrinology* 118:456, 1986b.

Adashi EY, Resnick CE, Hernandez ER et al: Characterization and regulation of a specific cell membrane receptor for somatomedin C/insulin-like growth factor I in cultured rat granulosa cells. *Endocrinology* 122:194–201, 1988.

Arceci RJ, Shanahan F, Stanley ER et al: Temporal expression and location of colony stimulating factor-1 (CSF-1) and its receptor in the female reproductive tract are consistent with CSF-1 regulated placental development. *Proc Natl Acad Sci USA* 86:8818–8822, 1989.

Austin CR: The "capacitation" of the mammalian sperm. *Nature* (*Lond*) 170:326, 1952.

Baker TG: A quantitative and cytological study of germ cells in human ovaries. *Proc R Soc Lond* (*Biol*) 158:417, 1963.

Baker TG: Oogenesis and ovulation, in Austin CR, Short RV (eds): *Germs Cells and Fertilization* (2d ed; vol 1). Cambridge, Cambridge University Press, 1982, p 17.

Batista MC, Cartledge T, Merriam GR, Nieman LK: The antiprogesterone RU486 delays ovulation in spontaneous and GnRH-induced menstrual cycles (abstract). *Endocrinology* 124:431A, 1989.

Battalia DE, Yanagimachi R: Enhanced and coordinated movement of the hamster oviduct during the periovulatory period. *J Reprod Fertil* 56: 515–520, 1979.

Beers W, Strickland S: Studies on the role of plasminogen activator in ovulation. *J Biol Chem* 251:5694, 1976.

Blandau RJ: Gamete transport, comparative aspects, in Hafez ESE, Blandau RJ (eds): *The Mammalian Oviduct*. Chicago, University of Chicago Press, 1969, p 129.

Blandau RJ: Gamete transport in the female mammal, in Greep RO, Astwood EB (eds): *Handbook of Physiology* (sec 7: *Endocrinology II*). Washington, DC, American Physiological Society, 1973, pp 153–163.

Blandau RJ, Gaddum-Rosse P. Mechanism of sperm transport in pig oviducts. *Fertil Steril* 25:61–67, 1974.

Blumenfeld Z, Lunenfeld B: The potentiating effect of growth hormone on follicle stimulating with human menopausal gonadotropin in a pan hypopituitary patient. *Fertil Steril* 52:328–331, 1989.

Bongso A, Ng SC, Sathananthan H et al: Improved quality of human embryos when co-cultured with human ampullary cells. *Hum Reprod* 14:706, 1989.

Bonilla-Musoles F, Ferrer-Barriendos J, Pellicer A: Cyclical changes in the epithelium of the fallopian tube. Studies with scanning electron microscopy (SEM). *Clin Exp Obstet Gynecol* 10:79, 1983.

Borell U, Nilsson O, Westman A: Ciliary activity in the rabbit fallopian tube during oestrus and after copulation. *Acta Obstet Gynecol Scand* 36:22–28, 1957.

Borland RM, Biggers JD, Lechene CP et al: Elemental composition of fluid in the human fallopian tube. *J Reprod Fertil* 58:479–482, 1980.

Brody SA, McAtee NM, Wachter KW, Merriam GR, Loriaux DL: The progression of 24-hour plasma gonadotropin patterns in the normal menstrual cycle: A longitudinal study. *Clin Res* 39A:684, 1982.

Caffier P: Studien zum Eitransport beim Menschen: I. Mitteilung: Der Eiabnahmemechanismus. *Zentralb Gynakol* 60:1873, 1936.

Casper R, Yen SSC: Induction of luteolysis in the human with a long acting analog of luteinizing hormone–releasing factor. *Science* 205:408, 1979.

Catt KJ, Dufau ML, Vaitukaitis JL: Appearance of hCG in pregnancy plasma following the initiation of implantation of the blastocyst. *J Clin Endocrinology Metab* 40:537, 1975.

Chang MC: Fertilizing capacity of spermatozoa deposited into the fallopian tubes. *Nature* 168:697–698, 1951.

Clarke IJ, Cummins JT: The temporal relationship between gonadotropin releasing hormone (GnRH) and luteinizing hormone secretion in ovariectomized ewes. *Endocrinology* 111:1737–1739, 1982.

Colburn GL: The anatomy of the fallopian tube, in Siegler AM (ed): *The Fallopian Tube*. Mount Kisco, NY, Futura Publishing, 1986, pp 8–9.

Collins RC, Brody SA, Loriaux DL et al: Patterns of LH secretion during the midcycle surge in normal women. *Excerpta Med* 652:447, 1984.

Coutinho, Maia H, Mattos C: Contractility of the fallopian tube. *Gynecol Invest* 6:146–161, 1975.

Critoph F, Dennis K: Ciliary activity in the human oviduct. *Br J Obstet Gynecol* 84:216–218, 1977.

Crowley WF, McArthur JW: Stimulation of the normal menstrual cycle in Kallman's syndrome by pulsatile administration of luteinizing hormone–releasing hormone (LHRH). *J Clin Endocrinol Metab* 51:173, 1980.

Croxatto HB, Ortiz ME, Diaz S et al: Studies on the duration of egg transport by the human oviduct. *Am J Obstet Gynecol* 132(6):629–634, 1978.

Csapo AI, Pulkkinen MO, Ruttner B et al: The significance of the human corpus luteum in pregnancy maintenance. *Am J Obstet Gynecol* 112:1061, 1972.

David A, Serr DM, Czernobilsky B: Chemical composition of human oviductal fluid. *Fertil Steril* 24:435, 1973.

Davoren JB, Hsueh AJW, Li CH: Somatomedin C augments FSH-induced differentiation of cultured rat granulosa cells. *Am J Physiol* 249:E26, 1985.

Davoren JB, Kasson BG, Li CH et al: Specific insulin-like growth factor (IGF) I– and II–binding sites on rat granulosa cells: Relation to IGF action. *Endocrinology* 119:2155–2162, 1986.

diZerega GS, Hodgen GD: Folliculogenesis in the primate ovarian cycle. *Endocr Rev* 2:27, 1981.

Donnez J, Goenen E, Casanas-Roux F et al: Monoamine oxidase reactivity in the human fallopian tube. *Fertil Steril* 43:488–490, 1985.

Doyle JB: Exploratory culdotomy for observation of tubo-ovarian physiology at ovulation time. *Fertil Steril* 2:475, 1951.

Doyle JB: Tubo-ovarian mechanism: Observations at laparotomy. *Obstet Gynecol* 8:686, 1956.

Eden JA, Jones J, Carter GF et al: A comparison of follicular fluid levels of IGF-I in normal dominant and cohort follicles, polycystic and multicystic ovaries. *Clin Endocrinol* 29:327, 1988.

Edwards RG: Maturation in vitro of mouse, sheep, cow, pig, rhesus monkey, and human oocytes. *Nature* 208:349, 1965.

Enders AC, Nelson DM: Pinocytotic activity in the uterus of the rat. *Am J Anat* 138:277, 1973.

Erickson GF: Normal ovarian function. *Clin Obstet Gynecol* 21:31, 1978.

Erickson GF, Garzo VG, Magoffin DA: Insulin like growth factor-I regulates aromatase activity in human granulosa and granulosa luteal cells. *J Clin Endocrinol Metab* 69:716, 1989.

Ferenczy A, Richart RM, Agate FJ et al: Scanning electron microscopy of the human fallopian tube. *Science* 175:783, 1972.

Filicori M, Santoro N, Merriam GR et al: Characterization of the physiological pattern of episodic gonadotropin secretion throughout the human menstrual cycle. *J Clin Endocrinol Metab* 62:1136, 1986.

Forman A, Anderson K-E, Maigaard S: Concentrations and contractile effects of substance P in the human ampullary-isthmic junction. *Acta Physiol Scand* 124:17–23, 1985.

Fowler RE, Edwards RG: The genetics of early human development. *Prog Med Genet* 9:49, 1973.

Fritz MA, Speroff L: The endocrinology of the menstrual cycle: The interaction of folliculogenesis and neuroendocrine mechanisms. *Fertil Steril* 38:509, 1982.

Garcia JE, Jones GS, Wright GL: Prediction of the time of ovulation. *Fertil Steril* 36:308, 1981.

Gates GS, Bayer S, Seibel M et al: Characterization of insulin-like growth factor binding to human granulosa cells obtained during in vitro fertilization. *J Receptor Res* 7:885, 1987.

Geisthovel F, Moretti-Rojas I, Asch RH et al: Expression of insulin-like growth factor-II (IGF-II) messenger ribonucleic acid (mRNA) but not IGF-I mRNA in human preovulatory granulosa cells. *Human Reprod* 4:899–902, 1989.

Ghahary A, Murphy LJ: Uterine insulin-like growth factor-1 receptors: Regulation by estrogen and variation throughout the estrous cycle. *Endocrinology* 125:597–604, 1989.

Gindoff PR, Jewelewicz R, Hembree W et al: Sustained effects of opioid antagonism during the normal human luteal phase. *J Clin Endocrinol Metab* 66:1000, 1988.

Giudice LC, Farrell EM, Pham H et al: Identification of insulin-like growth factor binding protein-3 (IGFBP-3) and IGF BP-2 in human follicular fluid. *J Clin Endocrinol Metab* 71:1330–1338, 1990.

Giudice LC, Milki AA, Milkowski DA et al: Human granulosa contain messenger ribonucleic acids encoding insulin-like growth factor–binding proteins (IGFBPs) and secrete IGFBPs in culture. *Fertil Steril* 56(3):475–480, 1991.

Goebelsmann U: The menstrual cycle, in Mishell DR, Davajan V (eds): *Infertility, Contraception and Reproductive Endocrinology* (2d ed). Oradell, NJ, Medical Economics, 1986a, p 71.

Goebelsmann U: The menstrual cycle, in Mishell DR, Davajan V (eds): *Infertility, Contraception and Reproductive Endocrinology* (2d ed). Oradell, NJ, Medical Economics, 1986b, p 73.

Good RG, Moyer DL: Estrogen-progesterone relationships in the development of secretory endometrium. *Fertil Steril* 19:37, 1968.

Gupta DN, Karkun J, Kar AB: Biochemical changes in different parts of the rabbit fallopian tube during passage of ova. *Am J Obstet Gynecol* 106:833–837, 1970.

Harper MJK: Stimulation of sperm movement from the isthmus to the site of fertilization in the rabbit oviduct. *Biol Reprod* 8:369–377, 1973a.

Harper MJK: Relationship between sperm transport and penetration of eggs in the rabbit oviduct. *Biol Reprod* 8:441–450, 1973b.

Helm G, Ekman R, Owman C: Cyclic fluctuations of vasoactive intestinal polypeptide measured radioimmunologically in various regions of the human fallopian tube. *Int J Fertil* 32:467–471, 1987.

Hodgson BJ, Eddy CA: The autonomic nervous system and its relationship to tubal ovum transport—A reappraisal. *Gynecol Invest* 6:162, 1975.

Hoff JD, Quigley ME, Yen SSC: Hormonal dynamics at midcycle: A reevaluation. *J Clin Endocrinol Metab* 57:792, 1983.

Hofmann GE, Anderson TL: Immunohistochemical localization of epidermal growth factor receptor during implantation in the rabbit. *Am J Obstet Gynecol* 162(3):837–841, 1990.

Homburg R, Eshel A, Abdalla HI et al: Growth hormone facilitates ovulation induction by gonadotropins. *Clin Endocrinol* 29:113–117, 1988.

Hunter RH: Observations on the development of the human female genital tract. *Contrib Embryol* 22(129):91–107, 1930.

Hunter RHF: Sperm transport and reservoirs in the pig oviduct in relation to the time of ovulation. *J Reprod Fertil* 63:109–117, 1981.

Hunter RHF: Human fertilisation in vivo, with special reference to progression, storage and release of competent spermatozoa. *Human Reprod* 2(4):329–332, 1987.

Hunter RHF: *The Fallopian Tubes*. Berlin, Springer Verlag, 1988, p 61.

Jalkanen J, Suikkari A-M, Koistinen R et al: Regulation of insulin-like growth factor binding protein-1 production in human granulosa luteal cells. *J Clin Endocrinol Metab* 69:1174–1179, 1989.

Jones HW: Surgery for anomalies of the müllerian ducts, in Mattingly RF, Thompson, JD (eds): *TeLinde's Operative Gynecology* (6th ed). Philadelphia, JB Lippincott, 1985, pp 345–348.

Jost A: Hormonal factors in the sex differentiation of the mammalian foetus. *Philos Trans R Soc Lond (Biol)* 259:119, 1970.

Kannisto P, Ekblad E, Helm G et al: Existence and coexistence of peptides in nerves of the mammalian ovary and oviduct demonstrated by immunocytochemistry. *Histochemistry* 86:25–34, 1986.

Kliman HJ, Coutifaris C, Feinberg RF et al: Implantation: In vitro models utilizing human tissues, in Yoshinaga K (ed): *Blastocyst Implantation (Serono Symposia)*. Boston, Adams Publishing, 1989, pp 83–91.

Knobil E: The neuroendocrine control of the menstrual cycle. *Recent Prog Horm Res* 36:53, 1980.

Leese HJ: Formation and function of oviduct fluid: A review. *J Reprod Fertil* 82:843–856, 1988.

LeMaire WJ, Leidner R, Marsh JM: Pre- and post-ovulatory changes in the concentration of prostaglandins in rat graafian follicles. *Prostaglandins* 9:221, 1975.

Lenton EA, Landgren B, Sexton L: Normal variation in the length of the luteal phase of the menstrual cycle: Identification of the short luteal phase. *Br J Obstet Gynaecol* 91:685, 1984.

Lindblom B, Wilhelmsson Wikland M et al: Prostaglandins and oviductal function. *Acta Obstet Gynecol Scand* (suppl) 113:43–46, 1983.

Lobo RA: The menstrual cycle, in Mishell DR, Davajan V, Lobo RA (eds): *Infertility, Contraception and Reproductive Endocrinology*. Cambridge, MA, Blackwell Scientific, 1991, p 116.

Lumsden MA, Kelly RW, Templeton AA et al: Changes in the concentrations of prostaglandins in pre-ovulatory human follicles after administration of hCG. *J Reprod Fertil* 77:119, 1986.

McLachlan JA, Nelson KG, Takahashi T et al: Estrogens and growth factors in the development, growth, and function of the female reproductive tract, in Schomberg DW (ed): *Growth Factors In Reproduction* (*Serono Symposia*). New York, Springer-Verlag, 1991, pp 197–203.

McLachlan RI, Cohen NL, Dahl KD et al: Serum inhibin levels during the periovulatory interval in normal women: Relationships with sex steroid and gonadotropin levels. *Clin Endocrinol* 32:39, 1990.

McNatty KP, Smith DM, Makris A et al: The microenvironment of the human antral follicle: Interrelationships among the steroid levels in antral fluid, the population of granulosa cells, and the status of the oocyte in vivo and in vitro. *J Clin Endocrinol Metab* 49:851–860, 1979.

March CM, Goebelsmann U, Nakamura RM et al: Roles of estradiol and progesterone in eliciting the mid cycle luteinizing hormone and follicle-stimulating hormone surges. *J Clin Endocrinol Metab* 49:507, 1979.

Markee JE: Menstruation in intraocular endometrial transplants in the rhesus monkey. *Contrib Embryol* 28:219–308, 1940.

Martel D, Frydman R, Sarantis L, Roche D, Psychoyos A: Scanning electron microscopy of the uterine luminal epithelium as a marker of the implantation window, in Yoshinaga K (ed): *Blastocyst Implantation* (*Serono Symposia*), Boston, Adams Publishing, 1989, pp 225–230.

Martin O, Arias-Stella F: Plasminogen activator production by trophoblast cells in vitro. Effect of steroid hormones on protein synthesis inhibitors. *Am J Obstet Gynecol* 142:402–409, 1982.

Mastroianni L, Shah U, Abdul-Karim R: Prolonged volumetric collection of oviduct fluid in the rhesus monkey. *Fertil Steril* 12:417, 1961.

Menezo YJR, Guerin JF, Czyba JC: Improvement of human early embryo development in vitro by co-culture in monolayers of Vero cells. *Biol Reprod* 42:301, 1990.

Mishell DR Jr, Nakamura RM, Barbiera JM et al: Initial detection of human chorionic gonadotropin in serum in normal human gestation. *Am J Obstet Gynecol* 118:990, 1974.

Moszkowski E, Woodruff JD, Jones GS: The inadequate luteal phase. *Am J Obstet Gynecol* 83:363, 1962.

Necklaws EC, LaQuaglia MP, Maclaughlin D et al: Detection of Mullerian inhibiting substance in biological samples by a solid phase sandwich radioimmunoassay. *Endocrinology* 118:791, 1986.

Neill JD, Patton JM, Dailey RA et al: Luteinizing hormone releasing hormone (LHRH) in pituitary stalk blood of rhesus monkeys: Relationship to level of LH release. *Endocrinology* 101:430, 1977.

Nexo E, Hollenberg MD, Figuerva A et al: Detection of epidermal growth factor–urogastrone and its receptor during fetal mouse development. *Proc Natl Acad Sci USA* 77:2782, 1980.

Nippoldt TB, Reame NE, Kelch RP et al: The roles of estradiol and progesterone in decreasing luteinizing hormone pulse frequency in the luteal phase of the menstrual cycle. *J Clin Endocrinol Metab* 69:67, 1989.

Noyes RW, Hertig AT, Rock J: Dating the endometrial biopsy. *Fertil Steril* 1:3–25, 1950.

O'Gray JP, Kohorn EI, Glass RH et al: Inhibition of progesterone synthesis in vitro by prostaglandin $F_{2\ alpha}$. *J Reprod Fertil* 30:153, 1972.

Oliver AE, Atman TJ, Powel JF et al: Insulin like growth factor-I gene expression in the rat ovary is confined to granulosa cells of developing follicles. *Endocrinology* 124:2671, 1989.

Overstreet JW, Cooper GW: Reduced sperm motility in the isthmus of the rabbit ovuduct. *Nature* (*Lond*) 258:718–719, 1975.

Overstreet JW, Katz DF: Sperm transport and selection in the female genital tract, in Johnson MH (ed): *Development in Mammals* (vol 2). Amsterdam, Elsevier, 1977, pp 31–65.

Patek E: The epithelium of the human fallopian tube. *Acta Obstet Gynecol Scand* (53 Suppl) 31:1–28, 1974.

Patek E, Nilsson L, Johannisson E: Scanning electron microscopy of the human fallopian tube: Report I. The proliferative and secretory stages. *Fertil Steril* 23:459, 1972.

Pauerstein CJ, Eddy CA, Croxatto HD et al: Temporal relationship of estrogen, progesterone, luteinizing hormone levels to ovulation in women and infra human primates. *Am J Obstet Gynecol* 130:876, 1978.

Pincus G, Enzmann EV: The comparative behaviors of mammalian eggs in vivo and in vitro. *J Exp Med* 62:655, 1935.

Plant TM: Gonadal regulation of hypothalamic gonadotropin-releasing hormone release in primates. *Endocr Rev* 7(1):75, 1986.

Queenan JT Jr, Kao L-C, Arboleda CE et al: Regulation of urokinase type plasminogen activator production by cultured human cytotrophoblasts. *J Biol Chem* 262:10903–10906, 1987.

Ralt D, Goldenberg M, Fetterolf P et al: Sperm attraction to a follicular factor(s) correlates with human egg fertilizability. *Proc Natl Acad Sci USA* 88:2840–2844, 1991.

Rappolee DA, Brenner CA, Schultz R et al: Developmental expression of PDGF, TGF-alpha, and TGF-beta genes in preimplantation mouse embryos. *Science* 241:1823, 1988.

Rosenfeld RG, Lamson GL, Oh YM et al: Insulin-like growth factor binding proteins. *Recent Prog Horm Res* 46:99–159, 1990.

Ryan KJ, Smith OW: Biogenesis of steroid hormones in the human ovary. *Recent Prog Horm Res* 21:367–403, 1965.

Sachy A, Jukosowski A, Bernet F et al: Electromyographic activity and noradrenaline content of the rabbit oviduct under different hormonal states. *Reprod Nutr Dev* 29:171–183, 1989.

Santen RJ, Bardin CW: Episodic luteinizing hormone secretion in man. *J Clin Invest* 52:2617–2628, 1973.

Sathananthan AH, Trouson AO, Wood C et al: Ultrastructural observations on the penetration of human sperm into the zona pellucida of the human egg in vitro. *J Androl* 3:356, 1982.

Sayegh R, Mastroianni L Jr: Recent advances in our understanding of tubal function. *Ann NY Acad Sci* 626:266–275, 1991.

Schoonmaker JN, Victery W, Karsch F: A receptive period for estradiol-induced luteolysis in the rhesus monkey. *Endocrinology* 108:1874, 1981.

Settlage DSF, Motoshima M, Tredway DR: Sperm transport from the vagina to the fallopian tubes in women, in Hafez ESE, Thibault CG (eds): *The Biology of Spermatozoa*. Basel, Karger, 1975, pp 74–82.

Sherman BM, Korenman SG: Measurement of plasma LH, FSH, estradiol, and progesterone in disorders of the human menstrual cycle: The short luteal phase. *J Clin Endocrinol Metab* 38:89, 1974.

Sjoberg JT, Wahlstrom JT, Grudzinskas JG et al: Demonstration of pregnancy-associated plasma protein-A (PAPP-A) like material in the fallopian tube. *Fertil Steril* 46:517–521, 1986.

Soules M, Steiner R, Clefton D, Cohen D, Aksel S, Bremner WJ: Progesterone modulation of pulsatile luteinizing hormone secretion in normal women. *J Clin Endocrinol Metab* 58:378, 1984.

Stone SL, Pomerantz SH, Schwartz-Kripner A et al: Inhibition of oocyte maturation from porcine follicular fluid: Further purification and evidence for reversible action. *Biol Reprod* 19:585, 1978.

Strott CA, Cargille CM, Ross GT, Lipsett MB: The short luteal phase. *J Clin Endocrinol Metab* 30:246, 1970.

Suikkari A-M, Jalkanen J, Koistinen R et al: Human granulosa cells synthesize low molecular weight insulin-like growth factor binding protein. *Endocrinology* 124:1088–1090, 1989.

Tabibzadeh S: Human endometrium: An active site of cytokine production and action. *Endocr Rev* 12(3):272–290, 1991.

Takahashi M, Koide SS, Donahoe PK: Müllerian inhibiting substance as oocyle meiosis inhibitor. *Mol Cell Endocrinol* 47:225, 1986.

Talo A, Pulkkinen M: Electrical activity in the human oviduct during the menstrual cycle. *Am J Obstet Gynecol* 142:135–147, 1982.

Tamada H, McMaster MT, Flanders KC et al: Cell type-specific expression of TGF-beta in the mouse uterus during preimplantation period. *Mol Endocrinol* 4:965, 1990.

Thibault C: Sperm transport and storage in vertebrates. *J Reprod Fertil* (suppl) 18:39–53, 1973.

Tiitinen A, Wahlstrom T, Julkunen M et al: The content and immunohistochemical localization of placental 10 (PP10) in the fallopian tube. *Br J Obstet Gynecol* 93:924, 1986.

Tsafiri A, Channing CP: Inhibitory influence of granulosa cells and follicular fluid upon porcine oocyte meiosis in vitro. *Endocrinology* 96:922, 1975.

Tsafiri A, Pomerantz SH: Oocyte maturation inhibitor. *J Clin Endocrinol Metab* 15:157, 1986.

Ui M, Shimonaka M, Shimasaki S, Ling N: An insulin-like growth factor–binding protein in ovarian follicular fluid blocks follicle-stimulating hormone stimulated steroid production by ovarian granulosa cells. *Endocrinology* 125:912–916, 1989.

Vande Wiele RL, Bogumil J, Dyrenfurth I et al: Mechanisms regulating the menstrual cycle in women. *Recent Prog Horm Res* 26:63, 1970.

Van Vugt DA, Lam NY, Ferin M: Reduced frequency of pulsatile luteinizing hormone secretion in the luteal phase of the rhesus monkey: Involvement of endogenous opiates. *Endocrinology* 115:1095–1101, 1984.

Verhage HG, Bareither ML, Jaffe RC et al: Cyclic changes in ciliation, secretion, and cell height of the oviductal epithelium in women. *Am J Anat* 156:505, 1979.

Vigier, B, Picard J-Y, Tran D et al: Production of anti-Müllerian hormone: Another homology between Sertoli and granulosa cells. *Endocrinology* 114:1315, 1984.

Volpe A, Coukos G, Bareca A, Artini PG, Minute F, Giordano G, Genazzani Ar: Ovarian response to combined growth hormone–gonadotropin treatment in patients resistant to induction of superovulation. *Gynecol Endocrinol* 3(2):125–133, 1989.

Voutilainen R, Miller WL: Coordinate trophic hormone regulation of mRNAs for insulin-like growth factor II and the cholesterol side chain cleavage enzyme P450scc in human steroidogenic tissues. *Proc Natl Acad Sci USA* 84:1590, 1987.

Wassarman PM: The biology and chemistry of fertilization. *Science* 235:553, 1987.

Waugh PV, Lippes J: Human oviductal fluid proteins: III. Identification and partial purification. *Fertil Steril* 51:81–88, 1989.

Wentz AC: Physiologic and clinical considerations in luteal phase defects. *Clin Obstet Gynecol* 22:169, 1979.

Westman A: Transit of ovum from ovary to uterus in rabbits. *Acta Obstet Gynecol Scand* 5:7, 1926.

Wiemer KE, Cohen J, Amborski GF et al: In vitro development and implantation of human embryos following culture on fetal bovine uterine fibroblast cells. *Hum Reprod* 4:595, 1989.

Wilhemsson L, Lindblom B: Adrenergic responses of the various smooth muscle layers at the human utero tubal junction. *Fertil Steril* 33:280–282, 1980.

Witschi E: Migration of the germ cells of human embryos from the yolk sac to the primitive gonadal folds. *Contrib Embryol Carnegie Instit* 32:67, 1948.

Yanagimachi R, Noda YD: Ultrastructural changes in the hamster sperm head during fertilization. *J Ultrastruct Res* 31:465–485, 1970.

Ying S-H: Inhibins, activins and follistatins: Gonadal proteins modulating the secretion of follicle-stimulating hormone. *Endocr Rev* 9:267, 1988.

Yoshimi T, Strott C, Marshall J et al: Corpus luteum function in early pregnancy. *J Clin Endocrinol Metab* 29:225, 1969.

Yoshimura Y, Wallach EE: Studies on the mechanism(s) of mammalian ovulation. *Fertil Steril* 47:22, 1987.

Zimmermann RC, Westhof G, Hoedemaker M et al: Is follicular fluid steroid content representative of follicular steroid secretion? *Hum Reprod* 4:29–32, 1989.

CHAPTER 3

The Early Conceptus: Fertilization to Implantation

Koji Yoshinaga

FERTILIZATION

Maturation of Gametes

Gamete Maturation: Sex Differences

Mature germ cells (sperm and ova) are haploid, that is, with $1n$ DNA. Fertilization results in the formation of the zygote with $2n$ DNA, and subsequent cells are diploid, as are somatic cells. *Maturation* of the ovum is often used to mean *meiosis,* which is completed at the time of extrusion of the second polar body following sperm penetration. In the male the second meiotic division of germ cells results in production of spermatids in the testis. However, spermatozoa, the end products of transformation of spermatids (spermiogenesis), have to go through additional processes outside the testis to gain fertilizability. Thus the term *maturation* is used not only for the completion of meiosis, but also to imply, in the male, acquisition of fertilizability of the germ cells. The fertilizability of oocytes is acquired *before* completion of meiosis while that of sperm is acquired *after* completion of meiosis.

Other striking sex differences are the timing of initiation of meiotic division, the stage of arrest in cell division, and the end products of meiosis. In fetal life, germ cells of both sexes divide mitotically until the time of gonadal sex differentiation. Male germ cells enter a long resting stage with $2n$ DNA before the initiation of meiosis; female germ cells enter the early stages of meiosis in fetal life. They go through the initial stages of meiosis—the leptotene, zygotene, and pachytene—and reach the diplotene stage with $4n$ DNA in late prophase of the first meiotic phase. The oocytes then enter a long resting stage that lasts until the time of ovulation during adulthood.

In contrast to the early initiation of meiosis in the oocyte, the meiosis of spermatocytes is initiated around the time of puberty. For a detailed account of the embryology of mammalian gonads, see Byskov and Høyer (1988). In the female fetus, primordial germ cells, which originated in the yolk sac, migrate to reach the gonadal ridge around the third month of fetal life. These primordial germ cells penetrate the coelomic epithelium of the gonadal ridge and enter the underlying mesenchymal tissue. Cells from the mesonephros also migrate to this region. Initiation of meiosis of the migrated germ cells is closely related to the timing of the germ cells' association with the mesonephric cells. The mesonephric cells play a crucial role in the initiation of germ cell meiosis because meiosis is prevented if the mesonephros is separated from undifferentiated ovaries. The mesonephric cells invade the ovary to situate themselves in the medulla. Those oocytes located close to the medulla initiate their meiosis around 3 months of fetal age, but those located in the periphery of the ovary initiate meiosis about the eighth month. As the oocytes reach the diplotene (dictyate) stage, the germinal vesicle stage of development, they are surrounded by presumptive granulosa cells, which arise from mesonephric-derived rete cells. Oocyte meiosis appears to be controlled by paracrine mechanisms. Meiosis-inducing substance (MIS), meiosis-preventing substance (MPS), and meiosis-stimulating factor have been proposed as factors involved in the initial stages of meiosis. One should not confuse meiosis-inducing substance (with a molecular weight of less than 2000) with müllerian inhibiting substance (a homodimer of 72,000-dalton subunits) although both are abbreviated in the same way.

Recent studies have revealed the importance of maturation-promoting factor (MPF) in controlling meiosis and mitosis in general (see a concise review by Murray and Kirschner, 1991). MPF is a protein found in the cytoplasm; it activates cell division. MPF is composed of two proteins: cdc2 and cyclin. The name *cdc* is derived from *cell-division cycle*. Cdc2 is a protein kinase and thus phosphorylates proteins. Cyclin is a regulator of cdc2 activity: the greater the level of cyclin in the cell, the greater the activity of cdc2 and, therefore, of MPF. Cyclin regulates its own level by stimulating the activity of the enzyme that breaks down cyclin itself. The cyclic fluctuation of the cyclin level is closely related to the cycle of cell division. The breakdown of the germinal vesicle (nuclear envelope) of the oocyte is due to the phosphorylation of lamin, a major constituent of the nuclear envelope. The nuclear envelope is formed by dephosphorylation of lamin by phosphorylase. MPF appears to control all the cellular events observed in cell division.

The size of cells produced by meiotic division shows a clear sex difference. In the male, one large-sized primary spermatocyte ($2n$ chromosome, $2n$ DNA) divides once to produce two small-sized secondary spermatocytes ($1n$ chromosome, $2n$ DNA), which, in turn, each produce two equal-sized spermatids ($1n$ chromosome, $1n$ DNA). Thus, one primary spermatocyte produces four equal-sized spermatids. In the female, on the other hand, the first meiotic division results in one large oocyte and one tiny cell (the first polar body). The second meiotic division is triggered by sperm penetration into the oocyte cytoplasm. The presence of a second polar body thus signifies the occurrence of fertilization. The second polar body may or may not divide subsequently. Since the polar bodies contain the same genetic information as the oocyte, genetic defects may be detected by analyzing the polar body DNA multiplied by the polymerase chain reaction.

Maturation of Male Gametes

The seminiferous tubules of the testis are lined with epithelium (seminiferous epithelium) composed of spermatogonia and their evolving and maturing cells at various stages of their differentiation up to spermatozoa, and of Sertoli cells that nurture these germ cells. During spermatogenesis the germ cells gradually move from the basal portion of Sertoli cells to the apical portion, and finally, when morphogenesis is complete, spermatozoa are released from the apical surface of Sertoli cells into the lumen of the seminiferous tubule. The spermatozoa released from Sertoli cells are barely motile and still too immature for fertilization. The spermatozoa released into the tubal lumen are transported to the rete testis, to the ductuli efferentes, and to the caput portion of the epididymis.

Since a large proportion (more than 90 percent) of the fluid secreted by the testis is absorbed by the ductuli efferentes and the epididymis, the spermatozoa released into the lumen are considered to be carried away in the direction of the secretion flow. Seminiferous tubules in laboratory and domestic animals are surrounded by peritubular contractile cells (myoid cells). Although the contractility of the seminiferous tubules in these animals may serve the release and transport of spermatozoa from the site of production, it may not be the case in humans, as primate seminiferous tubules are not contractile.

Spermatozoa mature during their stay in the epididymis. This organ is a duct lined with pseudostratified epithelium composed of tall columnar cells and rounded basal cells. The surface of the epithelium is covered with characteristically long microvilli (stereocilia). The physiologic significance of stereocilia has not been clarified. The local environment provided by the epididymis has been considered to be important for sperm maturation. One of the environmental factors is a variety of (glyco)protein secretion. Synthesis and secretion of some of the proteins are androgen-dependent. The secreted proteins interact with spermatozoa as they migrate and mature in the epididymis. It remains to be clarified which of the proteins added on the sperm surface are associated with maturation of the sperm. Some of the sperm surface proteins may be important for functional maturation of sperm, because immunoneutralization of a specific protein has been reported to result in a loss of their fertilizability in experimental animals. The spermatozoa in the caudal portion of the epididymis are considered to have completed their maturation process. One obvious sign of maturation is their ability to move. After ejaculation, however, the spermatozoa are considered to gain "real" fertilizability, that is, they are *capacitated* in the female genital tract. The capacitation phenomenon has been clearly demonstrated in some species of experimental animals in which the time period required for the capacitation process is several hours. Human spermatozoa seem to require little time for capacitation.

Maturation of Female Gametes

For a discussion of the oocyte maturation, refer to the articles by Wassarman (1988) and Haseltine (1989). The initial stages of maturation of oocytes are completed during the fetal period, but the oocytes remain in the dictyotene stage, that is, the resting stage of the meiosis, until the follicle is stimulated to grow into a graafian follicle. The primordial follicles are found in clusters in the cortex of the ovary. These follicles contain a large oocyte surrounded by one layer of flat follicle (granulosa) cells. The oocyte of a primordial follicle initiates its growth by enlargement of the oocyte, and the surrounding follicle cells become cuboidal and multilayered. The growing oocyte in the primary follicle synthesizes a large amount of mRNA. The mRNA that encodes the zona pellucida protein is mainly in the growing oocyte, not in the surrounding granulosa cells, which suggests that the zona pellucida is secreted by the oocyte and laid down on its plasma membrane. The cellular junctions between the oocyte and surrounding granulosa cells are maintained despite a thick accumulation of the zona between them. The plasma membrane of the oocyte extends numerous microvilli into the zona; the granulosa cells also extend their irregularly shaped processes into the zona, and at times extend deep into the ooplasm. This communication between the oocyte and granulosa cells is important to maintain the oocyte at the dictyotene stage via adenylate cyclase activity. The granulosa cells make special junctional complexes (gap junctions) that allow passage of small molecules. It is well established that a certain level of cAMP must be available to maintain the meiotic arrest. Besides cAMP, hypoxanthine and other purines have been reported to play important roles in the meiotic arrest.

Steroids, prostaglandins, oocyte maturation inhibitor (OMI), and a host of growth factors, particularly epidermal growth factor (EGF), are considered to play roles in the regulation of meiotic arrest (see Racowsky, 1991). In a large preovulatory follicle, the nucleus of the oocyte moves to beneath the plasma membrane and the nuclear envelope breaks down. This is followed by appearance of fine spindle fibers (microfilaments), and the first meiotic division takes place. This cell division is asymmetric and produces the oocyte and the first polar body. After ovulation the stimulation associated with fertilization triggers the expulsion of the second polar body, namely, the second meiotic division.

The growing follicle (the primary follicle) becomes the secondary follicle when it accumulates follicular fluid (liquor folliculi) in the antrum, the space surrounded by granulosa cells. Under the influence of pituitary gonadotropins [follicle-stimulating hormone (FSH) and luteinizing hormone (LH)], selected follicles grow and the antrum enlarges to occupy the major portion of the follicle (graafian follicle). Hormonal interplay in the hypothalamo–pituitary–ovarian axis leading to the preovulatory surge of gonadotropins results in ovulation. Finding inhibin, activin, follistatin, and growth factors in follicular fluid and neuropeptides in the brain implies existence of a regulatory mechanism much more complicated than a simple ovarian steroid–gonadotropin feedback system for the preovulatory gonadotropin surge (Knobil and Hotchkiss, 1988).

Encounter of the Gametes

Soon after ovulation the oocyte is covered with cumulus cells. These cumulus cells are joined together with amorphous matrix that is rich in hyaluronic acid. The cumulus mass containing the oocyte is picked up by the fimbria of the fallopian tube, and with the aid of cilial and muscular movements the mass is transported to the ampulla portion of the tube. When the cumulus mass is removed from the tube and treated with hyaluronidase, the cumulus cells are readily dispersed to expose the oocyte. Thus it appears that penetration of spermatozoa into the cumulus mass and later into the oocyte is facilitated by enzymes released from the acrosome. The acrosome is a caplike structure covering the anterior portion of the sperm head; it contains hydrolytic enzymes: acrosin, hyaluronidase, and many others. The ampulla of the fallopian tube (oviduct) is the site of fertilization. Spermatozoa are considered to reach the ampulla by their own movement aided by contraction of the female genital tract muscles.

Penetration of Spermatozoa through the Zona Pellucida

Detailed accounts of the process of fertilization are found in Yanagimachi (1988) and Austin (1972). The initial step of fertilization is attachment of spermatozoa to the zona pellucida of the oocyte. This attachment is a cell–cell adhesion process that involves species-specific sperm head protein(s) and its (their) receptor protein (ZP3) in the zona pellucida. Of the three major proteins (ZP1, ZP2, and ZP3) found in the zona pellucida, ZP1 is considered to serve as a framework component of the zona, and ZP3 is the primary sperm receptor, ZP2 being the secondary receptor that binds to spermatozoa after the acrosome reaction (see Wassarman, 1987). Since the sperm head protein is located on the acrosome membrane as well as on the sperm plasma membrane, a spermatozoon attaches to the zona pellucida irrespective of whether it is acrosome-reacted or acrosome-intact.

When an oocyte is fertilized, the entry of other sperm into the oocyte is inhibited at two separate levels: the zona pellucida and the oocyte plasma membrane. In human fertilization, prevention of entry of other sperm appears to depend mainly on the change in the zona pellucida (the zona reaction). In lower animals, especially in invertebrate animals, polyspermy is blocked by an alteration of the nature of the oocyte plasma membrane due to the release of cortical granules located just under the oocyte membrane. How the entry of the first spermatozoon induces a block to polyspermy is not clearly understood. It may be mediated, at least partly, by the release of contents of the cortical granules such as proteinase. Cortical granules are indeed observed in unfertilized mammalian oocytes, but they disappear after fertilization. Enzymes released from the granules may be considered to convert crucial components of the zona pellucida, that is, ZP2 and ZP3, in such way that they no longer serve as sperm receptors.

When the blocking mechanism does not work, polyspermy takes place. Polyspermy is an abnormal fertilization in which two or occasionally more spermatozoa enter the oocyte and the multiple spermatozoa take part in fertilization. These two spermatozoa become two male pronuclei that become enlarged. They are, however, much smaller than those observed in normal fertilization. The two male and one female pronuclei cluster, shrink, and assemble together. The zygote thus formed has three sets ($3n$) of chromosomes, instead of the regular diploid set ($2n$), leading to a triploid embryo. Triploid embryos may continue to develop, but they fail either to implant into the uterine endometrium or to continue their development through midgestation.

The species-specific ligand–receptor interaction between the components of the zona pellucida and the surface proteins of spermatozoa is a useful mechanism to avoid interspecific fertilization. The use of zona-free hamster eggs to test the fertilizability of human sperm is based on the idea of eliminating this barrier. Recently sperm microinjection, the so-called zona drilling technique, has been used to increase the success rate of fertilization in vitro. Although a recent report indicates that sperm microinjection did not increase the incidence of chromosomally abnormal eggs using a small number of eggs (18 eggs fertilized by microinjection and 30 with conventional in vitro fertilization), further studies are needed to ascertain that the avoidance of this natural selection barrier in fertilization does indeed not result in any defective outcome.

Entry of Spermatozoon into the Oocyte

Entrance of a spermatozoon into an oocyte is initiated by apposition of the sperm head to the plasma membrane of the oocyte in such a way that the long axis of the head is in parallel with the oocyte surface. Attachment of the sperm head with the oocyte is made at the equatorial segment of the sperm head plasma membrane; then the membranes fuse. The membrane fusion results in formation of a continuous membrane that includes the sperm head within the oocyte cytoplasm. As soon as the sperm head is exposed to the oocyte cytoplasm, the head swells up (decondenses) and transforms into the male pronucleus. The decondensed male pronucleus expands further coincidentally with the formation of female pronucleus, prior to which the oocyte nucleus was arrested at the metaphase of the second meiotic division. With the development of the pronuclei, duplication of chromosomes proceeds. Fully developed male and female pronuclei come together near the center of the oocyte. The envelopes of the pronuclei break down and chromosomes mingle, signifying the completion of fertilization.

DEVELOPMENT OF PREIMPLANTATION EMBRYO

From Zygote to Blastocyst

The mingling of chromosomes from the male and the female pronuclei is the beginning of the first mitotic division of the zygote. The term *cleavage* is used to imply rapid cell divisions of the fertilized egg before cell differentiation begins. A zygote divides to produce a preimplantation embryo at the 2-cell stage, which, in turn, divides again, and, at about the 16-cell stage the embryo becomes a morula. At this stage the embryo enters the uterus, and formation of the blastocoele cavity with concomitant fluid accumulation results in a blastocyst. For detailed accounts of mechanisms involved in the development of preimplantation embryos refer to recent publications on this topic (Heyner and Wiley, 1990; Kimber, 1990; Yoshinaga and Mori, 1989; Pedersen, 1988). One of the noticeable events in early development is transition (from maternal to embryonic) of protein synthesis control. In the mouse, zygote protein synthesis is mainly dependent on the maternal mRNAs. At the 2-cell stage, however, protein synthesis depends on embryonic mRNAs. In comparison with this early transition in the mouse, for most species studied, including the human, the transition is observed at later stages and protein synthesis becomes dependent on embryonic transcription by the 8- to 16-cell stage.

Blastomeres of the human embryo at the 8-cell stage have not yet committed to differentiate into particular tissues or organs. A recent report (Hardy et al, 1990) indicates that removal of one or two cells at the 8-cell stage, while reducing the cellular mass, does not adversely affect the preimplantation development of biopsied embryos in vitro. The cells that are removed at the 8-cell stage can be used for diagnosis of genetic defects.

Requirement for Preimplantation Development

When fertilized eggs are cultured in vitro, their development is blocked at certain stages of development. The stage at which the development is blocked is species-specific. Although the cause of the block is not clear, it may be due to lack of factor(s) essential for development and/or to substances harmful in the culture medium. In vitro blocks, however, may be overcome by coculturing embryos with oviduct tissues. Media conditioned by oviductal tissue also facilitate overcoming the developmental block. Thus, the oviduct appears to provide substance(s) essential for embryonic development, particularly at certain stages. It has been suggested that ovarian hormone-dependent and stage-specific proteins secreted by the oviduct may play beneficial roles in embryonic development because they bind to the zona pellucidae of eggs in the oviduct, fertilized eggs, and embryos in primates, but not to ovarian oocytes. Further elucidation of the proteins is awaited.

It has been shown that early stages of embryonic development in vitro coincide with the onset of embryonic transcription and an abrupt increase in cell cycle length (Eyestone and First, 1990). As described in the section "Gamete Maturation" in this chapter, MPF plays an important role in controlling the cell cycle. Embryonic block in culture may be associated with abnormal conditions tampering with appropriate operation of MPF.

Initial Differentiation

Rodent embryos exhibit a peculiar behavior at the 8-cell stage. The blastomeres polarize and flatten on each other to maximize intercellular contact (compaction). This process is calcium-dependent, and the blastomere contact is so close that it becomes difficult to distinguish the outline of each blastomere. Adhesion of blastomeres depends mainly on a cell surface glycoprotein, uvomorulin. The apical region of each blastomere differs from the basolateral region in that it is rich in microfilaments, microtubules, pericentriolar material, and many associated organelles (Johnson, 1989). One hypothesis of the first tissue differentiation in the embryo is based on these regional differences in blastomere cytoplasm. The blastomere cytoplasm is divided unequally in the daughter cells: the cells that contain the cytocortical cytoplasm are polarized and become trophoblast cells, and those cells that do not contain the cytocortical cytoplasm do not polarize, and they become inner cell mass cells. Another hypothesis is based on the site or position of a blastomere within the early embryo; at the 10- to 16-cell stage embryo, a small number of blastomeres are completely surrounded by other blastomeres, and consequently these cells have no exposed outer surfaces and become inner cell mass cells. Human embryos do not seem to exhibit compaction clearly enough to be recognized, as rodent embryos do.

As tissue differentiation takes place a fluid-accumulating cavity forms. The formation involves active fluid transport across the trophectoderm. The sodium pump (Na^+, K^+-ATPase) plays an important role in the fluid transport, and the timing of blastocoele formation coincides with expression of the gene encoding this enzyme.

Following cavity formation, the blastocyst shows a number of differentiations (maturation) before implantation. According to Enders (1989) morphologic maturation of the blastocyst includes maturation of cellular organelles:

diminution of storage materials, an increase in the protein synthetic apparatus, and mitochondrial changes. Trophoblast cell junctions change from primitive ones between the blastomeres to apical junctions, trophoblast cells increase endocytotic activity, and the first appearance of basal lamina of the organism is formed under the trophectoderm, and also within the inner cell mass. Endoderm formation takes place prior to implantation in those mammals whose implantation occurs later than 5 days after ovulation. These morphologic manifestations of blastocyst maturation are considered to be regulated endogenously by autocrine–paracrine mechanisms. However, uterine factors may also participate in regulating maturational and functional changes of the blastocyst. When a mouse preimplantation embryo is cultured by itself, the embryonic development is poor. However, 5 to 10 embryos cultured together developed better than a single embryo in culture. Addition of EGF or transforming growth factors (TGFs) also improved the development of embryos into the blastocyst stage (Paria and Dey, 1990). Cultured mouse blastocysts produce growth factors that exhibit the *biological* properties of fibroblast growth factor–like (FGF-like) growth factor (Rizzino et al, 1990). These results suggest the possibility of interaction between the blastocyst and the uterus using paracrine mechanisms, because EGF is produced by the endometrial epithelium under the influence of ovarian hormones and because blastocysts have EGF receptors.

Prior to implantation into the uterine endometrium, the mature blastocyst "hatches" out of the zona pellucida. Hatching can be observed in vitro. Empty zonae pellucidae together with hatched blastocysts may be occasionally observed in vivo in rodent uteri prior to implantation. Although the precise mechanism of the zona hatching is not clear, involvement of prostaglandins and enzymatic digestion by trophoblastic cells have been suggested.

IMPLANTATION

Implantation Processes

Human blastocyst implantation takes place during its seventh day of development (Hertig, 1975). Implantation usually takes place at the fundus of the uterus. The process of implantation may be divided into three steps: apposition, adhesion, and invasion. In the youngest-known human blastocyst specimen, the blastocyst has penetrated the uterine epithelium. The epithelial tissue has disappeared at the site of blastocyst attachment. This site of invasion has been closed by a coagulum of fibrin. Differentiation of cytotrophoblasts into a syncytium is observed (Larsen, 1980). There is a gap in our information between the time of adhesion of the blastocyst to the epithelium and this stage. By extrapolation of the results of studies on rhesus monkeys and other laboratory animals, the following process may be considered to take place: the blastocyst adheres on the epithelial surface, probably with stage-specific and hormone-dependent glycoproteins. The microvilli of the trophoblast cells and those of the uterine epithelial cells interdigitate, and desmosomes are formed between them. Trophoblast cells penetrate between the columnar epithelial cells, which degenerate and are then phagocytosed by the trophoblastic cells. Fusion of trophoblast cells with the uterine epithelial cells (symplasm) has been reported in the sheep, but this has not been observed in the human. Penetration of the basal lamina by trophoblast cells takes some time. If the hypothesis for cancer metastasis proposed by Liotta (1986) is applied to the invasion of the uterine epithelium by trophoblast cells, the process of the basal lamina penetration would involve (1) anchoring of the trophoblast cells by means of the ligand–receptor binding, that is, laminin receptors on the trophoblast membrane bind laminin molecules in the basal lamina, (2) secretion of hydrolytic enzymes such as collagenase IV by the trophoblast cells to break the lamina fibers, and (3) invasion into the uterine stroma. The basal lamina is broken at many places by club-shaped processes; then trophoblast cells invade into the extracellular matrix of the highly edematous endometrium. Stromal cells have decidualized by this time, but trophoblast cells do not appear to "attack" maternal cells (Larsen, 1980). Blocking of oxygen supply caused by invading cells is considered to result in degeneration of maternal cells. By the 11th or 12th day, the human blastocyst is completely embedded in the stromal tissue and the uterine epithelium regenerates to cover the blastocyst. Capillaries adjacent to the trophoblast become congested and become intercommunicating sinusoids. Invading syncytia tap these sinusoids, and maternal blood flows into the developing trophoblastic lacunae. When the trophoblastic invasion reaches the stage at which trophoblastic lacunae are in direct communication with arterioles and venules, the difference in blood pressure induces directional flow of maternal blood, establishing the interlacunar circulation and leading to formation of the future placenta (Hamilton et al, 1972).

The process of implantation outlined above is a presumptive description of human implantation. One has to recognize diverse species variations in this process. If results obtained in experimental animals are to be extrapolated to human implantation, one should choose an animal model based on the similarity of one or a few aspect(s) of the entire process between the experimental animal and humans. Major species differences to be considered include (1) the relation of the implanted blastocyst to the uterine lumen (central, eccentric, or interstitial implantation), (2) the developmental stage of the blastocyst at the time of implantation (whether or not the endoderm and other tissues are formed), (3) the placental type—the relation between the fetal and maternal blood (epitheliochorial, endotheliochorial, or hemochorial), (4) the faculty of the blastocyst to undergo a dormant state (delayed implantation), (5) the hormonal requirement for uterine preparation for implantation, and (6) the uterine tissue response to the implanting blastocyst (epithelial plaque formation, decidualization of stromal cells, etc.). The reader is referred to the literature dealing with the comparative aspect of implantation (morphology: Boyd and Hamilton, 1961; endocrinology: Yoshinaga, 1977).

Embryonic Signals

The postovulatory increase in progesterone secretion is essential for the establishment of pregnancy. The presence of the developing embryo in the genital tract subsequent to mating alters the maternal endocrine function from the cyclic pattern to the pregnancy pattern characterized by prolonged luteal function. In the human, human chorionic gonadotropin (hCG) secreted by trophoblast cells "rescues" the corpus luteum destined to degenerate, and thus maintains high levels of progesterone secretion. Because human embryos cultured in vitro have been reported to secrete hCG (Fishel et al, 1984), it is possible that the blastocyst starts secreting hCG prior to implantation. Measurable hCG is detectable in circulating blood soon after implantation. The molecular weight of hCG is much greater than that of LH because of a large carbohydrate moiety in the hCG molecule. When hCG binds the LH/CG receptors on the luteal membrane, the hCG molecule remains bound much longer than the LH molecule. These gonadotropins are presumed to stimulate progesterone secretion by stimulating adenylate cyclase via a G-protein-mediated signal transduction system in luteal membranes.

Another type of embryonic signal that leads to maintenance of luteal function has been extensively studied in domestic animals, particularly in the sheep. Attachment of the sheep conceptus to the uterine wall takes place around 16 days after fertilization. During preimplantation development between 13 and 20 days after fertilization, the blastocyst secretes a specific protein, oTP-1, an interferon of the alpha family. This protein binds its receptors on the uterine epithelium and attenuates secretion of prostaglandin $F_{2\alpha}$, which is a potent luteolytic agent in the sheep. Thus the embryonic signal in the form of interferon acts through a paracrine mechanism to prevent decline of progesterone secretion by attenuation of luteolytic secretion by the uterus, but not by direct stimulation of luteal cells in this species of animals. When sheep embryos are removed from the uterus before the secretion of oTP-1, luteal function is not maintained and the next cycle ensues (Roberts, 1989).

The antiviral activity and immunologic implication of interferons and the presence of interferons in the human placenta make one wonder if there may be other types of embryonic signal besides hCG that affect the maternal system to secure the lease on its symbiotic existence in the uterus for the duration of pregnancy. It is conceivable that an early conceptus is capable of secreting a host of factors, and, in the human, the secretion of hCG is one of the factors that stimulates progesterone secretion. It would not be surprising if other factors secreted by the human blastocyst are found to have antiviral or immunosuppressive activities such as oTP-1 in sheep. The mechanism of blastocyst implantation continues to demand the exploration of many problems of endocrinologic, immunologic, cellular, and molecular bases.

Uterine Receptivity

The uterus is usually not a very accommodating organ. In the human, the uterus accepts a blastocyst to implant on about the seventh day after ovulation, and this receptive period may last for 4 days. The uterus rejects embryos transferred to it during the nonreceptive period of the cycle. The receptive period is often referred to as an *implantation window*. Although the hormonal requirements for rendering the uterus receptive or refractory to blastocysts are well documented in laboratory animals, particularly in the rat (Psychoyos, 1973; Yoshinaga, 1988), there is little information available about the biochemical, cellular, and molecular requirements for establishing uterine receptivity. In order to improve the success of embryo transfer, much research is urgently needed.

REFERENCES

Austin CR: Fertilization, in Austin CR, Short RV (eds): *Reproduction in Mammals* (book 1: *Germ Cells and Fertilization*). Cambridge, Cambridge University Press, 1972, pp 103–133.

Boyd JD, Hamilton WJ: Early development and implantation of the egg, in Parks AS (ed): *Marshall's Physiology of Reproduction* (vol 2). London, Longmans, Green, 1961, chap 14, pp 1–126.

Byskov AG, Høyer PE: Embryology of mammalian gonads and ducts, in Knobil E, Neill JD et al (eds): *The Physiology of Reproduction* (vol 1). New York, Raven Press, 1988, pp 265–302.

Enders AC: Morphological manifestations of maturation of the blastocyst, in Yoshinaga K, Mori T (eds): *Development of Preimplantation Embryos and Their Environment.* New York, Alan R Liss, 1989, pp 211–223.

Eyestone WH, First NL: Bovine oocyte maturation and embryo development, in Hegner S, Wiley LM (eds): *Early Embryo Development and Paracrine Relationships* (vol 117: *UCLA Symposium on Molecular and Cellular Biology. New Series*). New York, Wiley-Liss, 1990, pp 1–10.

Fishel SB, Edwards RG, Evans CJ: Human chorionic gonadotropin secreted by preimplantation embryos cultured in vitro. *Science* 223:816–818, 1984.

Hamilton WJ, Boyd JD, Mossman HW: The implantation of the blastocyst and the development of the fetal membranes, placenta and decidua, in *Human Embryology.* Cambridge, Heffer, 1972, pp 83–131.

Hardy K, Martin KL, Leese HJ, Winston RML, Handyside AH: Human preimplantation development in vitro is not adversely affected by biopsy at the 8-cell stage. *Hum Reprod* 5:708–714, 1990.

Haseltine FP: New concepts for old problems in meiosis, in Yoshinaga K, Mori T (eds): *Development of Preimplantation Embryos and Their Environment.* New York, Alan R Liss, 1989, pp 3–19.

Hertig AT: Implantation of the human ovum: The histogenesis of some aspects of spontaneous abortion, in Behrman SJ, Kistner RW (eds): *Progress in Infertility* (2d ed). Boston, Little, Brown, 1975, pp 411–438.

Heyner S, Wiley LM (eds.): *Early Embryo Development and Paracrine Relationships* (vol 117: *UCLA Symposia on Molecular and Cellular Biology. New Series*). New York, Wiley-Liss, 1990.

Johnson MH: How are two lineages established in early mouse development? in Yoshinaga K, Mori T (eds): *Development of Preimplantation Embryos and Their Environment.* New York, Alan R Liss, 1989, pp 189–198.

Kimber SJ: Glycoconjugates and cell surface interactions in pre- and peri-implantation mammalian embryonic development. *Int Rev Cytol* 120:53–167, 1990.

Knobil E, Hotchkiss J: The menstrual cycle and its neuroendocrine control, in Knobil E, Neill JD et al (eds): *The Physiology of Reproduction* (vol 2). New York, Raven Press, 1988, pp 1971–1994.

Larsen JF: Human implantation and clinical aspects. *Prog Reprod Biol* 7:284–296, 1980.

Liotta L: Tumor invasion and metastases—Role of the extracellular matrix: Rhodes Memorial Award Lecture. *Cancer Res* 46:1–7, 1986.

Murray, AW, Kirschner MW: What controls the cell cycle? *Sci Am* 264(3):56–63, 1991.

Paria BC, Dey SK: Preimplantation embryo development in vitro: Cooperative interactions among embryos and role of growth factors. *Proc Natl Acad Sci USA* 87:4756–4760, 1990.

Pedersen RA: Early mammalian embryology, in Knobil E, Neill JD et al (eds): *The Physiology of Reproduction* (vol 1). New York, Raven Press, 1988, pp 187–230.

Psychoyos A: Endocrine control of egg implantation, in Greep RO, Astwood EB (eds): *Handbook of Physiology* (sec 7: *Endocrinology;* vol 2 Part 2). American Physiological Society, Washington, DC, 1973, pp 187–215.

Racowsky C: Gamete resources: Origin and production of oocytes, in *Animal Applications of Research in Mammalian Development. Current Communications in Cell and Molecular Biology* (vol 4). Cold Spring Harbor, NY, Cold Spring Harbor Laboratory Press, 1991, pp 23–82.

Rizzino A, Tiesman J, Hines R, Kelly D: Developmental regulation of the KS-FGF oncogene by embryonal carcinoma cells and early mouse embryos, in Heyner S, Wiley

LM (eds): *Early Embryo Development and Paracrine Relationships.* New York, Wiley-Liss, 1990, pp 53–66.

Roberts RM: Minireview. Conceptus interferons and maternal recognition of pregnancy. *Biol Reprod* 40:449–452, 1989.

Wassarman PM: The biology and chemistry of fertilization. *Science* 235:553–554, 1987.

Wassarman PM: The mammalian ovum, in Knobil E, Neill JD et al (eds): *The Physiology of Reproduction* (vol 1). New York, Raven Press, 1988, pp 69–102.

Yanagimachi R: Mammalian fertilization, in Knobil E, Neill JD (eds): *The Physiology of Reproduction* (vol 1). New York, Raven Press, 1988, pp 135–185.

Yoshinaga K: Hormonal interplay in the establishment of pregnancy, in Greep RO (ed): *International Review of Physiology. Reproductive Physiology II* (vol 13). Baltimore, University Park Press, 1977, pp 201–223.

Yoshinaga K: Uterine receptivity for blastocyst implantation. *Ann NY Acad Sci* 541:424–431, 1988.

Yoshinaga K, Mori T (eds): *Development of Preimplantation Embryos and Their Environment.* New York, Alan R Liss, 1989.

PART

II

MECHANISMS OF REPRODUCTIVE FAILURE

CHAPTER 4

The Testis, Spermatogenesis, and the Excurrent Duct System

Peter K. Working and Gary J. Chellman

The reproductive process in males is dependent on a complex series of biological interactions, involving multiple organs, multiple cell types, and multiple types of molecules, as well as precise temporal and spatial coordination of events. Not surprisingly, this intricate biological system is readily vulnerable to derangement by environmental factors, both physical and chemical. Previous chapters in this volume have presented an introduction to reproductive physiology and discussed mechanisms of reproductive failure that act via effects on reproductive hormones in the male. In this chapter, we will discuss primarily mechanisms of reproductive failure mediated via direct cytotoxic effects on the testis, spermatogenesis, the epididymis, the excurrent duct system, and the accessory sex glands.

The clinician's primary concern is, of course, human reproductive failure. The male rat, however, is the species most completely described in terms of its basic reproductive physiology and its general pharmacologic and toxicologic response to xenobiotic exposure. Although there are anatomic and physiologic differences between rats and humans that may affect their individual responses to toxic substances, the use of a higher primate as a model for human response would not necessarily prove more salutary. This is because testicular anatomy and spermatogenesis are no more similar among higher primates than between rats and men (for review, see Amann, 1986; Working, 1988). Accordingly, in this presentation we will focus on the activity of several well-known reproductive toxicants in the male rat, with particular emphasis on the cellular sites of action of each within the reproductive organs. As will be seen, reproductive toxicants often seem to specifically target particular testicular cell types or individual reproductive organs (Table 4-1). However, this apparent specificity may be dependent on both the dose of chemical and the duration of treatment. A chemical that affects only one cell

Table 4-1. Targets in the male reproductive tract

Target	Chemical	Reference
Testis		
Spermatogonia	Busulfan	Jackson et al (1962)
	Procarbazine	Meistrich et al (1982) Gould et al (1983) Ward et al (1989)
Spermatocytes	2-Methoxyethanol	Foster et al (1983) Chapin et al (1984a)
	Procarbazine	Glode et al (1982) Delic et al (1986)
Spermatids	Methyl chloride	Chapin et al (1984b) Working et al (1985a)
Sertoli cells	1,3-Dinitrobenzene	Blackburn et al (1988)
	2,5-Hexanedione	Chapin et al (1983)
Leydig cells	Ethane dimethanesulfonate	Jackson and Morris (1977) Kerr et al (1987)
Epididymis	α-Chlorohydrin	Ericsson (1970)
	Ethane dimethanesulfonate	Klinefelter et al (1990)
	Methyl chloride	Chapin et al (1984b) Chellman et al (1986a)
	Sulfasalazine	Wong et al (1987)
Accessory sex glands	Imidazoles	Vickery et al (1985, 1986)

type or organ under one treatment regimen will frequently be toxic to more than one stage of spermatogenic cell (e.g., procarbazine) or to more than one reproductive organ (e.g., methyl chloride) if the dose is increased or the treatment duration lengthened.

TESTICULAR ANATOMY AND SPERMATOGENESIS

The testicular parenchyma comprises two functional compartments, the seminiferous epithelium and the interstitial tissue. The seminiferous epithelium is formed into long convoluted tubules connected at boths ends to the beginning of the excurrent duct system, the rete testis (Fig. 4-1). Within the tubules are Sertoli cells, which are somatic cells that extend from the basement membrane of the tubule to its lumen, and germ cells of various stages of development. The Sertoli cells surround and support the germ cells (for review, see Chap. 1 and Russell, 1980), and tight junctions between the Sertoli cells form the blood–testis barrier. This anatomic barrier effectively

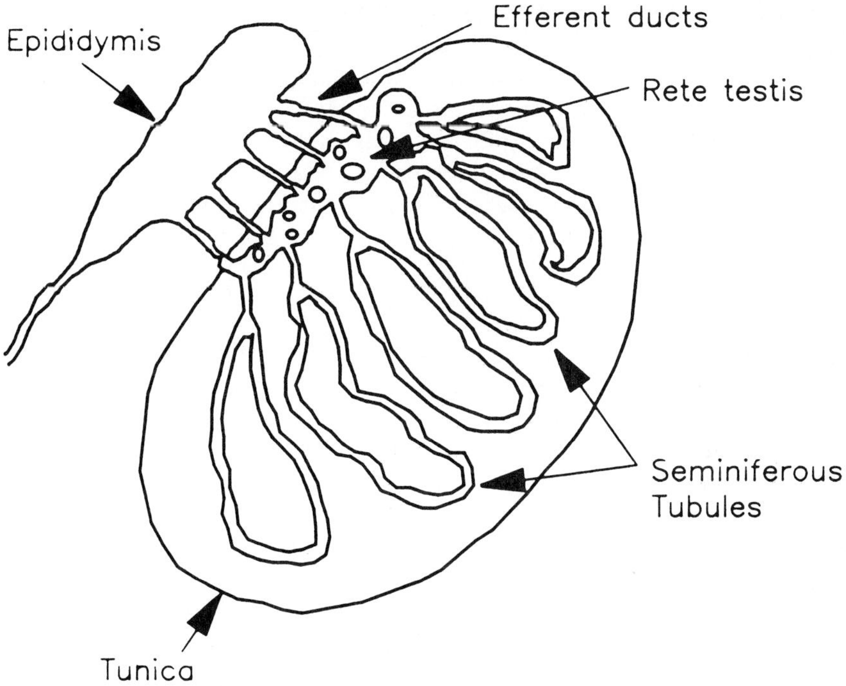

Figure 4-1. Rat testis, showing seminiferous tubules, efferent ducts and the caput epididymis. [*Reprinted from Foster (1988), with permission.*]

divides the tubules into basal and adluminal compartments that have differential exposure to bloodborne xenobiotics. In addition, Sertoli cells fulfill a number of other important functions, including secretion of fluid into the seminiferous tubule, coordination of the movement of germ cells from the periphery of the tubule to its lumen, spermiation (the release of mature spermatids at the end of spermatogenesis), and the secretion of a number of important proteins, including androgen-binding protein (Fritz et al, 1976), transferrin (Skinner and Griswold, 1980), and inhibin (Verhoeven and Franchimont, 1983).

The highly vascularized tissue of the interstitial compartment contains the peritubular cells, which are located at the seminiferous tubule periphery, macrophages, smooth muscle cells, and Leydig cells, the primary source of testosterone in the testis (see Fawcett et al, 1973; Connell and Connell, 1977; Ewing, 1983). Toxicants that target the interstitial cells may indirectly affect spermatogenesis via their disturbance of the homeostatic mechanisms of the testis.

Spermatogenesis is the gradual process by which spermatogonial cells divide and differentiate into spermatozoa. The continued production of spermatogenic cells is dependent on the existence of stem cell spermatogonia, which have been identified as isolated, undifferentiated A-type spermatogonia (Huckins, 1971). Beginning with the stem cells, the spermatogonia undergo a series of mitotic divisions (six in the rat), the final division yielding preleptotene primary spermatocytes and initiating meiosis. After a prolonged prophase, each tetraploid primary spermatocyte divides to two secondary spermatocytes, which quickly undergo the final meiotic division to yield a total of four haploid spermatids. These spermatids divide no further and, round at formation, are subsequently transformed into the typical elongated shape of mammalian sperm in a process known as spermiogenesis.

Each step of spermatogenesis is precisely timed, and complex associations of cells mature in synchrony (Clermont, 1972). As the cycle of the seminiferous epithelium progresses, older cells continually become associated with younger ones, so that in any given area of the tubule, five or six generations of cells appear together, forming cell associations of definite and fixed composition (Leblond and Clermont, 1952; Hess, 1990). Specifically, one or two generations of spermatids are always associated with one or two generations of spermatocytes, as well as with spermatogonia, each in a particular developmental stage. The cycle of the seminiferous epithelium in the rat has been divided into 14 specific stages (Leblond and Clermont, 1952), each consisting of a specific mixture of cells (Fig. 4-2). The duration of spermatogenesis corresponds to the time required for a given stem cell to produce mature spermatozoa (released into the excurrent duct system), which, in the rat, takes about 60 days.

The seminiferous tubules are folded and refolded within the testis; thus, a cross-section of rat testis reveals several hundred tubules, each identifiable as one of the 14 different spermatogenic stages. Classification of the tubule is based primarily on the morphologic appearance and location of the spermatids within the seminiferous epithelium (Leblond and Clermont, 1952).

EXCURRENT DUCTS, EPIDIDYMIS, AND ACCESSORY SEX GLANDS

After completion of the spermiogenesis phase of spermatogenesis, spermatids move out of the testis into an extensive duct system that is critical to their final maturation into spermatozoa and to their acquisition of fertilizing capacity. This duct system consists of the efferent ducts (ductuli efferentes), epididymis, ductus deferens, and ejaculatory duct (Fig. 4-3). Associated with this duct system are the accessory sex glands, which in humans consist of the seminal vesicles, prostate, and bulbourethral (Cowper's) glands. In rodents, coagulating glands and preputial glands are also present. The duct system and most accessory sex glands in humans and laboratory rodents (rats and mice) are quite similar in structure and also have many common functional properties (Eddy, 1988).

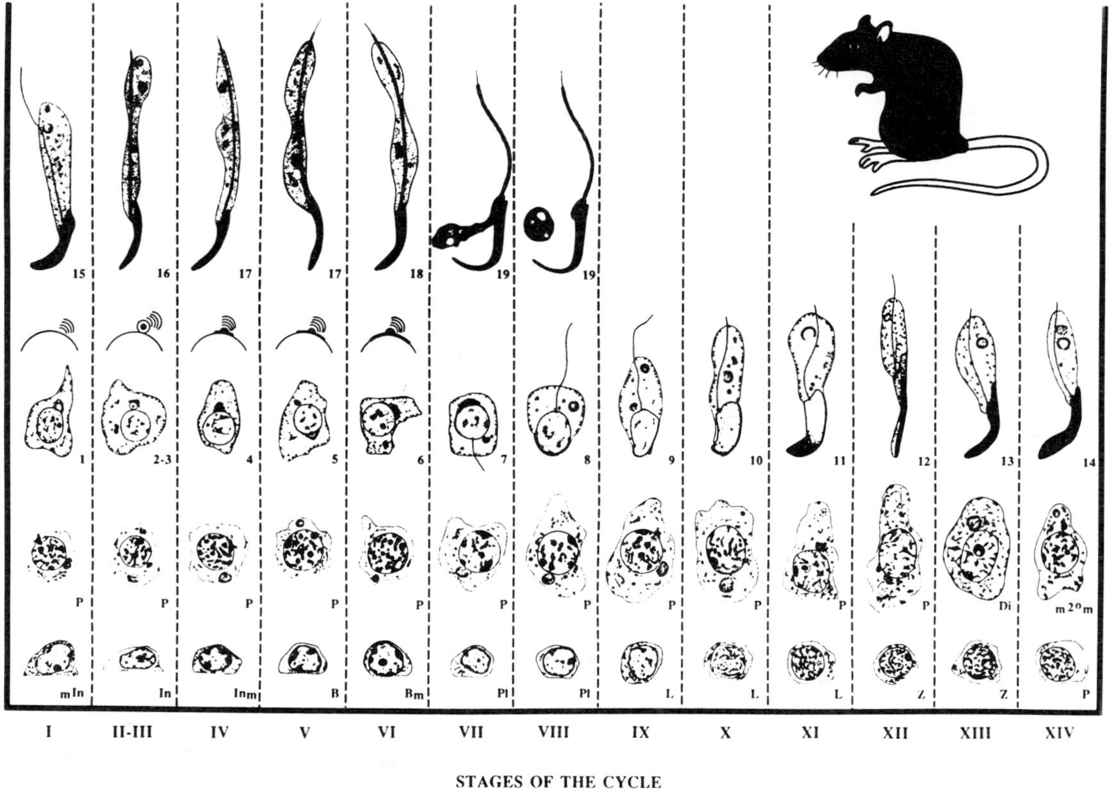

Figure 4-2. Spermatogenesis in the rat. The vertical columns (Roman numerals) represent cell stages or associations, according to Leblond and Clermont (1952). Abbreviations: In, intermediate spermatogonium; B, type B spermatogonium; Pl, preleptotene primary spermatocyte; L, leptotene primary spermatocyte; Z, zygotene primary spermatocyte; P, pachytene primary spermatocyte; Di, diplotene primary spermatocyte; 2°, secondary spermatocyte. Arabic numbers refer to steps of spermiogenesis. The subscript "m" indicates that a mitotic or meiotic division occurs prior to or after the indicated cellular stage. [*Reprinted from Russell et al. (1990), with permission.*]

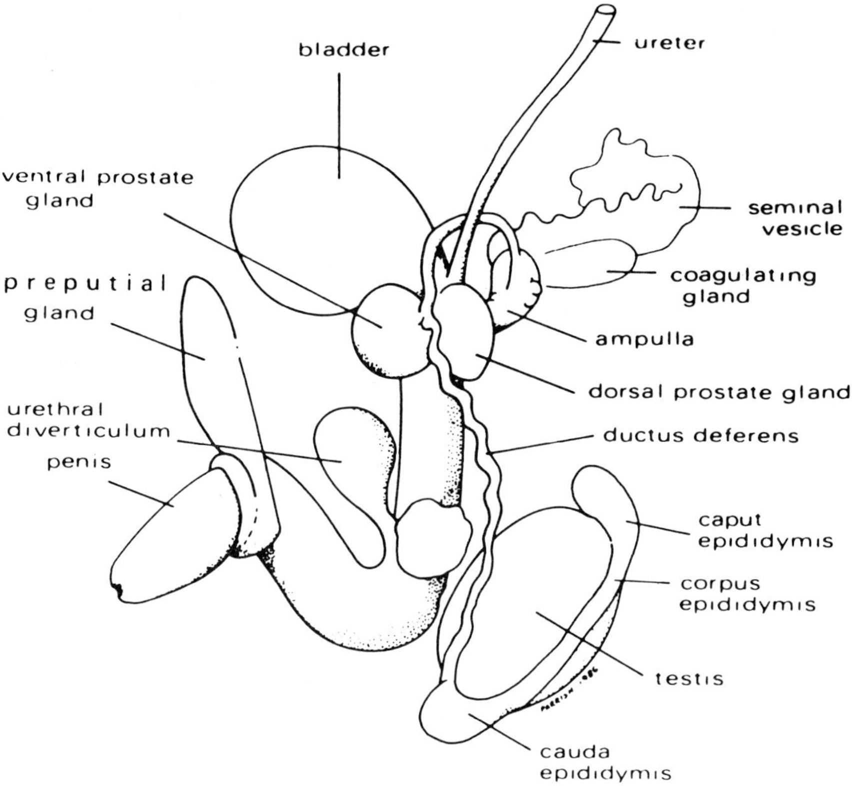

Figure 4-3. Lateral view of the urogenital system in the male mouse. The duct system includes the epididymis, ductus deferens, and ejaculatory duct. Accessory glands in the rodent include the seminal vesicles, prostate, coagulating glands, and preputial glands. [*Reprinted from Eddy (1988), with permission.*]

The epididymis is the principal site of sperm maturation and is also a reservoir for the storage of spermatozoa until ejaculation. It consists of three main regions: the head (caput epididymis), the body (corpus epididymis), and the tail (cauda epididymis). Secretory and absorptive functions of the epididymal epithelium are critical for sperm transport, modification of epididymal fluid composition, and changes in sperm morphology, composition, and function that occur during sperm maturation (Bedford, 1975).

The accessory sex glands produce a variety of secretory products that contribute most of the volume of the ejaculate. Fertility is reduced in vivo in the absence of accessory sex gland products, indicating that they contribute to the overall process of reproduction (Eddy, 1988). Sex gland secretions are thought to facilitate transport of spermatozoa within the female reproductive tract after ejaculation and to modify the environment for optimum sperm survival and function.

Less is known about the effects of toxic substances on the duct system and accessory sex glands of the male than is known about their effects on the testis, although it is clear that both direct and indirect effects are possible. Examples are discussed below. The androgen dependence of the duct system and sex glands makes them susceptible to secondary effects from agents that primarily affect the testis, since testicular damage often results in altered androgen production.

TARGETS IN THE TESTIS

Specific chemicals and their cellular targets in the testis are summarized in Table 4-1. Using the criteria of Leblond and Clermont (1952), the stages of spermatogenesis can easily be identified by microscopic examination of testicular cross-sections (Fig. 4-4). We have utilized data from such histopathologic analyses to categorize a variety of reproductive toxicants according to their cellular target(s) in the testis. Target cells can readily be identified by light microscopy in properly fixed and processed testicular cross-sections

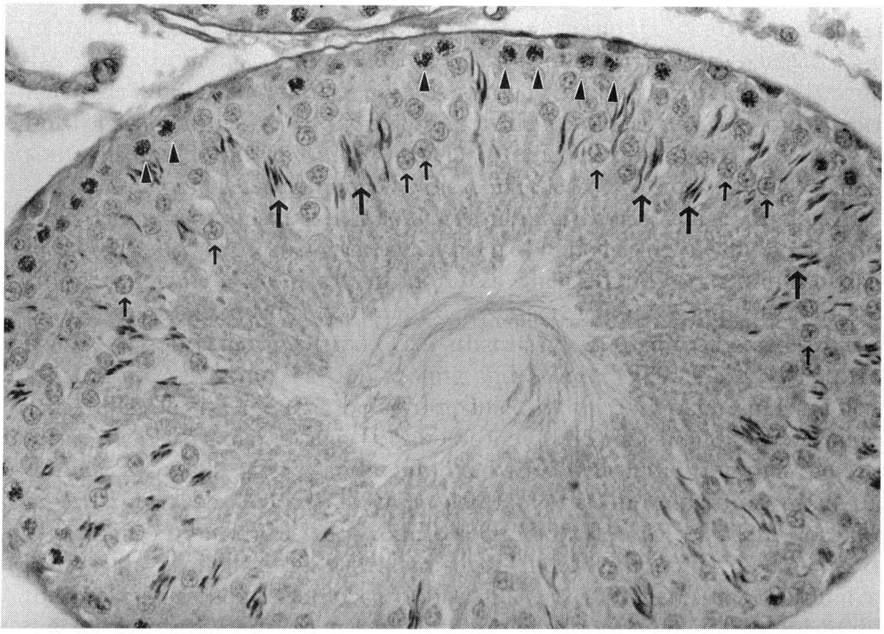

Figure 4-4. Stage V of spermatogenesis in an untreated rat. Note pachytene spermatocytes (arrowheads), step 5 round spermatids (small arrows), and bundles of step 17 elongated spermatids (large arrows). Periodic acid–Schiff/ hematoxylin, × 700.

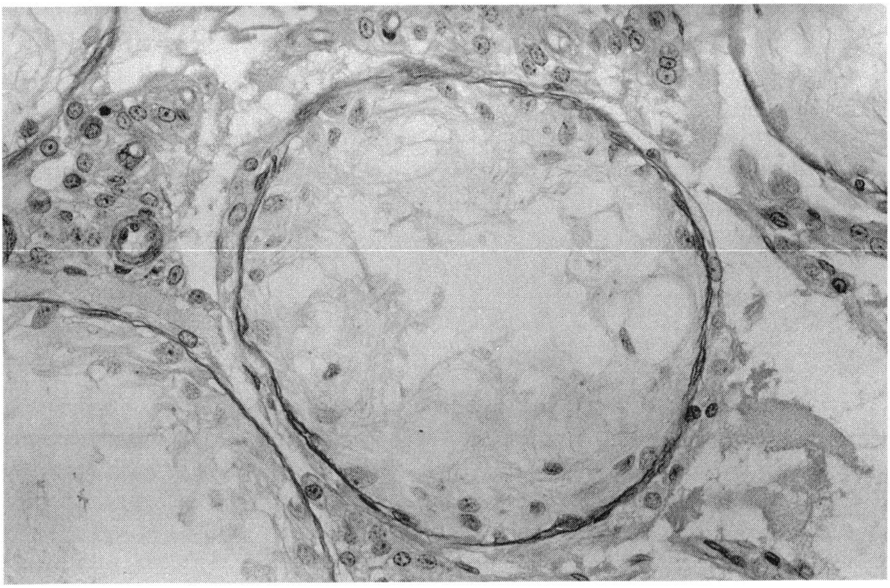

Figure 4-5. Seminiferous tubules from rat that received four weekly intraperitoneal injections of procarbazine (150 mg/kg). No germ cells are evident in the shrunken tubules, which contain only Sertoli cells and debris. Periodic acid–Schiff/hematoxylin, × 700.

(Chapin, 1988). However, to effectively determine the initial cellular target, one must examine sections of testis soon after exposure to the toxicant, often within hours, since damage to many of the cell types in the testis can ultimately result in the same end-stage lesion: tubules that contain Sertoli cells and little else (Fig. 4-5). Little useful information can be realized from examination of testes that have reached this final stage.

Spermatogonia

The mitotically active cells in the seminiferous epithelium are the spermatogonia, found only at the basement membrane of the seminiferous tubules (de Rooij, 1983). The progenitors of all spermatogonia are the stem cells, which both maintain their own number and produce more-differentiated stages of spermatogonia that eventually mature to spermatozoa. Damage to spermatogonia, and particularly to stem cells, is likely to produce severe effects on spermatogenesis. Decreased fertility that results only from spermatogonial damage will take 6 to 8 weeks to appear in the rat, as undamaged cells complete their progression through spermatogenesis; however, it may persist for a prolonged period, or even be permanent, if the undifferentiated A-type spermatogonia are destroyed.

Busulfan

Sterility, ranging from short-term to permanent depending on dosage and duration of treatment, can be induced in rats by busulfan (Myleran), a methanesulfonic diester that selectively targets spermatogonia (Jackson et al, 1962). Early investigators studied this and other diester methanesulfonates as possible means of fertility control in males, but because of the equally suppressive action of the diesters on bone marrow stem cells, work was abandoned.

Procarbazine

Procarbazine, an alkylating anticancer drug used primarily in the treatment of Hodgkin disease (Spivak, 1974), is a spermatogonial toxicant in rats. Treatment with procarbazine will destroy all stages of germ cells in the testes of male rats (Glode et al, 1982; Delic et al, 1986). It is well known that procarbazine is toxic to type A spermatogonia, the stem cells (Meistrich et al, 1982; Gould et al, 1983), producing permanent sterility in some strains of rat (Ward et al, 1989; Velez de la Calle and Jegou, 1990). Examination of the testes of $LBNF_1$ rats that received four weekly intraperitoneal injections of the chemical revealed atrophied seminiferous tubules containing only Sertoli cells (Fig. 4-5); this condition persisted for at least 84 days after the last treatment (Working, unpublished observations).

Spermatocytes

Destruction of spermatocytes, while not rendering a male permanently infertile, can cause decreased fertility via maturation depletion of sperm number. In other words, if one stage of spermatocyte is damaged or destroyed, the sperm number will decrease a specific number of weeks later (as with spermatogonial damage, the exact timing of the decrease will correspond to the precise stage of spermatocyte destroyed) but will recover once undamaged stages have completed their maturation to spermatozoa. Such an outcome can be detected both in sperm counts and in breeding studies. Spermatocytes are the initial target of a number of toxicants.

2-Methoxyethanol

The glycol ether 2-methoxyethanol (2-ME) is a testicular toxicant in rats (Foster et al, 1983; Chapin et al, 1984a). Just 24 h after a single dose of 2-ME, the testicular lesion is confined to late-stage pachytene spermatocytes

(Foster et al, 1983; Chapin et al, 1984a). Continued dosing results in the progressive depletion of other stages of spermatocytes, and, with time, the maturation depletion of early spermatids. Treated rats show a profound decrease in fertility approximately 5 weeks after treatment with 2-ME, which corresponds to the length of time necessary for pachytene spermatocytes to reach the epididymis and to be available for insemination (Fig. 4-6a; Chapin et al, 1985). An oxidized metabolite of 2-ME, 2-methoxyacetic acid (2-MAA), has similar toxic effects in the testis in vivo (Miller et al, 1982); however, only 2-MAA, and not the parent compound 2-ME, is toxic to

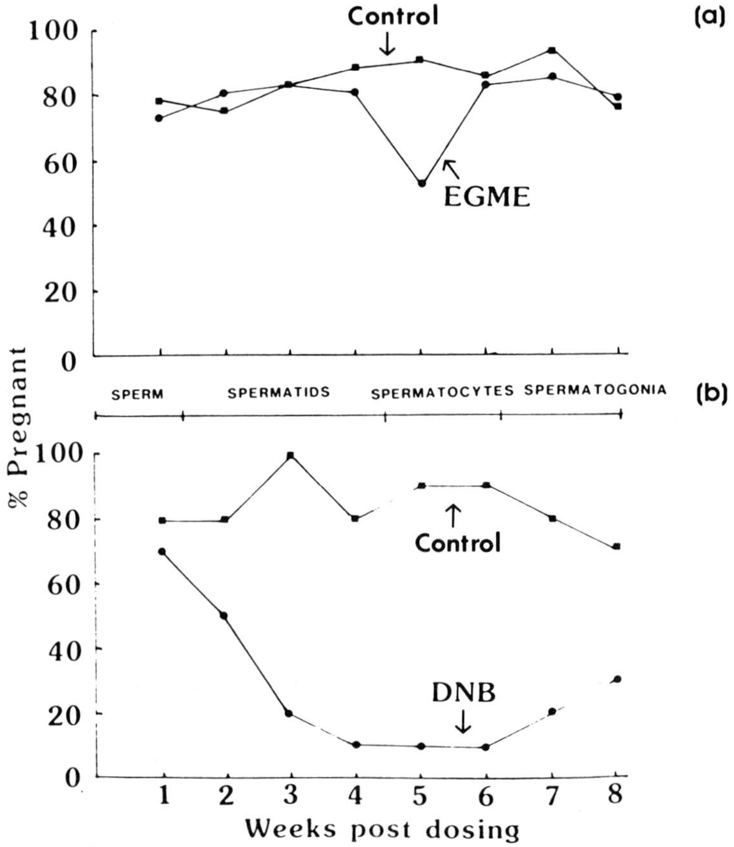

Figure 4-6. **Effects of (*a*) 2-methoxyethanol (ethylene glycol monomethyl ether, EGME; 100 mg/kg per day for 5 days), and (*b*) 1,3-dinitrobenzene (DNB; 10 mg/kg per day for 5 days) on fertility of male rats in a multiweek breeding study. The timing of the decreases in fertility is related to the specific cell stage of spermatogenesis damaged. [(*a*) *Reprinted from Chapin et al (1985), with permission. Entire figure reprinted from Foster (1988), with permission.*]**

spermatocytes in vitro (Gray et al, 1985). These data suggest that metabolism of 2-ME is required for it to demonstrate testicular toxicity, and, further, that the germ cells themselves are not capable of performing the necessary oxidative steps.

Procarbazine

Pachytene spermatocytes are damaged or missing in seminiferous tubules at stages V to IX from rats that received a single intraperitoneal injection of procarbazine 1 week earlier (Fig. 4-7). After three more weekly injections, the lesion proceeds to the end-stage lesion pictured in Fig. 4-5 and discussed above (Working, unpublished observations). Other investigators have reported that decreases in sperm counts first occur just 16 days after treatment with procarbazine, suggesting that not only spermatocytes but also spermatids have been damaged by the treatment (Russell et al, 1983; Velez de la Calle and Jegou, 1990).

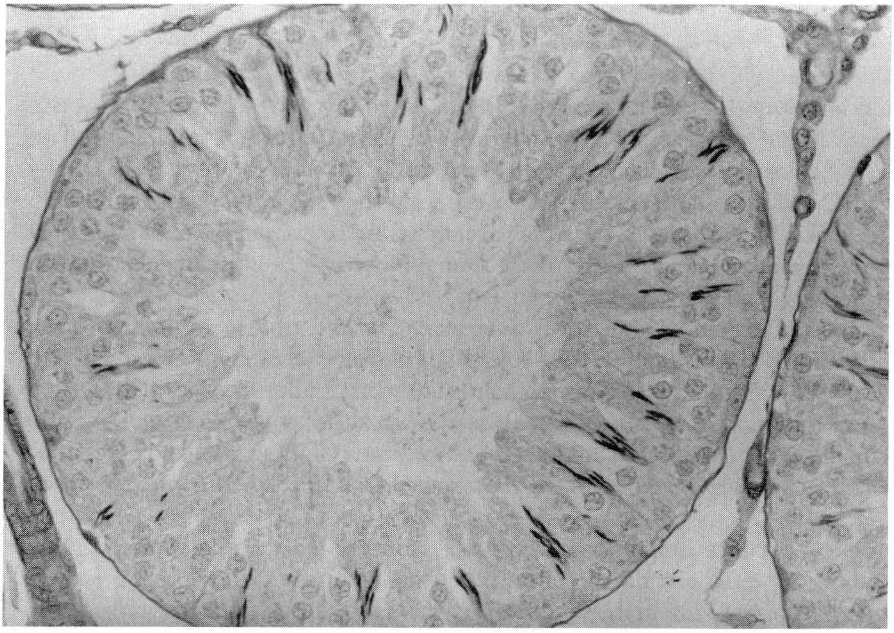

Figure 4-7. Cell-specific lesion in the testis. Stage V of spermatogenesis in a rat that received a single intraperitoneal injection of procarbazine (150 mg/kg) 1 week earlier. Note normal-appearing step 5 round spermatids and step 17 elongated spermatids, but absence of pachytene spermatocytes (see Fig. 4-4 for comparison). Periodic acid–Schiff/hematoxylin, × 700.

Spermatids

Spermatids, which undergo no further division, during their last 3 weeks in the testis undergo extensive morphologic modifications in a process known as spermiogenesis. At the completion of spermiogenesis, the spermatids are released into the tubule lumen and swiftly move into the excurrent ducts and on into the epididymis. Because spermatids are the final developmental stage of spermatogenic cells, the effects of spermatid-specific toxicants will quickly be evident as decreased sperm counts and decreased fertility.

Methyl Chloride

Methyl chloride is a halogenated hydrocarbon once used extensively for fumigation and as a chemical intermediate in multiple industrial applications. Exposure of male rats to methyl chloride causes testicular degeneration, epididymal inflammation, and sperm granuloma formation (Chapin et al, 1984b; Working et al, 1985a). The initial lesion in the testis appears to be delayed spermiation of mature spermatids at stage VIII; instead of being released, elongated spermatids are still present through at least stage XI (Chapin et al, 1984b). This lesion is followed by a more generalized disruption of the seminiferous epithelium, including some spermatogonial toxicity, although this is a minor component of methyl chloride's action in the testis (Working et al, 1985b). Sperm counts decline soon after exposure, as does fertility; however, these latter effects may be partially attributable to the chemical's epididymal effects (Chellman and Working, 1989), which will be discussed below. Interestingly, the degenerative effects of methyl chloride in the testis and in other organs of the rat are effectively blocked by pretreatment with the anti-inflammatory agent BW755C (Chellman et al, 1986a), which inhibits the cyclooxygenase and lipoxygenase enzymes (Higgs et al, 1979; Egan et al, 1983). The precise mechanism of this effect is not known, but these observations suggest that methyl chloride toxicity may be mediated by the action of prostaglandins and/or leukotrienes.

Sertoli Cells

Because of the multifunctional role of Sertoli cells in the testis (see discussion above), Sertoli cell toxicants will have marked and often immediate effects on spermatogenesis. Damage to the Sertoli cells that results in significant cell death may cause permanent infertility, since Sertoli cells normally do not divide in the mature adult testis.

1,3-Dinitrobenzene

Dinitrobenzene (DNB) is a nitroaromatic compound used widely as an intermediate in the synthesis of dyes and various plastics. Commercial DNB is a mixture of the 1,2-, 1,3-, and 1,4- isomers, with 1,3-DNB predominating. Although all three isomers cause methemoglobinemia in rats (Goldstein and Rickert, 1985), only 1,3-DNB is a testicular toxicant in male rats, inducing testicular atrophy and decreased fertility (Cody et al, 1981; Linder et al, 1986). After a single dose, the testicular lesions are limited to stages VIII to XIII of the cycle and are characterized by Sertoli cell damage, followed by degeneration of pachytene spermatocytes (Blackburn et al, 1988; Cossum et al, 1986). Ultrastructural examinations confirm that the initial target of 1,3-DNB is the Sertoli cell and that the germ cell damage is secondary to the Sertoli cell toxicity (Blackburn et al, 1988). The immediacy of the DNB effect on spermatogenesis is exemplified by the immediate and prolonged decrease in fertility seen in treated males (Fig. 4-6b; Foster, 1988), which have decreased fertility from weeks 3 to 8 posttreatment, time points that correspond to spermatogenic stages ranging from late spermatids (week 3) to A-type spermatogonia (week 8). Fertility was not decreased significantly in weeks 1 and 2, when sperm located in the epididymis during DNB exposure would be the fertilizing sperm. 1,3-DNB is metabolized in vivo via nitroreduction or glutathione conjugation (Nystrom and Rickert, 1987), and small amounts of both nitroreduced and glutathione-conjugated metabolites are found in the testis after exposure (Nystrom et al, 1988). However, none of the major reduced metabolites of 1,3-DNB are testicular toxicants in vivo, suggesting that either the parent compound or some transient intermediate metabolite is the toxic species (Cossum et al, 1986).

2,5-Hexanedione

A similar type of toxicity to Sertoli cells is seen in male rats treated with 2,5-hexanedione (2,5-HD). After 6 weeks of exposure to 2,5-HD in drinking water, the testes of rats were azoospermic (Chapin et al, 1982). In this study, Sertoli cell–specific enzymes were found to be decreased even when testicular morphology appeared normal, that is, changes in Sertoli cell biochemistry preceded distinguishable morphologic changes in the seminiferous tubules. Subsequent ultrastructural studies confirmed that the Sertoli cell is the initial testicular target of 2,5-HD (Chapin et al, 1983).

Leydig Cells

Unlike the cell types discussed above, the Leydig cells are located in the interstitial compartment of the testis. They are relatively small cells, generally found clustered around the blood vessels (Clark, 1976; Connell and Connell, 1977), and seem to be particularly associated with tubules at spermatogenic stages during which testosterone is most needed (Bergh, 1985). The effects of compounds that have Leydig cells as their testicular target are primarily mediated via effects on the production of testosterone and other steroids.

Ethane Dimethanesulfonate

Ethane dimethanesulfonate (EDS) is an alkylating sulfonic ester that exerts a primary cytotoxic effect on testicular Leydig cells. This has been demonstrated in multiple species, including rat, guinea pig, and hamster (Jackson and Morris, 1977; Kerr et al, 1987). The toxicity of EDS to Leydig cells results secondarily in decreased levels of intratesticular testosterone (disrupting spermatogenesis and causing infertility), and decreased levels of circulating testosterone (disrupting accessory sex gland function) (Jackson, 1973). The mechanism of action of EDS appears to be unique, since the structural analogs busulfan, methylene dimethanesulfonate, and propylene dimethanesulfonate selectively damage spermatogonia in testis irrespective of hormonal status (Jackson, 1971). The testosterone depletion induced by EDS also causes secondary epididymal effects (see discussion below).

EXTRATESTICULAR TARGETS

Epididymis

The epididymis consists of a single, highly coiled tubule and provides the environment in which spermatozoa undergo morphologic compositional, and functional changes (including development of motility) that impart the ability to fertilize. It is a highly vascularized organ, and its epithelium is active in a variety of secretory and absorptive functions critical for normal maturation of spermatozoa (Bedford, 1975; Hamilton, 1977). The epididymis is also highly androgen-dependent (Orgebin-Crist and Jahad, 1978). Agents that disrupt the normal functions of the epididymal vasculature or epithelium, or that alter androgen production, can have profound effects on sperm maturation, motility, or fertilizing ability.

α-Chlorohydrin

α-Chlorohydrin at lower doses in rats causes reversible infertility by decreasing the motility and respiratory activity of epididymal spermatozoa (Waller et al, 1985). Higher doses damage the epididymal vasculature, with resultant epithelial desquamation, which blocks the caput epididymis (Ericsson, 1970). Spermatoceles and sperm granulomata form in the caput epididymis, leading to irreversible infertility due to back pressure on the testis and subsequent cessation of spermatogenesis. α-Chlorohydrin also inhibits binding of specific epididymal secretory proteins to the sperm surface (Tsang et al, 1981). Interestingly, α-chlorohydrin may in part be responsible for the reported posttesticular toxic effects of the nematocide 2-dibromo-3-chloropropane (DBCP) (Kluwe et al, 1983), since DBCP is metabolized to epichlorohydrin and in turn to α-chlorohydrin (Kluwe, 1983).

Ethane Dimethanesulfonate

Although Leydig cells in the testis are the primary target of EDS (discussed above), the resulting testosterone depletion causes secondary epididymal effects in rats, including decreased weight of the caput epididymis, diminution of several sperm membrane proteins, and decreased motility and progressive motility of sperm (Klinefelter et al, 1990). Implantation of testosterone-filled Silastic capsules does not ameliorate all the toxic effects of EDS on the epididymis, however, including altered morphology of the corpus epididymis, dedifferentiation of clear cells from the cauda epididymis, and perforation and loss of the plasma membrane of caudal sperm (Klinefelter et al, 1990). Thus, EDS appears to exert both androgen-dependent and androgen-independent effects on the epididymis.

Methyl Chloride

In the rat, methyl chloride is toxic not only to the testis (discussed above), but also to the epididymis, where it induces a unique inflammatory response in the cauda epididymis. The initial response is an acute inflammatory response against the tubular epithelium (Chapin et al, 1984b). This is followed within 1 to 2 weeks by formation of sperm granulomata secondary to chronic inflammation. Sperm passing through the cauda epididymis during the acute inflammatory phase (weeks 1 and 2 postexposure) can be exposed to reactive metabolic intermediates produced by the inflammatory cells and thereby sustain genotoxic and/or cytotoxic damage. The genotoxic damage to sperm manifests as increased postimplantation loss in dominant lethal studies in which male rats exposed to methyl chloride are mated to nonexposed females (Working et al, 1985a). However, this dominant lethal effect appears to be secondary to the epididymal inflammation rather than a direct effect of methyl chloride, since coadministration of the anti-inflammatory agent BW755C, which prevents the epididymal inflammation, also prevents the postimplantation loss (Chellman et al, 1986b).

Sulfasalazine

Sulfasalazine is used for treatment of inflammatory bowel disease, principally ulcerative colitis but also Crohn disease. Male infertility is a frequent clinical complication of sulfasalazine therapy (Giwercman and Skakkebæk, 1986). Chronic administration to men causes reduced sperm count, compromised sperm motility and fertilizing capability, and altered sperm morphology (enlarged "megalo" sperm heads). Dose-related antifertility also occurs in rats (O'Morain et al, 1984). It has been proposed that the antifertility effect may be mediated through a direct effect of sulfasalazine or its metabolite sulphapyridine on spermatozoa stored in the epididymis, hence compromising some processes vital for fertilization (Wong et al, 1987). Sperm from sulfasalazine-treated hamsters are unable to undergo the acrosome reaction or penetrate hamster eggs in vitro (O'Morain et al, 1985).

Accessory Sex Glands

The accessory sex glands produce a variety of secretory products that constitute most of the volume of the ejaculate. The precise roles of most of these products are unknown, but in general the accessory sex gland secretions are thought to be important for carrying spermatozoa into the female reproductive tract and for creating an environment in which the spermatozoa can survive and function. There are few reports of agents causing toxic effects specifically on the accessory sex glands, but it is not clear whether this is due to lack of occurrence or lack of detection. Agents could exert toxic effects on the accessory sex glands either directly or indirectly (secondary to perturbed androgen metabolism). The androgen dependence of the accessory sex glands and the degenerative effects of exogenous hormones such as diethylstilbestrol and estradiol have been studied extensively (Cavazos, 1977; Feagans et al, 1961). In addition, numerous drugs and toxic agents have been found in semen, and these not only might affect the male reproductive tract (including sperm, as discussed below) but also might gain access to the female reproductive tract and possibly the conceptus (Hales et al, 1986).

Imidazoles

Substituted imidazoles have been investigated as antimicrobial drugs, and a number of them are toxic either to the testis (de Kretser, 1974; Patanelli, 1975) or to epididymal sperm (McClain and Downing, 1988). However, certain 1-substituted imidazoles enter the accessory sex glands and are ejaculated into the semen, with subsequent toxic effects on sperm. Ketoconazole, for example, accumulates in the accessory sex gland secretions of dogs within several hours after oral administration, and when mixed with sperm upon ejaculation, exerts a spermicidal or spermatostatic effect (Vickery et al, 1985). Analogs such as RS-29984 and RS-29636 are more potent spermicides and less toxic (Vickery et al, 1986). The unique mechanism of action of these agents suggests that male contraceptive agents with a posttesticular site of action may ultimately be developed.

CONCLUSIONS

The reproductive process in the male involves a complex biological system, vulnerable at multiple sites to the effects of toxic agents. Some reproductive toxicants exhibit a surprising specificity for particular cell types or organs, whereas others have a broader effect, acting at diverse locations. The specificity of many toxicants is a function of both dose level and treatment duration; chemicals that damage only one cell type or reproductive organ often cause more widespread damage if the dose or duration of treatment is increased.

Such dose and duration effects may be one reason for the observation that substances that are toxic to the testis are frequently also toxic to the epididymis or the accessory sex glands. The similar susceptibility of different parts of the reproductive tract also may be because they are similar both metabolically and physiologically, or because the entire reproductive tract is highly dependent on androgens. Thus, agents that directly damage the testis and alter androgen production may indirectly affect the remainder of the tract.

Clearly, to understand the long-term effects of exposure to reproductive toxicants, their mechanisms of action in the reproductive tract must first be understood. To accomplish this goal, it will be necessary to employ more specific approaches to elucidate specific sites and modes of action of reproductive toxicants in the male. Use of such information may permit development of more quantitative ways to monitor changes in the reproductive system of men exposed to potentially toxic agents in the environment.

REFERENCES

Amann RP: Detection of alterations in testicular and epididymal function in laboratory animals. *Environ Health Perspect* 70:149–158, 1986.

Bedford JM: Maturation, transport, and fate of spermatozoa in the epididymis, in Hamilton DW, Greep RO (eds): *Handbook of Physiology: Endocrinology, Male Reproductive System* (sec 7; vol 5). Washington, DC, American Physiological Society, 1975, pp. 303–317.

Bergh A: Development of stage-specific paracrine regulation of Leydig cells by the seminiferous tubules. *Int J Androl* 8:80–85, 1985.

Blackburn DM, Gray AJ, Lloyd SC, Sheard DM, Foster PMD: A comparison of the effects of the three isomers of dinitrobenzene on the testis of the rat. *Toxicol Appl Pharmacol* 92:54–64, 1988.

Cavazos LF: The mammalian accessory sex glands: A morphological and functional analysis, in Greep RO, Koblinsky MA (eds): *Frontiers in Reproduction and Fertility Control: A Review of the Reproductive Sciences and Contraceptive Development.* Cambridge, MA, MIT Press, 1977, pp 402–410.

Chapin RE: Morphologic evaluation of the seminiferous epithelium of the testis, in Lamb JC IV, Foster PMD (eds): *Physiology and Toxicology of Male Reproduction.* New York, Academic Press, 1988, pp 155–177.

Chapin RE, Norton RM, Popp JA, Bus JS: The effects of 2,5-hexanedione on reproductive hormones and testicular enzyme activities in the F-344 rat. *Toxicol Appl Pharmacol* 62:262–272, 1982.

Chapin RE, Morgan KT, Bus JS: The morphogenesis of testicular degeneration induced in rats orally administered 2,5-hexanedione. *Exp Mol Pathol* 38:149–169, 1983.

Chapin RE, Dutton SL, Ross MD, Sumrell BM, Lamb JC: The effects of ethylene glycol monomethyl ether on testicular histology in F-344 rats. *J Androl* 5:369–380, 1984a.

Chapin RE, White RD, Morgan KT, Bus JS: Studies of the lesions induced in the testis and epididymis of F-344 rats by inhaled methyl chloride. *Toxicol Appl Pharmacol* 76:328–343, 1984b.

Chapin RE, Dutton SL, Ross MD, Lamb JC: Effects of ethylene glycol monomethyl ether on mating and epididymal sperm parameters in F-344 rats. *Fundam Appl Toxicol* 5:182–189, 1985.

Chellman GJ, Working PK: Importance of assessing multiple endpoints in reproductive toxicity studies, in Working PK (ed): *Toxicology of the Male and Female Reproductive Systems*. New York, Hemisphere, 1989, pp 257–271.

Chellman GJ, Morgan KT, Bus JS, Working PK: Inhibition of methyl chloride toxicity in male F-344 rats by the anti-inflammatory agent BW755C. *Toxicol Appl Pharmacol* 85:367–379, 1986a.

Chellman GJ, Bus JS, Working PK: Role of epididymal inflammation in the induction of dominant lethal mutations in Fischer-344 rat sperm by methyl chloride. *Proc Natl Acad Sci USA* 83:8087–8091, 1986b.

Clark RV: Three-dimensional organisation of testicular interstitial tissue and lymphatic space in the rat. *Anat Rec* 184:203–218, 1976.

Clermont Y: Kinetics of spermatogenesis in mammals: Seminiferous epithelium cycle and spermatogonial renewal. *Physiol Rev* 52:198–236, 1972.

Cody TE, Witherup S, Hastings L, Stemmer K, Christian RT: 1,3-Dinitrobenzene: Toxic effects in vivo and in vitro. *J Toxicol Environ Health* 7:829–847, 1981.

Connell CJ, Connell GM: The interstitial tissue of the testis, in Johnson AD, Groves ER (eds): *The Testis* (vol 4). New York, Academic Press, 1977, pp 333–369.

Cossum PA, Rickert DE, Working PK: Comparison of the testicular toxicity of the dinitrobenzenes and their major reduced metabolites. *Pharmacologist* 28:178, 1986.

de Kretser DM: The regulation of male fertility: The state of the art and future possibilities. *Contraception* 9:561–600, 1974.

Delic JI, Bush C, Peckham MJ: Protection from procarbazine-induced damage of spermatogenesis in the rat by androgen. *Cancer Res* 46:1909–1914, 1986.

de Rooij DG: Proliferation and differentiation of undifferentiated spermatogonia in the mammalian testis, in Potten CS (ed): *Stem Cells. Their Identification and Characterization*. New York, Churchill, 1983, pp 89–117.

Eddy EM: Duct system and accessory glands of the male reproductive tract, in Lamb JC, Foster PM (eds): *Physiology and Toxicology of Male Reproduction*. New York, Academic Press, 1988, pp 35–69.

Egan RW, Tischler AN, Baptist EM, Ham EA, Soderman DE, Gale PH: Specific inhibition and oxidative regulation of lipoxygenase, in Samuelsson B, Paoletti P, Ramwell P (eds): *Advances in Prostaglandin, Thromboxane and Leukotriene Research* (vol. 11). New York, Raven Press, 1983, pp 151–157.

Ericsson RJ: Male antifertility compounds: U-5897 as a rat chemosterilant. *J Reprod Fertil* 22:213–222, 1970.

Ewing LL: Leydig cell, in Lipschultz LI, Howards, SS (eds): *Infertility in the Male.* New York, Churchill, 1983, pp 43–69.

Fawcett DW, Neaves WB, Flores MN: Comparative observations on the intertubular lymphatics and the organization of the interstitial tissue of the mammalian testis. *Biol Reprod* 9:500–532, 1973.

Feagans WM, Cavazos LF, Ewald AT: A morphological and histochemical study of estrogen-induced lesions in the hamster male reproductive tract. *Am J Anat* 108:31–45, 1961.

Foster PMD: Testicular organization and biochemical function, in Lamb JC, Foster PMD (eds): *Physiology and Toxicology of Male Reproduction.* New York, Academic Press, 1988, pp 7–34.

Foster PMD, Creasy DM, Foster JR, Thomas LJ, Cook MW, Gangolli SD: Testicular toxicity of ethylene glycol monomethyl and monoethyl ethers in the rat. *Toxicol Appl Pharmacol* 69:385–399, 1983.

Fritz IB, Rommerts FFG, Louis BG, Dorrington JH: Regulation by FSH and cAMP of the formation of cAMP in Sertoli-cell enriched cultures. *J Reprod Fertil* 46:17–24, 1976.

Giwercman A, Skakkebæk NE: The effect of salicylazosulphapyridine (sulphasalazine) on male fertility. A review. *Int J Androl* 9:38–52, 1986.

Glode LM, Robinson J, Gould SF, Nett TM: Protection of spermatogenesis during chemotherapy. *Drugs Under Exp Clin Res* 8:367–379, 1982.

Goldstein RS, Rickert DE: Relationship between red blood cell uptake and methemoglobin production by nitrobenzene and dinitrobenzene in vitro. *Life Sci* 36:121–125, 1985.

Gould SF, Powell D, Nett T, Glode LM: A rat model for chemotherapy-induced male infertility. *Arch Androl* 11:141–150, 1983.

Gray TJB, Moss EJ, Creasy DM, Gangolli SD: Studies on the toxicity of some glycol ethers and alkoxyacetic acids in primary testicular cell culture. *Toxicol Appl Pharmacol* 79:490–501, 1985.

Hales BF, Smith S, Robaire B: Cyclophosphamide in the seminal fluid of treated males: Transmission to females by mating and effect on pregnancy outcome. *Toxicol Appl Pharmacol* 84:423–430, 1986.

Hamilton DW: The epididymis, in Greep RO, Koblinsky MA (eds): *Frontiers in Reproduction and Fertility Control: A Review of the Reproductive Sciences and Contraceptive Development.* Cambridge, MA, MIT Press, 1977, pp 411–426.

Hess RA: Quantitative and qualitative characteristics of the stages and transitions in the cycle of the rat seminiferous epithelium: Light microscope observations of perfusion-fixed and plastic-embedded testes. *Biol Reprod* 43:525–542, 1990.

Higgs GA, Flower RJ, Vane JR: A new approach to anti-inflammatory drugs. *Biochem Pharmacol* 28:1959–1961, 1979.

Huckins C: Cell cycle properties of differentiating spermatogonia in adult Sprague–Dawley rats. *Cell Tissue Kinet* 4:139–154, 1971.

Jackson CM, Morris ID: Gonadotropin levels in male rats following impairment of Leydig cell function by ethylene dimethanesulphonate. *Andrologia* 9:29–35, 1977.

Jackson H: Chemical methods of male contraception. *Am Sci* 61:188–193, 1973.

Jackson H: Non-steroidal antifertility agents in the male, in Diczfalusy E, Borell U (eds): *Control of Human Fertility, Nobel Symposium* (no 15). New York, Wiley, 1971, pp 119–136.

Jackson H, Partington M, Fox BW: Effects of busulphan (Myleran) on the spermatogenic population of the rat testis. *Nature* 194:1184–1185, 1962.

Kerr JB, Knell CM, Abbott M, Donachie K: Ultrastructural analysis of the effect of ethane dimethanesulphonate on the testis of the rat, guinea pig, hamster, and mouse. *Cell Tissue Res* 249:451–457, 1987.

Klinefelter GR, Laskey JW, Roberts NR, Slott V, Suarez JD: Multiple effects of ethane dimethanesulfonate on the epididymis of adult rats. *Toxicol Appl Pharmacol* 105:271–287, 1990.

Kluwe WM: Chemical modulation of 1,2-dibromo-3-chloropropane toxicity. *Toxicology* 27:287–299, 1983.

Kluwe WM, Lamb JC IV, Greenwell AE, Harrington FW: 1,2-dibromo-3-chloropropane (DBCP)-induced infertility in male rats mediated by a post-testicular effect. *Toxicol Appl Pharmacol* 71:294–298, 1983.

Leblond CP, Clermont Y: Definition of the stages of the cycle of the seminiferous epithelium of the rat: I. Spermatogenesis and sperm maturation. *Ann NY Acad Sci* 55:548–573, 1952.

Linder RE, Hess RA, and Strader LF: Testicular toxicity and infertility in male rats treated with 1,3-dinitrobenzene. *J Toxicol Environ Health* 19:477–489, 1986.

McClain RM, Downing JC: The effect of ornidazole on fertility and epididymal sperm function in rats. *Toxicol Appl Pharmacol* 92:488–496, 1988.

Meistrich ML, Finch M, da Cunha MF, Hacker U, Au W: Damaging effects of fourteen chemotherapeutic drugs on mouse testis. *Cancer Res* 42:122–131, 1982.

Miller RR, Carreon RE, Young JT, McKenna MJ: Toxicity of methoxyacetic acid in rats. *Fundam Appl Toxicol* 2:158–160, 1982.

Nystrom DD, Rickert DE: Metabolism and excretion of dinitrobenzenes by male Fischer 344 rats. *Drug Metab Dispos* 15:192–205, 1987.

Nystrom DD, Working PK, Bentley KS, Rickert DE: Testicular metabolism of dinitrobenzenes. *Toxicologist* 8:182, 1988.

O'Morain C, Smethurst P, Dore CJ, Levi AJ: Reversible male infertility due to sulphasalazine: Studies in man and rat. *Gut* 25:1078–1084, 1984.

O'Morain C, Smethurst P, Levi AJ: Reversible male infertility induced by sulphasalazine, in Lobl TJ, Hafez ESE (eds): *Male Fertility and its Regulation*. Lancaster, MTP Press, 1985, pp 49–54.

Orgebin-Crist M-C, Jahad N: The maturation of rabbit epididymal spermatozoa in organ culture: Inhibition by antiandrogens and inhibitors of ribonucleic acid and protein synthesis. *Endocrinology* 103:46–53, 1978.

Patanelli DJ: Suppression of fertility in the male, in Hamilton DW, Greep RO (eds): *Handbook of Physiology: Endocrinology, Male Reproductive System* (sec 7; vol 5). Washington, DC, American Physiological Society, 1975, pp 245–258.

Russell LD: Sertoli–germ cell interactions: A review. *Gamete Res* 3:99–112, 1980.

Russell LD, Lee IP, Ettlin R, Malone JP: Morphological pattern of response after administration of procarbazine: Alteration of specific associations during the cell cycle of the seminiferous epithelium of the rat. *Tissue Cell* 15:391–404, 1983.

Russell LD, Ettlin RA, Sinha Hikim AP, Clegg ED: *Histological and Histopathological Evaluation of the Testis*. Clearwater, FL, Cache River Press, 1990.

Skinner MK, Griswold MD: Sertoli cells synthesize and secrete a transferrin-like protein. *J Biol Chem* 255:9523–9525, 1980.

Spivak SD: Procarbazine, diagnosis and treatment-drugs five years later. *Ann Intern Med* 81:795–800, 1974.

Tsang AYF, Lee WM, Wong PYD: Effects of antifertility drugs on epididymal protein secretion, acquisition of sperm surface proteins and fertility in male rats. *Int J Androl* 4:703–712, 1981.

Velez de la Calle JF, Jegou B: Protection by steroid contraceptives against procarbazine-induced sterility and genotoxicity in male rats. *Cancer Res* 50:1308–1315, 1990.

Verhoeven G, Franchimont P: Regulation of inhibin secretion by Sertoli-cell enriched cultures. *Acta Endocrinol* 102:136–143, 1983.

Vickery BH, Burns J, Zaneveld LJD, Goodpasture JC, Bergström K: Orally administered ketoconazole rapidly appears in seminal plasma and suppresses sperm motility. *Adv Contracept* 1:341–353, 1985.

Vickery BH, Grigg MB, Goodpasture JC, Bergström KK, Walker KAM: Toward a same-day, orally administered male contraceptive, in Zatuchni GI, Goldsmith A, Spieler JM, Sciarra JJ (eds): *Male Contraception: Advances and Future Prospects*. Philadelphia, Harper & Row, 1986, pp 271–292.

Waller DP, Killinger JM, Zaneveld LJD: Physiology and toxicology of the male reproductive tract, in Thomas JA, Korach KS, McLachlan JA (eds): *Endocrine Toxicology*. New York, Raven Press, 1985, pp 269–333.

Ward JA, Robinson J, Morris IA: Strain-dependency of procarbazine-induced testicular toxicity. *Reprod Toxicol* 3:43–50, 1989.

Wong PYD, Lau SKD, Fu WO: Antifertility effects of some sulphonamides and related compounds and their accumulation in the epididymides of male rats. *J Reprod Fertil* 81:259–267, 1987.

Working PK: Male reproductive toxicology: Comparison of the human to animal models. *Environ Health Perspect* 77:37–44, 1988.

Working PK, Bus JS, Hamm TE: Reproductive effects of inhaled methyl chloride in the male Fischer 344 rat: I. Mating performance and dominant lethal assay. *Toxicol Appl Pharmacol* 77:133–143, 1985a.

Working PK, Bus JS, Hamm TE: Reproductive effects of inhaled methyl chloride in the male Fischer 344 rat: II. Spermatogonial toxicity and sperm quality. *Toxicol Appl Pharmacol* 77:134–157, 1985b.

CHAPTER 5

Capacitation and Acrosome Reaction

Erma Z. Drobnis

During its residence in the female reproductive tract, the fertilizing sperm undergoes profound cellular changes culminating in sperm fusion with the oocyte. In general, mammalian sperm are not capable of fertilization immediately following ejaculation. Additional maturation of the sperm is required, either in the female reproductive tract or under appropriate culture conditions in vitro. The process by which sperm gain the ability to complete fertilization is termed capacitation. Once sperm are capacitated, they display hyperactivated motility, and appropriate stimulation will induce the acrosome reaction, an event involving exocytosis of the acrosome and release of the acrosomal contents. In vivo, the fertilizing sperm probably undergoes hyperactivation and completes the acrosome reaction in the vicinity of the oocyte. It is generally agreed that both hyperactivated motility and the acrosome reaction are required for penetration of the zona pellucida and that the acrosome reaction is required for sperm fusion with the oolemma. There is an extensive literature describing research into these cellular events and the mechanisms that support them. However, despite the effort devoted to these studies, the processes of capacitation and induction of the acrosome reaction are remarkably complex, and their underlying mechanisms have not been well characterized in any species.

Because sperm from the in vivo environment are not immediately accessible for study, capacitation and the acrosome reaction have been studied primarily in vitro. We know very little about these events in the female reproductive tract. Modifications in the structure and function of the sperm plasma membrane and sperm surface are central to capacitation and the acrosome reaction, and it is reasonable to accept as a working hypothesis that these changes are profoundly affected by the epithelial surfaces and secretions of the female reproductive tract (Amann, 1989). These relationships, which remain poorly characterized, are likely to be important considerations when investigating the cell biology of sperm dysfunction. Sperm dysfunction may result from naturally occurring subfertility, environmental conditions (including toxicologic exposure), or sperm preparation in vitro. In most cases, the only good measure of sperm function is reduced fertility. Using fertility as an endpoint requires large numbers of subjects or experimental animals and cannot determine the functional or cellular bases of sperm dysfunction. Reliable in vitro measures of sperm quality are required in order to make efficient progress at diagnosing sperm dysfunction and proposing therapies. Although capacitation and the acrosome reaction have remained major foci of research by both basic and applied sperm biologists, progress has been hampered by a lack of measures of normal sperm function. It remains a challenge to sperm biologists to choose endpoints that assess the ability of sperm to undergo normal capacitation and acrosome reactions in vivo. Better characterization of these processes at the cellular and molecular levels will be useful in developing better in vitro assays to assess potential sperm function in vivo.

Because the in vivo environment is important to sperm function, I have started this chapter with an overview of what we know about sperm interaction with the female reproductive tract and with the oocyte during sperm capacitation and the acrosome reaction. Capacitation is often viewed as the cellular changes that enable sperm to undergo the acrosome reaction. For this reason, I have described the process of the acrosome reaction first before returning to discuss the cellular modifications that prepare the sperm for this event. Finally, I have described some cellular lesions that may lead to abnormal capacitation and acrosome reactions. Very little toxicology research has been conducted in this area, so examples of sperm dysfunction come largely from studies of male infertility. These cases illustrate the difficulty of detecting diminished sperm function in vivo using in vitro assays. There are many recent reviews that contain more-detailed information of sperm physiology in the female reproductive tract (Hunter, 1980, 1988; Bedford, 1983; Moore and Bedford, 1983; Overstreet, 1983; Mortimer, 1983; Suarez et al, 1990; Overstreet and VandeVoort, 1990; Drobnis and Overstreet, 1992), sperm capacitation (Yanagimachi, 1981, 1988; O'Rand, 1982; Clegg, 1983; Bedford, 1983; Moore and Bedford, 1983; Fraser, 1984, 1990; Oliphant et al, 1985; Langlais and Roberts, 1985; Eddy and O'Brien, 1989; Parks and Ehrenwald, 1990; Myles et al, 1990; Roldan and Harrison, 1990; Florman and Babcock, 1991), penetration of the oocyte investing coats (Yanagimachi, 1981, 1988; Moore and Bedford, 1983; Talbot, 1985; Katz et al, 1989; Hunter, 1988; Saling, 1989), and the acrosome reaction in mammalian sperm (Yanagimachi, 1981, 1988; Meizel, 1985; Wassarman, 1987, 1988; Eddy and O'Brien, 1989; Kopf and Gerton, 1991; Meizel et al, 1990; Saling, 1989; Saling et al, 1990; Kopf, 1990a,b). Throughout this chapter I have primarily referenced those articles in which more detailed information is contained. The possible effects of toxicologic exposure on capacitation and the acrosome reaction have been discussed in the reviews by Gwatkin (1985) and Perreault (1989).

SPERM TRANSPORT AND FERTILIZATION IN VIVO

The requirement for capacitation of mammalian sperm was first recognized by Chang (1951) and Austin (1951). By the strictest definition, capacitation is the process by which sperm gain the ability to fertilize oocytes in vivo. Because this functional capacity is difficult to evaluate, various systems have been devised to detect sperm capacitation in vitro. Although these in vitro endpoints can be valuable clinical and research tools, they may not be capable of predicting normal sperm function in vivo. For example, in the majority of male factor subfertility cases in humans, the success rate of intrauterine insemination (which concentrates sperm near the oviduct) is low but that of in vitro fertilization is high (e.g., Rogers, 1989). This implies that the ability of sperm to fertilize oocytes in vitro may not predict fertility in vivo. We do know that capacitation occurs during sperm transport to the site of fertilization. The process may be completed during the initial stages of interaction with the oocyte investments. The following description of sperm transport and penetration of the oocyte investments is intended to provide an in vivo framework for the subsequent consideration of information that has been derived largely from experiments in vitro.

Sperm Transport to the Site of Fertilization and the Establishment of Sperm Reservoirs

Studies of sperm transport in the female reproductive tract of humans has been limited because invasive procedures are required to collect sperm for study. With the exception of sperm physiology in cervical mucus, most of our information in this field comes from experiments with model species and food animal species. Humans and other primates can be grouped with rabbits and ruminants as having significant cervical barriers that prevent whole semen from entering the uterus after coitus (Overstreet, 1983). In other species, including rodents, pigs, dogs, and horses, the bulk of the semen is deposited in the uterus or is drawn into the uterine lumen within minutes of coitus. In primates and ruminants, semen is deposited in the anterior vagina, and sperm migrate into the cervical mucus over several hours. Although, in all species studied to date, a few sperm are transported to the oviduct by visceral contractions within minutes of coitus, studies in a number of species strongly suggest that these sperm are not the fertilizing population (Overstreet, 1983; Hunter, 1988).

Functionally competent sperm migrate through the cervical mucus and into the uterus, where they are moved by contractions of the female reproductive tract to the uterotubal junction. As a consequence of the uterine ability to move immotile sperm and inert materials along its length (Egli and Newton, 1961; Mattner and Braden, 1963), sperm motility is not considered a requirement for this phase of transport. Nonetheless, some sperm may migrate along epithelial surfaces of the uterus, and it is not known if these sperm, rather than the population present in the uterine lumen, represent the fertilizing population. The uterotubal junction acts as a physiologic sphincter (Hunter, 1980, 1988), and vigorous motility is probably a requirement for sperm entry into the oviduct (Gaddum-Rosse, 1981; Mortimer, 1983).

Sperm that enter the initial segment of the oviduct, termed the isthmus, are retained within 2 cm of the uterotubal junction until ovulation occurs. In the species studied to date, this portion of the oviduct appears to serve as a storage reservoir for sperm because ligation of the oviduct above this region before the occurrence of ovulation prevents fertilization (Hunter and Nichol, 1983; Hunter and Wilmut, 1983; Hunter, 1984). Ligation of the uterotubal junction at various times after coitus in ruminants has demonstrated that a population of sperm capable of completing fertilization is established in the isthmus within 6 to 10 h (Hunter and Nichol, 1983; Hunter and Wilmut, 1983). These experiments have been interpreted as evidence that sperm remaining in the lower regions of the reproductive tract (i.e., the uterus, cervix, and vagina) do not participate in fertilization (Hunter, 1987). However, animals of normal fertility were used in these experiments, and mating was timed to be optimal with respect to ovulation. In consequence, these results do not preclude the possibility that sperm from the lower tract may continue to be an important source of sperm under less optimal conditions. The cervix has long been considered a possible sperm reservoir (Overstreet, 1983; Mortimer, 1983; Katz et al, 1989). Consistent with this view, human sperm collected from cervical mucus for up to 72 h after coitus or artificial insemination are physiologically similar to sperm collected at 1 h (Gould et al, 1984; Zinaman et al, 1989).

Sperm within the oviduct have been observed directly in rodents, which have transparent oviductal walls (Katz and Yanagimachi, 1980; Suarez, 1987; Smith and Yanagimachi, 1989). In the isthmus, sperm heads are closely adherent to the oviductal epithelium, and the flagella move only sluggishly. Similarly, sperm flushed from the isthmus of mated rabbits are sluggishly motile, but they rapidly become hyperactivated in culture medium (Cooper et al, 1979; Johnson et al, 1981; Burkman et al, 1984). It is not known if attributes of the isthmic environment contribute to these changes in sperm physiology, but the composition of isthmic fluid or interactions of sperm with epithelial cell surfaces may be involved. Changes in the surface properties of sperm may enable them to form this interaction with the epithelial surface, and reduction of flagellar motility may prolong their life span and promote their retention at this site. Retention of sperm in the proximal isthmus may also be promoted by fluid currents generated by the cilia of the epithelium or by coordinated smooth muscle movements of the oviductal wall.

Near the time of ovulation, a small number of sperm are released or escape from the isthmic storage site and ascend to the oviductal ampulla, where fertilization occurs. The mechanism(s) underlying this well-coordinated migration, is (are) unknown. Three possible contributing factors have been proposed, and there is supporting evidence for each (Blandau, 1969; Overstreet, 1983; Hunter, 1988; Suarez et al, 1990). Changes in oviductal fluid currents resulting from muscular contractions and/or ciliary activity, perhaps under endocrine control, may be responsible for this release. Alternatively, changes in the surface properties of sperm may allow some to dissociate from the isthmic epithelium and resume their migration. Finally, sperm viewed within or collected from the ampulla of the oviduct are vigorously motile, displaying hyperactivated motility (Yanagimachi, 1981, 1988), which may be at least partly responsible for their escape from the isthmic reservoir (Katz and Yanagimachi, 1980; Suarez et al, 1990). This vigorous motility pattern is characterized by high curvature of the flagellar bends, asymmetric beat shapes, and nonlinear swimming trajectories (Burkman, 1990; Drobnis and Katz, 1991). Although these erratic trajectories inhibit progression in vitro, this motility pattern may prevent entrapment of the hyperactivated sperm in the extensive folds of the oviductal epithelium (Yanagimachi, 1981; 1988; Suarez et al, 1990). Oviductal fluid currents may also favor the movement of sperm toward the oocyte. Recent evidence suggests that sperm may migrate in response to chemical signals from the oocyte (Ralt et al, 1991). At the time of fertilization, very few sperm are present in the ampulla, as few as 1 to 2 per oocyte (Zamboni, 1972; Yanagimachi and Mahi, 1976; Shalgi and Kraicer, 1978).

Because so few sperm are present at the site of fertilization, this population undoubtedly represents capacitated sperm that are capable of fertilization. But in what regions in the female reproductive tract is capacitation complete in the fertilizing sperm? The site at which capacitation is complete under natural conditions is not known (Yanagimachi, 1981, 1988; Bedford, 1983; Meizel, 1985; Talbot, 1985). Various regions of the female tract probably contribute to capacitation in vivo, but almost nothing is known about the contribution of specific factors of the in vivo environment to this process. In the rabbit, sperm become capacitated during incubation in either the uterus or oviduct, but sequential exposure to these sites is more effective than either alone. (Austin, 1952; Bedford, 1969; Hunter and Hall, 1974). Human sperm rapidly gain the ability to adhere to and penetrate the zona pellucida after migrating through a column of human cervical mucus in vitro, much more quickly than this functional capacity is gained by control, washed sperm (Overstreet et al, 1980a; Gould et al, 1985). It has been proposed that migration through the mucus microstructure imparts upon sperm some of the surface characteristics that accompany capacitation, perhaps removal of surface proteins (see the section "Modification of the Sperm Surface"). Although sperm gain the ability to fertilize oocytes during incubation at various sites within the female reproductive tract, or during culture in vitro, the capacitation process may not be the same under these various experimental conditions (Clegg, 1983; Moore and Bedford, 1983; Yanagimachi, 1981, 1988). Sperm that bypass the uterus, for instance, may not form the appropriate associations with the oviductal isthmus that ensure a prolonged life span. An indication that conditions in vitro differ significantly from those in vivo is that millions of sperm per oocyte are required to achieve optimal fertilization rates in vitro. There is more than one explanation for this phenomenon. The small proportion of the total sperm that enter the ampulla during the periovulatory period may represent a population selected during the process of sperm transport and having the required attributes to achieve fertilization. In other words, only sperm capable of fertilization may be capable of attaining this site in a timely manner (Mortimer, 1978; Bedford, 1983). Alternatively, the process of sperm transport may more effectively confer upon sperm the physiologic attributes required for fertilization than does incubation in vitro. It is also possible that oocyte physiology has been altered by the manipulations used to prepare it for insemination in vitro. The processed oocyte may no longer provide the optimal chemical and physical environment to support fertilization.

Sperm Interaction with the Oocyte

In most species, oocytes are surrounded during fertilization in vivo by the relatively rigid zona pellucida (hereafter referred to as the zona), composed of several glycoproteins, and by the cumulus oophorus (hereafter referred to as the cumulus), composed of cumulus cells and their extracellular matrices. Except perhaps in ruminants, the cumulus matrix, a viscoelastic gel composed primarily of the glycosaminoglycan hyaluronic acid, remains intact during fertilization and for several hours thereafter (Talbot, 1985). The location at which the fertilizing sperm completes the acrosome reaction is not known, but, for several reasons, it is believed to occur within the ampulla in close association with the oocyte and its vestments (Meizel, 1978, 1985; Yanagimachi, 1981, 1988; Talbot, 1985): (1) the majority of motile sperm collected from the ampulla after coitus have intact acrosomes, (2) survival following the acrosome reaction is limited, (3) conditions in the oviduct and/or molecules associated with the oocyte vestments are believed to induce the acrosome reaction, and (4) sperm enzymes associated with the acrosome are believed to participate in the acrosome reaction. The location at which the fertilizing sperm completes the acrosome reaction may seem to be of little practical consequence. However, without this information, it is difficult to devise assays to assess physiologic response to natural inducers of the acrosome reaction. Furthermore, sperm binding to the zona appears to involve specific receptors (discussed below), and these could be located on different membranes before and after the acrosome reaction. Without knowledge of the acrosomal status of the fertilizing sperm during initial contact with the zona, it is difficult to study the cell recognition elements involved. In other cell types, exocytotic events that occur rapidly, on demand, are triggered by specific inducers (Burgess and Kelley, 1987). This is also true in nonmammalian sperm that acrosome-react in response to molecules on the surface of specialized layers investing the oocyte (Macek and Shur, 1988; Garbers, 1989; Shapiro et al, 1990). It is not known what substance or combination of substances induce(s) the acrosome reaction in mammalian sperm in vivo. As is discussed below, a number of substances capable of inducing the acrosome reaction in vitro are present in oviductal fluid or are associated with the oocyte vestments. It is possible that the fertilizing sperm may respond to multiple signals, or that some inducers serve as backup systems to ensure that the acrosome reaction occurs under various conditions (Yanagimachi, 1981; Meizel, 1985; Boatman, 1990). In some invertebrates and amphibians, multiple components of the egg investments are required for maximal acrosomal response (Boatman, 1990). In mammals, the physiologic stimulus may depend upon the interval between coitus and ovulation, because this factor determines the duration of sperm storage in the female reproductive tract. Indeed, some sperm undergo the acrosome reaction spontaneously, and these may be capable of fertilization under some conditions. It is also possible that certain inducing substances actually serve to limit the number of sperm reaching the oolemma. Early induction of the acrosome reaction may be one of many strategies that limit the number of sperm reaching the oocyte, thus reducing the risk of polyspermic fertilization. It will be difficult to determine experimentally which molecules are necessary or sufficient to ensure induction of the acrosome reaction in vivo.

The relative contribution of sperm forces and enzymatic cleavage to penetration of the cumulus and zona is controversial (see Green, 1988; Katz and Drobnis, 1990). Two enzyme activities associated with sperm have been studied in this regard: hyaluronidase, which is capable of dispersing the cumulus cell matrix, and acrosin, which is able to disperse the zona (Talbot, 1985). These enzymes are localized primarily within the acrosomal interior and are released during the acrosome reaction. In general, the acrosome reaction is thought to occur either just prior to sperm penetration of the cumulus (initiated by the cumulus) or after sperm attachment to the zona (initiated by the zona). Because the sperm releases its hyaluronidase activity during the acrosome reaction (e.g., Zao et al, 1985), it is reasonable to hypothesize that the acrosome reaction occurs just before or during penetration of the cumulus. The first approach to testing this hypothesis has been to determine the acrosomal condition of sperm during cumulus penetration. When oocytes are flushed from the oviducts of mated hamsters after penetration of multiple sperm into the cumulus (i.e., after fertilization is complete), most of the sperm within the cumulus are acrosome-reacted or have "modified" acrosomes (Yanagimachi, 1981, 1988). However, it is unknown if these supernumerary sperm retain fertilization competence. A number of more recent experiments have confirmed that hamster and rabbit sperm having intact acrosomes are capable of penetrating the cumulus and adhering to the zona in vitro (Drobnis and Katz, 1991). Indeed, acrosome-reacted sperm stick to peripheral cumulus cells and are prevented from entering the matrix.

However, others have reported that closer inspection of hamster sperm penetrating the cumulus reveals that their acrosomes are modified (Cummins and Yanagimachi, 1986). These authors have suggested that the acrosome reaction is initiated before penetration of the cumulus, thereby allowing efflux of hyaluronidase activity from the modified acrosome to dissolve the cumulus matrix surrounding the sperm head. In contrast to this notion, transmission electron micrographs of sperm within the cumulus show the matrix material completely intact around the surface of penetrating hamster sperm (Yudin et al, 1988a). Scanning electron micrographs of sperm penetrating into the zona material show an acrosomal remnant associated with the sperm midpiece (Yanagimachi and Phillips, 1984) consistent with an acrosome reaction at the zona surface. Results of experiments using hyaluronidase inhibitors have given mixed results (Talbot, 1985). Overall, it remains unclear if the acrosome reaction is initiated within the cumulus and/or if acrosomal enzymes participate in sperm penetration of the cumulus in vivo.

Another approach used to determine the site of the acrosome reaction has been to establish whether or not acrosome-reacted sperm are capable of adhering to the zona (Saling, 1989). Sperm can attach to the zona before capacitation, but can they adhere to the zona and complete fertilization after the acrosome reaction? If sperm can attach to the zona only before the acrosome reaction, then this event must occur on the zona. This seems to be the case for mouse sperm (Saling et al, 1978, 1979; Saling, 1989). In contrast, the sperm of guinea pigs, rabbits, and humans can apparently adhere to the zona either before or after the acrosome reaction (Kuzan et al, 1984; Myles et al, 1987; Morales et al, 1989). In these species, if sperm acrosome-react within the cumulus, they may still be able to complete fertilization. One possible problem with the above studies is that cumulus-free oocytes were used, making the results difficult to interpret. The cumulus is a selective barrier that restricts the access of some highly motile sperm. For example, noncapacitated sperm cannot penetrate into the cumulus matrix, presumably due to their surface characteristics (e.g., Cummins and Yanagimachi, 1982, 1986; Cherr et al, 1986). Sperm that are capable of fertilizing cumulus-free oocytes may not be capable of cumulus penetration. The site of the acrosome reaction in the fertilizing sperm remains unclear.

The sperm acrosome contains the proenzyme form of acrosin, a trypsinlike protease that occurs in an inactive, zymogen form (Meizel and Mukerji, 1976; Parrish and Polakoski, 1979) that is activated during the acrosome reaction. Some acrosin activity may remain associated with the inner acrosomal membrane (Brown and Hartree, 1974, 1976; Morton, 1975). Thus, even if the acrosome reaction occurred in the cumulus at a distance from the zona, sufficient acrosin activity could remain with the sperm to function during zona penetration. Experiments using protease inhibitors have produced conflicting results (Talbot, 1985). The pronounced "penetration slit" left in the zona matrix following sperm penetration (e.g., Austin and Bishop, 1958; Bedford, 1972) has been taken as evidence both of enzymatic digestion (e.g., Stambaugh, 1978) and of mechanical cutting of the sperm head into the zona (e.g., Yanagimachi, 1988). The role of the sperm forces during penetration of the oocyte investing coats has been studied by direct observations and videomicrography (Gaddum-Rosse, 1985; Katz and Drobnis, 1990; Drobnis and Katz, 1991). Hamster sperm that enter the cumulus in vitro have motility qualitatively similar to hyperactivated sperm swimming in the culture medium outside the cumulus mass (Drobnis et al, 1988a). Quantitatively, sperm motions are damped by the viscous and elastic properties of the cumulus microstructure, and their velocity is considerably slower than in low-viscosity medium. Additional information about the physical properties of this biopolymer will be required before conclusions can be drawn about the contribution of sperm forces to cumulus penetration.

After penetrating the cumulus, hamster and rabbit sperm having intact acrosomes form loose associations with the zona surface, and the sperm head pivots around its point of attachment (e.g., Cummins and Yanagimachi, 1982; Suarez et al, 1984; Cherr et al, 1986). After the acrosome reaction, sperm head becomes more rigidly attached to the zona surface and initiates penetration into the zona material (Suarez et al, 1984; Cherr et al, 1986). These two steps in zona attachment may correspond to "loose" and "tight" binding, "attachment" and "binding," or "primary" and "secondary" binding that have been described based on the ease of removing adherent sperm from the zona, the temporal sequence involved (Saling, 1989), and/or the experimental conditions required for each type of adherence (sperm receptors for the zona are discussed in the section "Induction of the Acrosome Reaction"). By light and electron microscopy, the space immediately surrounding the point of zona penetration appears cleared of cumulus matrix (Drobnis et al, 1988; Yudin et al, 1988a). Hyaluronidase released during the acrosome reaction may dissolve the matrix locally, allowing the flagellum to produce higher-amplitude flagellar beats and produce greater forces against the zona material. Sperm display vigorous motility during zona penetration (Drobnis et al, 1988b). The flattened sperm head rocks within the zona material and appears to cut knifelike into the matrix. Calculations based on flagellar kinematics suggest that sperm forces are probably sufficient to enable sperm to penetrate the zona mechanically. It is likely that both enzymatic and mechanical mechanisms are required for sperm penetration of the zona.

Following penetration into the perivitelline space, most sperm fuse readily with the oocyte, and sperm motility ceases. The region of the sperm plasma membrane that fuses with the oolemma overlies the former equatorial segment of the acrosome. The equatorial segment is the most distal region of the acrosome (closest to the flagellum) adjacent to the boundary at which the outer and inner acrosomal membranes are continuous (Yanagimachi, 1981, 1988; Moore and Bedford, 1983). The acrosomal contents do not disperse completely from this region following the acrosome reaction and can be seen, using labeled antibodies or lectins that recognize acrosomal contents, as an equatorial band around the head of the acrosome-reacted sperm (Cross and

Meizel, 1989; Wolf, 1989; Bavister, 1990). Modification of this specialized region of the plasma membrane during the acrosome reaction is believed to prepare it for fusion with the oolemma (Moore and Bedford, 1983).

CELLULAR CHANGES THAT ACCOMPANY THE ACROSOME REACTION

The acrosome can be viewed as a large secretory organelle, positioned between the nuclear envelope and the plasma membrane in the anterior two-thirds of the sperm head. Although it is unique in some ways, the acrosome reaction is classified as an exocytotic event involving exocytosis of this secretory granule. In other cell types, exocytosis generally involves fusion of small, spherical vesicles with the plasma membrane such that the membrane of the exocytotic vesicle is incorporated into the plasma membrane and the vesicle contents are expelled into the extracellular fluid. The acrosome reaction differs from this normal pattern because the acrosome is irregularly shaped and is quite large, underlying the majority of the sperm head plasma membrane. The major intracellular compartments of the sperm head are the nucleus and the acrosome. Only a thin layer of cytoplasm is present (Yanagimachi, 1981). Fusion between the outer acrosomal membrane and the plasma membrane occurs at multiple locations, forming a "fenestrated shroud" or "acrosomal ghost" (Gwatkin et al, 1976). This acrosomal remnant, composed of fused outer acrosomal membrane and plasma membrane (Franklin et al, 1970), is lost from the cell following the acrosome reaction. In thin sections through the sperm head, the acrosomal ghost appears as individual vesicles (e.g., Yudin et al, 1988b). The complete shroud is visible by light microscopy in some species (e.g., Cummins and Yanagimachi, 1986), and the details can be seen in scanning electron micrographs (Gwatkin et al, 1976). Dispersal of the acrosomal contents may be similar to that in other exocytotic systems. Prior to the acrosome reaction, the acrosomal contents are condensed, having a highly ordered ultrastructural appearance that has been termed the acrosomal matrix. During the acrosome reaction, the acrosomal contents decondense (e.g., Yudin et al, 1988b), perhaps accompanied by conversion of zymogen enzymes, such as acrosin, to their active form (Meizel and Mukerji, 1976). Once membrane fusion has occurred, the acrosomal contents disperse rapidly.

The acrosome reaction can occur spontaneously both in vitro and in vivo. In general, as the proportion of sperm capable of induced acrosome reactions increases, so does the background occurrence of spontaneous acrosome reactions in the population. Although it is possible that sperm undergoing spontaneous reactions in vivo may become the fertilizing sperm, it is also possible that spontaneous acrosome reactions are associated with senescence or other damage to sperm. Although there is no strong experimental evidence for this view, it is generally agreed that the acrosome reaction in the fertilizing sperm is induced by specific substances. As is discussed in this section, the acrosome reaction is commonly viewed as a stimulus–secretion event analogous to those in other cell types.

Induction of the Acrosome Reaction

Many candidates for an inducer substance are present at the site of fertilization. Follicular fluid, the cumulus, and the zona can induce the acrosome reaction, as can various bioactive molecules associated with the oocyte or present in the oviduct, including progesterone, sulfated glycosaminoglycans (GAGs), serum albumin, serotonin, prostaglandins, unsaturated fatty acids, and a variety of enzymes (Meizel, 1978, 1985; Yanagimachi, 1981, 1988). Some of these substances may actually play a permissive rather than an inductive role. In other words, they may produce some physiologic change in sperm that is required for the acrosome reaction to occur. For example, if sperm are capacitated in the absence of calcium, the acrosome reaction will not occur, but subsequent addition of calcium results in a synchronized acrosome reaction (Meizel, 1978; Yanagimachi, 1981). This does not imply that calcium induces the acrosome reaction.

Stimulation of exocytosis in many cellular systems, such as hormone-sensitive cells, is initiated by the binding of an inducer substance or ligand to a specific receptor. Receptors are integral proteins in the plasma membrane of the target cell, having active sites exposed on the cell surface that recognize and bind strongly to specific ligands. Stimulus–secretion coupling involves transduction of this signal to the cell interior via intracellular second-messenger systems and/or changes in membrane permeability to ions. Exocytosis occurs as a consequence of these intracellular signals. Signal transduction elements involved in stimulus–secretion coupling in other cells are present in mammalian sperm, and two inducer substances have been partially characterized with respect to their interaction with these systems: zona glycoproteins and progesterone. These molecules are the best candidate ligands for receptor-mediated induction of the acrosome reaction because these molecules act at the sperm surface and elicit intracellular responses analogous to those seen in other exocytotic cells.

In many species, including humans (Cross et al, 1988), zona proteins or solubilized zonae are capable of inducing the acrosome reaction in capacitated sperm. In all species studied, there are three to four zona glycoproteins (Wassarman, 1988). The role of the specific zona proteins has been studied most extensively in the mouse, in which these proteins have been called ZP1, ZP2, and ZP3 (Wassarman, 1988). By distinct mechanisms that can be separated experimentally, ZP3 is the mouse zona protein responsible for both sperm adhesion to the zona and induction of the acrosome reaction (Kopf and Gerton, 1991). The term *adhesion* is commonly used to describe attachment of cells to extracellular matrices (e.g., Hortsch and Goodman, 1991). I will use this term to denote the attachment of sperm to the zona surface in order to differentiate this process from *binding* of ligand to a receptor.

Consistent with a receptor-mediated phenomenon, zona proteins bind to the sperm head (Bleil and Wassarman, 1986; O'Rand and Fisher, 1987; Gwatkin, 1988). However, in this regard, it is difficult to differentiate between the two different zona ligand–sperm receptor events: sperm adhesion to the zona and zona induction of the acrosome reaction. Each of these events appears to involve specific zona ligands and sperm receptors, and each involves at least two phases with separate controlling mechanisms (Saling et al, 1990; Kopf and Gerton, 1991). Although adhesion to the zona and induction of the acrosome reaction can be separated experimentally, it is not entirely clear whether this is due to a distinct zona ligand or due to other experimental factors.

As mentioned above, sperm adhesion to the zona involves two phases that I will refer to as *primary adhesion* (sometimes called loose binding or attachment) and *secondary adhesion* (sometimes called tight binding or simply binding). Primary sperm adhesion to the zona is not strictly species-specific [e.g., horse sperm adhere to the human zona (Liu et al, 1991)], suggesting considerable homology of the responsible adhesion zona ligand. Moreover, this phase of attachment does not require capacitation (e.g., Bedford, 1967; Gwatkin and Williams, 1977), and it can occur with zonae of embryos (Barros and Yanagimachi, 1972; Inoue and Wolf, 1975). In contrast to primary adhesion, secondary adhesion is species-specific, requires sperm capacitation, and does not occur on the zonae of fertilized eggs and embryos. There is now experimental evidence that primary adhesion occurs prior to the acrosome reaction and secondary adhesion occurs following the acrosome reaction (Wassarman, 1987; Saling, 1989): (1) direct observation of hamster and rabbit sperm on the zona have described a tighter association with the zona for acrosome-reacted sperm (see above), (2) the time interval between primary and secondary attachment is equivalent to that required for induction of the acrosome reaction with disaggregated (i.e., "solubilized") zona, (3) sperm tightly adherent to the zona are acrosome-reacted, and (4) tight adherence, like the acrosome reaction, requires extracellular calcium. Recent experiments have suggested a mechanism for this (Bleil and Wassarman, 1986). ZP3 is the principal zona protein binding to the sperm head before the acrosome reaction, while ZP2 has this property for acrosome-reacted sperm. One explanation of these results is that primary adhesion involves binding of a sperm receptor to ligand(s) on ZP3, and then, after the acrosome reaction, secondary adhesion involves binding of a different sperm receptor to ligand(s) on ZP2 (Saling, 1989).

In a number of species, proteins on the sperm surface bind to zona proteins and may act as receptors for zona ligands. These surface proteins are diverse (Fraser and Ahuja, 1988; Saling, 1989; Kopf and Gerton, 1991). Some of these proteins have been isolated and partially characterized. It is still not clear for any species which sperm proteins are involved in primary adhesion, secondary adhesion, and/or induction of the acrosome reaction. Research is active in this area, and the results of these efforts should provide answers to these questions in the near future. There is some evidence that multiple ligands, perhaps on the same zona molecule, must bind to and cross-link sperm receptors (Saling, 1989; Saling et al, 1990; Leyton and Saling, 1989). Proteolytic fragments of ZP3 that bind to sperm but do not induce the acrosome reaction, gain inducing activity under conditions in which multiple fragments are linked, for example, by anti-ZP3 antibodies (Saling et al, 1990). This evidence, as well as the results of ZP3-binding and acrosome reaction kinetics, suggests that the acrosome reaction is induced by aggregation of zona receptors on the sperm surface.

Progesterone is secreted by cumulus cells and is capable of inducing the acrosome reaction in human sperm (Meizel et al, 1990). This activity was first discovered in human follicular fluid, which contains significant amounts of progesterone; however, follicular fluid induces acrosome reactions in a higher proportion of sperm than does progesterone alone (Meizel et al, 1990). Some other component of follicular fluid, and presumably of the cumulus, may act together with progesterone to produce the full effect. Progesterone acts at the sperm surface, initiating a physiologic response (influx of calcium) within 2 s in capacitated human sperm (Meizel and Turner, 1991). Sperm surface receptors for progesterone have not been identified.

Signal Transduction Elements

Signal transduction pathways can be complex, involving multiple elements, each having multiple effects (Berridge, 1985). The best-studied signal transduction elements in other cell types also appear to be present in sperm. These include intracellular free calcium ion concentration ($[Ca^{2+}]_i$), pH_i, G-proteins, inositol phospholipid metabolism, and intracellular concentration of cAMP ($[cAMP]_i$). Appropriate manipulation of the concentrations or activities of transduction intermediaries affects the acrosome reaction, suggesting that these systems may be involved.

As in many other exocytotic systems, an increase in $[Ca^{2+}]_i$ is required for induction of the acrosome reaction (Yanagimachi, 1981; Fraser and Monks, 1990; Fraser, 1990; Florman and Babcock, 1991). Transport molecules in the plasma membrane maintain $[Ca^{2+}]_i$ levels of most cells in the nanomolar range. Elevation of $[Ca^{2+}]_i$ acts as an intracellular signal, modulating the activity of a variety of enzyme systems and promoting membrane fusion (Rasmussen, 1989). Influx of Ca^{2+} across the sperm plasma membrane is required for the increase in $[Ca^{2+}]_i$ that accompanies the acrosome reaction (Garbers and Kopf, 1980; Yanagimachi, 1981, 1988; Florman and Babcock, 1991). Sperm cells possess plasma membrane Ca^{2+}-ATPase activity that probably represents a transport protein (Ashraf et al, 1982; Breitbart et al, 1984). Most eucaryotic cells have similar "calcium pumps," which function to maintain low $[Ca^{2+}]_i$ by exporting calcium. A second transport protein may also be present that exchanges extracellular sodium ions (Na^+) for Ca^{2+} (see Bradley and Forrester, 1985; Roldan and Flemming, 1989; Florman and Babcock, 1991). If such an antiport system were active, the large gradient of Na^+ across the plasma membrane (maintained by the plasma membrane Na^+,K^+-ATPase) would drive the export of calcium. As is discussed below, $[Ca^{2+}]_i$ increases during capacitation. It is generally agreed that a threshold $[Ca^{2+}]_i$ level must be reached before the acrosome reaction can be induced. Induction of the acrosome reaction is then accompanied by a much larger increase in $[Ca^{2+}]_i$. An increase in $[Ca^{2+}]_i$ can occur by various mechanisms including (1) inhibition of Ca^{2+} export by the outwardly directed ATPase, by the Na^+, Ca^{2+} antiporter, or by other Ca^{2+} transport systems, (2) activation of a Ca^{2+} transport molecule that allows Ca^{2+} to enter the cytoplasm, and/or (3) influx of Ca^{2+} due to increased permeability of the lipid bilayer to Ca^{2+}. It is not known which of these mechanisms is responsible for the elevation of $[Ca^{2+}]_i$ during capacitation or the more extensive influx of Ca^{2+} during the acrosome reaction (Bradley and Forrester, 1985; Fraser and Monks, 1990; Florman and Babcock, 1991). There is some experimental evidence that mammalian sperm have voltage-gated calcium channels similar to those found in muscle cells, although other studies have produced conflicting results (see Roldan and Harrison, 1990; Florman and Babcock, 1991). Such channels respond to depolarization of the membrane potential by allowing Ca^{2+} influx. In some experiments, treatments expected to produce depolarization across the sperm plasma membrane resulted in increased $[Ca^{2+}]_i$, as would be expected if such channels exist, and this response was blocked by calcium channel blockers or by the absence of extracellular Ca^{2+} (Florman and Babcock, 1991). Also, the acrosome reaction in mouse sperm can be inhibited by appropriate calcium channel blockers (Fraser and Monks, 1990). Similar pharmacologic studies have suggested that this Ca^{2+} transport system is involved in the rapid increase in $[Ca^{2+}]_i$ induced by zona proteins but is not involved in spontaneous acrosome reactions (Florman and Babcock, 1991). Mammalian sperm may also have ligand-gated calcium channels that allow Ca^{2+} influx as a direct or indirect consequence of the binding of zona ligands to sperm receptors (Florman and Babcock, 1991).

If $[Ca^{2+}]_i$ is raised in capacitated sperm by increasing membrane permeability to Ca^{2+} with pharmacologic agents (e.g., the calcium ionophore A23187), the acrosome reaction occurs. This presumably bypasses membrane signal transduction events that normally produce increased $[Ca^{2+}]_i$ in response to extracellular inducers. Substances such as zona proteins and progesterone that induce the acrosome reaction by acting at the surface of capacitated sperm probably exert their effect by increasing membrane permeability to Ca^{2+} either directly or indirectly. Zona proteins are able to induce an increase in $[Ca^{2+}]_i$ in bovine sperm after a lag time of a few minutes (Florman et al, 1989). Once $[Ca^{2+}]_i$ is increased, the acrosome reaction requires up to 15 min for completion. In this regard, the acrosome reaction differs from other exocytotic events, which are often complete within seconds or even milliseconds. Progesterone induction of the acrosome reaction in human sperm comes closer to this normal time course: $[Ca^{2+}]_i$ is elevated within seconds (Thomas and Meizel, 1989; Blackmore et al, 1990), and the acrosome reaction is complete within 1 min (Meizel et al, 1990). There are several possible reasons for the temporal difference in events induced by zona proteins versus progesterone, including (1) differences among species, (2) differences in the efficacy of the in vitro capacitation and induction systems employed, or (3) a difference in the complexity of the signal transduction pathways for these inducers. It is not known how the kinetics of these induction systems seen experimentally compare with those occurring under natural conditions in vivo.

Increasing $[Ca^{2+}]_i$ alone will not always induce the acrosome reaction. If human sperm are treated with progesterone prior to capacitation, $[Ca^{2+}]_i$ increases without induction of the acrosome reaction (Thomas and Meizel, 1989). Similarly, treatment of human sperm with A23187 presumably increases the $[Ca^{2+}]_i$ but does not stimulate the acrosome reaction unless sperm are capacitated (e.g., Lee et al, 1987). Clearly there are elements of this system, downstream of Ca^{2+} influx, that gain the ability during capacitation to respond to the $[Ca^{2+}]_i$ increase. Calcium exerts its effects in other cells in part via calcium-binding proteins called calmodulins that are activated when bound to Ca^{2+} (Cheung, 1982). Calmodulin is present in mammalian sperm (Jones et al, 1980; Moore and Dedman, 1984; Weinman et al, 1986) and may be involved in induction of the acrosome reaction. Other calcium-sensitive systems that have been identified include the enzymes phospholipase A_2 and phospholipase C (Meizel, 1984).

A second change in ion distribution accompanies the acrosome reaction, alkalinization of the acrosome and of the cytosol. Normally the pH of the acrosome (pH_a) is maintained at approximately 4 to 5 (Meizel and Deamer, 1978; Meizel, 1984) and sperm pH_i at 6.3 to 6.7, compared with a pH of over 7 for physiologic media and secretions of the female tract. Secretion vesicles in some other cells are acidic until just before exocytosis, and this may reduce the activity of enzymes until secretion. Activation of acrosin is optimal at a pH greater than 5, so this may also be the functional significance of the acidic pH_a. The pH gradients across the acrosomal and plasma membranes are lost during the zona-induced acrosome reaction in mouse and bovine sperm, and proton efflux from the sperm occurs (Lee and Storey, 1985; Florman et al, 1989). Thus, proton transport occurs across both the acrosomal membrane and the plasma membrane. Presumably this ion efflux is coupled with transport of other ions across the two membranes to maintain electroneutral conditions. Exchange of potassium ions (K^+) for protons may occur across the acrosomal membrane (Mrsny and Meizel, 1981). Mechanisms involving K^+, Na^+, or bicarbonate ion (HCO_3^-) transport have been proposed, but it is not yet clear what transport processes are linked with the increases in pH_i and pH_a (Meizel, 1985; Yanagimachi, 1981, 1988; Florman and Babcock, 1991).

One mechanism by which binding of an inducer molecule to a surface receptor can be coupled to intracellular events is via GTP-binding regulatory proteins called G-proteins (Gilman, 1987; Levitzki, 1988). These mobile, integral membrane proteins can be activated by forming associations with the inducer-bound receptor. Subsequent interaction of the activated G-protein with other membrane proteins can modulate the membrane protein activity. This association activates enzymes or transport proteins, resulting in intracellular changes. In some cells, G-proteins mediate increases in $[Ca^{2+}]_i$ and pH_i, and there is recent evidence that they may also serve these functions in mammalian sperm (Kopf et al, 1986; Bentley et al, 1986; Endo et al, 1987a,b, 1988). G-proteins are present in mammalian sperm and have been localized in the acrosomal area. Specific inhibition of G-protein activity with pertussis toxin inhibits the ZP3-induced acrosome reaction in mouse sperm without affecting adhesion of sperm to the zona (Endo et al, 1987a,b, 1988). Pertussis toxin also inhibits the zona-induced acrosome reaction in bovine sperm and blocks the increases in $[Ca^{2+}]_i$ and pH_i (Florman et al, 1989).

A signal transduction pathway in many cellular systems that has received considerable recent attention involves polyphosphoinositides (Majerus, 1986). Activation of the membrane enzyme phospholipase C (PLC), via activation of G-proteins, results in the release of inositol phosphates and diacylglycerol (DAG) from membrane phosphatidylinositides (PPIs). The PLC involved is specific for PPIs, and it requires at least low levels of $[Ca^{2+}]_i$ for activity (Nishizuka, 1986). Two of the inositol phosphates, IP_3 and IP_4, stimulate increases in $[Ca^{2+}]_i$ via release from intracellular stores or influx from the extracellular fluid (Berridge, 1988). DAG is believed to activate protein kinase C, a key signal transduction protein that is involved in regulation of enzyme activity by phosphorylation. Some recent evidence indicates that this signal transduction system may be involved in induction of the acrosome reaction (Roldan and Harrison, 1990; Harrison and Roldan, 1990; Kopf and Gerton, 1991). All the requisite elements have been found in mammalian sperm. PPIs have been localized to the inner surface of the plasma membrane and to the outer surface of the underlying outer acrosomal membrane (Berruti et al, 1986). Calcium-dependent, PPI-specific PLC activity is present in sperm (Ribbes et al, 1987), and G-proteins have been detected as mentioned above. Induction of the acrosome reaction with calcium ionophore, follicular fluid, or progesterone in capacitated mammalian sperm from a number of species results in hydrolysis of PPIs and release of both IP_3 and DAG (Thomas and Meizel, 1989; Roldan and Harrison, 1989). This process is dependent upon influx of Ca^{2+} and is complete within 3 min of $[Ca^{2+}]_i$ elevation, after which completion of the acrosome reaction requires 5 to 15 min (Roldan and Harrison, 1990). These results suggest that the second messengers released during hydrolysis of PPIs may have a role in signal transduction downstream from the increase in $[Ca^{2+}]_i$ but prior to membrane fusion. Consistent with this model, specific inhibition of PPI breakdown with neomycin blocks acrosomal exocytosis in response to elevation of $[Ca^{2+}]_i$ with A23187, but only if the neomycin is added within 3 min of the $[Ca^{2+}]_i$ increase (Roldan and Harrison, 1989). In other cells, the DAG produced during hydrolysis of PPIs by PLC activates protein kinase C. However, it is not clear that protein kinase C is present in mammalian sperm (Roldan and Harrison, 1990). Nonetheless, application of exogenous DAG, or stimulation of DAG buildup in sperm, accelerates the acrosome reaction induced by A23187 (Roldan and Harrison, 1990). It has been suggested that this action may be via stimulation of phospholipase A_2 because DAG does have this activity in other cells (Roldan and Harrison, 1990) and because this enzyme is involved in the mammalian sperm acrosome reaction (see below).

Cyclic AMP serves as a second messenger in many cellular systems (Berridge, 1985). The $[cAMP]_i$ is controlled by the activity of two enzymes, adenylate cyclase, an integral membrane protein that catalyzes the conversion of ATP to cAMP, and cyclic nucleotide phosphodiesterase, a cytosolic enzyme that catalyzes the breakdown of cAMP to AMP. The activity of adenylate cyclase is often stimulated by the binding of hormones or other ligands to their specific receptors on the cell surface. This interaction may be mediated by G-proteins and may be calcium-dependent. The intracellular activity of cAMP involves the activation of various protein kinases that in turn activate or inactivate other enzymes and/or transport molecules by phosphorylation. In mammalian sperm, $[cAMP]_i$ increases during capacitation (see below), but the relationship of this to induction of the acrosome reaction is not clear (Fraser and Ahuja, 1988; Fraser and Monks, 1990). Elevation of $[cAMP]_i$ by pharmacologic means does not induce the acrosome reaction in the absence of elevated $[Ca^{2+}]_i$, but it has a permissive effect. The best evidence is that $[cAMP]_i$ must reach some threshold value in order for the acrosome reaction to occur.

The acrosome reaction may take place in at least two steps having separate controlling mechanisms. Evidence of an intermediate stage in the ZP3-induced mouse acrosome reaction has been obtained using the fluorescent dye chlortetracycline (CTC), which fluoresces when bound to Ca^{2+} in a hydrophobic environment. Just before the acrosome reaction is complete, a distinct change in fluorescence pattern occurs that has been used to characterize two steps in the acrosome reaction (Saling and Storey, 1979; Ward and Storey, 1984). The first stage may be accompanied by the loss of the pH gradient across the plasma membrane (Lee and Storey, 1985). Each of these two stages of the acrosome reaction can be inhibited separately, indicating that they are separate processes (Kopf and Gerton, 1991). For example, pertussis toxin, the G-protein inhibitor, does not reduce sperm adhesion to the zona, does not prevent the alkalinization of the acrosome, and does not alter the kinetics of the first stage of the acrosome reaction, characterized by increased CTC fluorescence. However, pertussis toxin–treated mouse sperm do not complete the acrosome reaction (Endo et al, 1987a, 1988). Similarly, the zona of activated oocytes induces only the first stage of the acrosome reaction in sperm bound to the zona or treated with solubilized zona. CTC fluorescence increases, but, in this case, the pH gradient across the sperm membranes is maintained (Endo et al, 1987b; Kligman et al, 1991). Agents that activate protein kinase C, such as DAG, accelerate the first stage of the reaction and slow down the second stage of the reaction, indicating that these stages may have independent control elements (Lee et al, 1987). This technique was first described for mouse sperm but has now been applied to other species. During in vitro capacitation and the acrosome reaction, human sperm treated with CTC show a progression through three labeling patterns similar to those seen in mouse sperm (Lee et al, 1987), and this may indicate that the acrosome reaction also occurs in discrete stages in human sperm.

Membrane Fusion and Dispersion of the Acrosomal Contents

In order for the acrosome reaction to occur, changes in the plasma membrane overlying the acrosome and the outer acrosomal membrane must occur in order to allow fusion of these membranes. Membrane fusion is highly unfavorable thermodynamically. Even once the membranes have been brought into close apposition, the hydrophobic interiors of the two membrane bilayers must somehow overcome the intervening hydrophilic surfaces in order to become a single, continuous membrane. Certain fusogenic lipids in the membrane are believed to destabilize the lipid bilayer, helping to overcome this barrier (Meizel, 1984; Langlais and Roberts, 1985; Yanagimachi, 1988). The gradual buildup of fusogenic lipids during capacitation is discussed below. However, a more rapid buildup of fusogenic lipids may participate in the membrane fusion events of the acrosome reaction (Meizel, 1984; Roldan and Harrison, 1990). The membrane enzyme phospholipase $A_2(PLA_2)$ has been implicated in this process. This membrane enzyme is calcium-dependent, so it may be activated by the increase in $[Ca^{2+}]_i$ during the acrosome reaction. Once activated by calcium, PLA_2 cleaves the fatty acid from the central carbon of the glycerol backbone of phospholipids. The resulting products, fatty acids and lysophospholipids (and perhaps derivatives of these), are fusogenic lipids. The addition of exogenous lysophosphatidylcholine (LPC), the most common membrane lysophospholipid, to capacitated sperm will induce the acrosome reaction in a number of species (Meizel, 1984; Yanagimachi, 1988). PLA_2 has been found in the sperm of many mammalian species, and specific inhibitors of its activity block the acrosome reaction (Meizel, 1984). Although PLA_2 is calcium-dependent, activation of this enzyme is not the only process during the acrosome reaction that requires increased $[Ca^{2+}]_i$. Stimulation of the acrosome reaction in vitro with LPC still requires extracellular Ca^{2+}, indicating that increased $[Ca^{2+}]_i$ is required for other events.

The argument can be made that induction of the acrosome reaction by exogenously applied fusogens is not physiologic because these agents can destabilize membranes nonspecifically. Membrane destabilization may simply increase membrane permeability to Ca^{2+} and protons. However, exogenous enzymes may be present at the site of fertilization and may be involved in producing LPC, which then acts on sperm membranes (Langlais and Roberts, 1985). PLA_2 activity is found in serum and follicular fluid, and some may remain associated with the cumulus matrix following ovulation (Langlais and Roberts, 1985). Similarly lecithin:cholesterol acyltransferase (LCAT) is present in serum and follicular fluid (Langlais and Roberts, 1985) as is discussed further below. This enzyme transfers a fatty acid from phosphatidylcholine to cholesterol, producing LPC and cholesterol ester. Thus, the induction of the acrosome reaction by exogenous LPC may not be entirely nonphysiologic.

At least one trypsinlike protease may be involved in the membrane fusion events of the acrosome reaction (Meizel, 1984; Meizel et al, 1990; Pillai and Meizel, 1991) as well as in exocytosis in other cell types. Permeant inhibitors of trypsin activity (that cross cell membranes) can inhibit or delay the acrosome reaction in sperm from a number of species. Impermeant inhibitors generally do not inhibit the acrosome reaction (Meizel and Lui, 1976; Lui and Meigel, 1979). This suggests that the requisite protease activity is intracellular. The protease may be involved in activation of PLA_2 or adenylate cyclase as it is in other cells (Meizel, 1984). There is some evidence that protease activity is involved in the lateral reorganization of membrane components to produce the protein-free areas of the plasma membrane believed to participate in membrane fusion (Langlais and Roberts, 1985).

In addition to protease activity in the membrane events of the acrosome reaction, a protease appears to be required for activation of acrosin and dispersion of the acrosomal contents (Meizel, 1984; Langlais and Roberts, 1985). At the ultrastructural level, the acrosomal contents are electron-dense, having an ordered appearance that has been termed the acrosomal matrix. The acrosomal matrix decondenses during the acrosome reaction (e.g., Yudin et al, 1988b). Decondensation and dispersal of this matrix is apparently controlled separately from the membrane fusion events of the acrosome reaction (Shams-Borhan and Harrison, 1981; Perreault et al, 1982; Fraser, 1982). Prevention of this process with protease inhibitors prevents the complete dispersal of the matrix following the membrane fusion events of the acrosome reaction and inhibits sperm fusion with the oolemma (Green, 1978; Shams-Borhan and Harrison, 1981; Perreault et al, 1982; Fraser, 1982), supporting a role for protein hydrolysis in this process.

CELLULAR CHANGES THAT ACCOMPANY CAPACITATION

The earliest work on sperm capacitation involved experiments in which sperm were capacitated in the female reproductive tract. The ability of sperm to fertilize oocytes was the experimental endpoint (Bedford, 1970a). At that time, the incubation conditions required for capacitation in vitro had not been determined. This early experimentation was time-consuming, technically challenging, and difficult to control. With the advent of in vitro capacitation and fertilization techniques (Yanagimachi and Chang, 1963) and the ability to capacitate sperm and fertilize oocytes in "defined" culture media (Toyoda et al, 1971), more rapid progress was possible toward characterizing the capacitation process. In general, media used for capacitation in vitro are modified Krebs–Ringer or Tyrode's balanced salt solutions, partially buffered with bicarbonate and including metabolites and serum albumin. Although no specific component of these media is an absolute requirement for capacitation, specific combinations of constituents are more effective for sperm of individual species (Rogers, 1978; Bavister, 1981; Oliphant et al, 1985). Specific chemical composition, gas phase, and temperatures enhance capacitation in various species. Although these in vitro systems have provided valuable tools for the study of sperm capacitation, it is important to remember that results obtained in vitro must be validated for physiologic relevance. The fact that certain experimental conditions are sufficient to permit the acrosome reaction in vitro does not demonstrate that similar conditions are either necessary or sufficient in vivo. Many pharmacologic conditions will support the acrosome reaction.

If, as is often the case, capacitation is viewed as the cellular changes that prepare sperm to respond to induction of the acrosome reaction by appropriate inducers, then at least the following must be true of capacitated sperm: (1) for induction of the acrosome reaction by a specific ligand, the receptor must be in its active form and accessible to the ligand, (2) the membrane elements of the transduction system must be active and must be mobile in the membrane as required, (3) $[Ca^{2+}]_i$ and $[cAMP]_i$ must have reached their required threshold levels, (4) signal transduction must result in a further increase in $[Ca^{2+}]_i$, and (5) the increase in $[Ca^{2+}]_i$ must result in initiation of the acrosome reaction. Each of these requirements can be viewed as occurring during capacitation because none of them, as far as we know, are true for freshly ejaculated sperm. Noncapacitated sperm are capable of primary adhesion to the zona (see "Induction of the Acrosome Reaction"), but they cannot penetrate the cumulus, and they will not undergo the acrosome reaction, even when $[Ca^{2+}]_i$ is elevated (e.g., Meizel and Turner, 1991).

Many cellular changes have been described that accompany capacitation in vitro. For the most part, it is modifications of the sperm plasma membrane that constitute capacitation. These include (1) modification of protein and saccharide components of the membrane surface, (2) altered activity and distribution of integral membrane proteins, and (3) changes in the composition and physical properties of the lipid bilayer.

Capacitation-Associated Changes in the Sperm Plasma Membrane

The basic structure of cellular membranes consists of a lipid bilayer in which proteins are embedded (Yeagle, 1987; Gennis, 1989). Some proteins have structural functions. Others are receptors, enzymes, and transport molecules involved in maintaining homeostasis and in transmitting extracellular signals across the membrane. The plasma membrane is usually about half lipid and half protein. In most animal cells, the plasma membrane contains significant amounts of cholesterol in addition to the phospholipid in the lipid bilayer (Gennis, 1989). Some membrane proteins are mobile and free to move laterally in the plane of the fluid membrane, at least within a given domain of the cell surface. Separate domains exist in the lateral plane of the membrane, characterized by specific lipid and protein composition and by specific physical and functional properties. Mammalian sperm are quite complex with respect to their membrane domains (Koehler, 1978; Myles and Primakoff, 1985) and are often used in textbooks as examples of cells having distinct membrane domains. Because the membrane is fluid, the existence of domains is puzzling. It is not known how membrane domains are maintained, although there is some evidence for several mechanisms (Houslay and Stanley, 1982): (1) cytoskeletal elements on the inner surface of the plasma membrane anchor some proteins and may even be active in moving membrane proteins to new surface locations, (2) coating substances on the cell exterior form associations with membrane proteins, (3) membrane proteins form associations and aggregate, and (4) lipids form associations due to their geometric shapes and thermodynamic properties. The outer surface of the plasma membrane can be quite complex. A saccharide layer, composed of oligosaccharide groups of membrane glycoproteins and the saccharide head groups of certain membrane lipids, forms a surface coat similar in thickness to the lipid bilayer (typically extending 0.1 nm from the membrane surface). These so-called glycoconjugates are at least partly responsible for the cell's surface charge. Often cells have a negative surface charge as the result of terminal sialic acid residues on membrane glycoconjugates. In the case of mammalian sperm, sulfated saccharides and/or sulfolipids may be more important than sialic acid in producing the negative surface charge (Bedford and Cooper, 1978). Membrane glycoconjugates are important in cell–cell recognition and adhesion events. In addition to these membrane elements, adsorbed proteins are often noncovalently associated with the cell surface (i.e., they can be removed by changing pH and/or ionic strength of the extracellular medium). Coordinated with these membrane surface elements, many cells have extracellular coats composed of GAGs and their associated protein.

Based on what we know about the composition and structure of the plasma membrane of mammalian sperm, the membrane does not differ significantly from that of other cells. As discussed in this section, sperm capacitation is complex, involving major alterations of each element of the plasma membrane. The composition, organization, and physical properties of the sperm surface and membrane bilayer are all modified in preparation for the acrosome reaction (Yanagimachi, 1981, 1988; Eddy and O'Brien, 1989).

Increased Permeability to Calcium

As mentioned above, $[Ca^{2+}]_i$ increases gradually during capacitation, and a threshold value may be required before the acrosome reaction can be induced. It appears that the ability to accumulate Ca^{2+}, which may be viewed as an increased membrane permeability to Ca^{2+}, develops during capacitation. Freshly ejaculated sperm maintain very low $[Ca^{2+}]_i$, on the order of 100 nM (e.g., Irvine and Aitken, 1982; Florman et al, 1989). The extracellular levels of free Ca^{2+} to which sperm are exposed in vivo can be difficult to determine because Ca^{2+} ions may be noncovalently associated with other molecules such that they are not free to participate in other processes. Nonetheless, at least in culture medium, $[Ca^{2+}]$ is on the order of 1 to 5 mM, but free levels may be considerably lower. There is a large difference in $[Ca^{2+}]$ across the sperm plasma membrane before capacitation. Cells maintain low $[Ca^{2+}]_i$ by means of plasma membrane transport mechanisms. As discussed above, mammalian sperm plasma membranes appear to possess at least two calcium pumps that export Ca^{2+} from the cell as well as voltage-gated and perhaps ligand-gated calcium channels believed to participate in the large Ca^{2+} influx that occurs just prior to the acrosome reaction. The gradual rise in $[Ca^{2+}]_i$ during capacitation is probably regulated via these or other calcium transport molecules (Fraser and Ahuja, 1988; Florman and Babcock, 1991).

One consequence of increased $[Ca^{2+}]_i$ during capacitation may be the increase in $[cAMP]_i$ seen in mammalian sperm. During capacitation, adenylate cyclase activity increases and cyclic nucleotide phosphodiesterase activity decreases (Morton and Albagli, 1973; Stein and Fraser, 1984). Calcium is involved in similar regulation of these enzymes in other types of cells. Increased $[cAMP]_i$ during capacitation, and its activation protein kinases, may in turn be involved in regulation of membrane Ca^{2+} transport. In general, the intracellular function of cAMP is protein phosphorylation via stimulation of cAMP-dependent protein kinases. In mouse sperm there is one protein that is only phosphorylated under capacitating conditions (Fraser, 1990), and this may be cAMP-dependent.

Modification of the Sperm Surface

During capacitation, the molecular composition and physical properties of the sperm surface change (Yanagimachi, 1981, 1988; Clegg, 1983; Eddy and O'Brien, 1989). Surface proteins and carbohydrates are removed, added, modified, and redistributed (Yanagimachi, 1981, 1988). There is also a loss of the negative surface charge that was acquired by the sperm during epididymal maturation. These surface modifications have been detected using a variety of techniques, and they may not be entirely independent. Changes in carbohydrates may involve glycoproteins, and decreased negative charge may reflect modification of terminal sugar groups. One intriguing aspect of capacitation is that it is reversible in vitro. Capacitated sperm can be reversibly "decapacitated," particularly by incubation with specific seminal plasma proteins and lipids called "decapacitation factors" (Chang, 1957; Dukelow et al, 1965). These factors become incorporated into the membrane surface elements and are later lost as the sperm capacitate again. The reversible nature of capacitation has been an important consideration in the formulation and testing of hypotheses regarding the cellular mechanism of sperm capacitation (Yanagimachi, 1981, 1988; Langlais and Roberts, 1985; Florman and Babcock, 1991).

Removal of adsorbed proteins from the sperm surface is generally recognized as a prerequisite for capacitation (Yanagimachi, 1981, 1988; O'Rand, 1982; Oliphant et al, 1985; Eddy and O'Brien, 1989). Certain proteins become strongly associated with the sperm surface during epididymal maturation or at ejaculation and are lost during capacitation. Some of these have been shown to act as decapacitation factors (Oliphant, 1976; Oliphant et al, 1985; Vernon et al, 1987). In vivo, these substances may prevent premature capacitation and ensure that hyperactivated motility and/or the acrosome reaction occur(s) in the vicinity of the oocyte. Even extensive washing does not remove from the sperm surface some of these proteins that are subsequently lost during capacitation (Okabe et al, 1986a,b). In vivo, removal of seminal fluids from sperm is initiated soon after coitus as sperm migrate rapidly out of the seminal plasma and into the secretions of the female genital tract. During incubation in the female reproductive tract or in tract fluids, sperm both lose proteins from their surfaces (e.g., Dravland and Joshi, 1981; Vernon et al, 1987) and perhaps adsorb new proteins (Volglmayr and Sawyer, 1986).

There is evidence for the presence of decapacitation factors in the semen of a variety of species. For example, one glycoprotein, called acrosome-stabilizing factor (ASF), has been isolated from rabbit epididymal fluid and seminal plasma. ASF has decapacitating activity, is found associated with ejaculated but not with epididymal sperm (Eng and Oliphant, 1978), and is lost (or modified) during sperm residence in the uterus. Moreover, caudal epididymal sperm from a number of species have been shown to capacitate more rapidly than ejaculated sperm, suggesting that sperm-associated seminal constituents inhibit capacitation.

It is not known how protein decapacitation factors exert their effects. Adsorbed proteins may form associations with membrane proteins and inhibit their mobility within the plane of the membrane. Inhibition of the migration or of the interaction of membrane receptors, transport molecules, and/or signal transduction elements may then prevent the acrosome reaction until the adsorbed proteins are removed (Yanagimachi, 1988). One mechanism for which there is some evidence is that decapacitation factors bind to membrane proteins and act as regulators of membrane enzymes and/or transport proteins, perhaps inhibiting the increase in $[Ca^{2+}]_i$ or $[cAMP]_i$ (Fraser, 1990; Florman and Babcock, 1991). Proteins that inhibit Ca^{2+} uptake by sperm, and therefore inhibit capacitation, are present in the seminal plasma of various species, including humans (Babcock et al, 1979). A decapacitation protein that can be removed from mouse sperm by gentle washing may be involved in sperm response to extracellular Ca^{2+} (Fraser et al, 1990). A calcium transport inhibitor in bovine seminal plasma that appears to act on the Na^{+}/Ca^{2+} exchange molecule has been partially characterized and has been called caltrin (Rufo et al, 1982, 1984; San Agustin et al, 1987). Another protein that stimulates Ca^{2+}-ATPase activity and may promote Ca^{2+} efflux has been isolated from the seminal plasma of sheep and humans (Bradley and Forrester, 1985). It has been suggested that capacitation may be reversible until $[Ca^{2+}]_i$ reaches a threshold level (Langlais and Roberts, 1985). According to this model, reversal of processes that increase membrane permeability to Ca^{2+} can then be viewed as decapacitating activity.

The role of seminal plasma in capacitation is an important consideration in sperm studies using some model species. In rodents, epididymal rather than ejaculated sperm are usually studied because ejaculated sperm are difficult to collect directly from the male. Because the time course of capacitation is different for epididymal sperm than for ejaculated sperm in at least some species, a change in sperm physiology mediated through seminal plasma, or changes in the seminal plasma itself, may affect sperm capacitation in vivo. These effects cannot be detected by studying epididymal sperm.

Changes in the sperm surface during capacitation in vitro have been detected using labeled antibodies and lectins. Lectins are proteins that recognize specific terminal saccharide residues and can be used to detect changes in glycoconjugates. In a number of species, including humans, the pattern of lectin binding on the sperm surface changes during capacitation, indicating that surface carbohydrates have been lost, modified, unmasked, and/or rearranged (e.g., Nicholson and Yanagimachi, 1974; Nicholson et al, 1977; Millette, 1977; Schwarz and Koeler, 1979; Koeler, 1981; Cross and Overstreet, 1987). These glycoconjugates could be associated either with extracellular glycoproteins and GAGs or with glycolipids and glycoproteins of the membrane bilayer.

Not only surface proteins are modified during capacitation. Integral membrane proteins of the lipid bilayer are also modified. Monoclonal antibodies that recognize specific integral membrane proteins change their degree of binding and their labeling pattern on the sperm surface. The best-studied membrane protein in this regard is PH20 of guinea pig sperm (Myles et al, 1990). Monoclonal antibodies directed against this protein inhibit sperm adhesion to the zona, suggesting that PH20 may be a zona receptor. PH20 migrates from the anterior sperm head to the posterior sperm head during epididymal maturation, and then migrates to the anterior head after the acrosome reaction. This lateral reorganization of PH20 distribution in the membrane may prevent loss of this protein during the acrosome reaction and may account for the observation that guinea pig sperm can adhere to the zona either before or after the acrosome reaction. These observations are consistent with alterations of surface molecules during capacitation in vitro. Much less work has been done to examine the sperm surface during capacitation in vivo.

It is likely that the female reproductive tract participates in modification of the sperm surface by removing, altering, or adding substances.

The sperm acquires an increasingly negative surface charge during epididymal maturation (e.g., Bedford, 1963) due to an increase in negatively charged sialic acid residues (Nicholson et al, 1977) and sulfate groups (e.g., Farooqui, 1978; Langlais et al, 1981). Reduction of the negative surface charge during capacitation is believed to be due to loss, removal, or modification of sialoglycoproteins, sulfated glycoproteins, sulfated GAGs, and sulfated lipids in the membrane or on its surface (Yanagimachi, 1981; Langlais and Roberts, 1985). It has been proposed that specific enzymes present in sperm and in the female reproductive tract have the ability to effect these modifications. One example is arylsulfatase, which catalyzes the removal of sulfate groups from lipids, GAGs, and glycoproteins. Enzymes having arylsulfatase activity are found in relatively high concentrations in seminal plasma, associated with sperm, in the secretions of the female reproductive tract, and associated with the cumulus (Langlais and Roberts, 1985). The functional significance of the change in sperm surface charge is not known. Surface charge may be involved in the interaction of sperm with the secretions and cells of the female reproductive tract. Also, membranes having a negatively charged surface are less permeable to cations. The reduction of the negative surface charge may be involved in increasing the permeability of the membrane to Ca^{2+} and promoting membrane fusion.

Membrane Lipid Composition and Membrane Fluidity

The lipid composition of sperm membranes changes significantly during capacitation. Over the years, these changes have usually been studied by extracting the membrane lipids from whole sperm or from plasma membrane preparations. Reviewing the results of this work, one could envision overall changes in the lipid composition of sperm membranes. However, in order for the acrosome reaction to occur, it is not necessary that all sperm membranes become fusogenic. In fact, a general increase in membrane fluidity and permeability would reduce the ability of the sperm to maintain homeostasis. The establishment of apposed fusogenic membrane domains in the plasma and outer acrosomal membranes should be sufficient (Yanagimachi, 1988; Parks and Ehrenwald, 1990). Indeed, there is evidence that this is what occurs. During capacitation, the plasma membrane undergoes major physical changes. Studies using fluorescence recovery after photobleaching have revealed that the mobility of sperm membrane constituents increases during capacitation, indicating greater membrane fluidity (Wolf et al, 1986; Wolf, 1987). As seen using freeze-fracture electron microscopy, areas form over the sperm head during capacitation that are free of intramembranous particles (IMPs) (Koehler and Gaddum-Rosse, 1975; Friend et al, 1977; Tesarik, 1984). These IMPs are thought to represent intrinsic membrane proteins and are present in all plasma membranes except certain specialized areas in some cells. The clear areas that form during sperm capacitation are believed to be the points at which membrane fusion is initiated. This is because removal of protein would allow the close apposition of the plasma and outer acrosomal membranes required for membrane fusion (Friend et al, 1977). Experiments using filipin, which binds to membrane cholesterol and can be detected by freeze-fracture, have indicated that the IMP-free domains appearing on the plasma membrane over the acrosome during capacitation are also free of cholesterol (Bearer and Friend, 1982; Tesarik and Flechon, 1986). The absence of appreciable cholesterol in these domains is thought to enhance the fusogenic properties of the membrane.

There are two concomitant changes in lipid composition during capacitation that are believed to prepare sperm membranes (or domains of sperm membranes) for the acrosome reaction: a decrease in the ratio of cholesterol to phospholipid (C/P) in the plasma membrane, and an increase in fusogenic lipids. Cholesterol can act to stabilize membranes, making them less permeable, less fluid, and less susceptible to fusion (Chapman and Benga, 1984; Shinitzky, 1984; Bloch, 1985; Yeagle, 1985, 1989). In addition to these direct effects on the lipid bilayer, alteration of membrane physical properties can influence the mobility, conformation, and activity of membrane enzymes and transport molecules (Gordon and Mobley, 1985; Yeagle, 1989). In many animal cells, the alteration of the C/P is an important mechanism for regulating membrane fluidity and permeability; these membrane properties normally increase with lower C/P. The plasma membrane of uncapacitated sperm from most mammalian species has a high C/P (Watson, 1981). Human sperm have particularly high cholesterol levels, approaching 50 percent (mole fraction) of the plasma membrane lipid. Sulfated sterols have even stronger stabilizing effects on membranes than cholesterol does (Langlais and Roberts, 1985). Cholesteryl sulfate is secreted by the epididymis and incorporated into the sperm membrane during epididymal maturation (Legault et al, 1979) and is also present in seminal plasma (Lalumiere et al, 1976).

Recently, evidence has been accumulating that loss of cholesterol from sperm membranes is required for sperm capacitation (Langlais and Roberts, 1985; Parks and Ehrenwald, 1990). C/P does decrease in mammalian sperm membranes (Go and Wolf, 1983; Ehrenwald et al, 1988a). Furthermore, cholesterol is able to reversibly decapacitate rabbit sperm, and exposure of sperm to cholesterol inhibits sperm capacitation in a number of species (Davis and Hungund, 1976; Davis, 1976, 1980; Fleming and Yanagimachi, 1981; Go and Wolf, 1985; Ehrenwald et al, 1988b). The membrane content of sulfated sterols also decreases during capacitation, and these membrane lipids are potent inhibitors of capacitation and fertilization in vitro and in vivo (Bleau et al, 1975; Fayrer-Hosken et al, 1987).

Because cholesterol has low solubility in aqueous media, extracellular cholesterol is usually found associated with transport molecules. Acceptor molecules can take up cholesterol from cell surfaces, promoting further efflux of cholesterol from the plasma membrane. Diffusion from the membrane is the rate-limiting step in cholesterol efflux (Levy, 1981). A complex system of acceptor proteins is involved in cholesterol transport in the blood and regulation of free cholesterol levels at cell surfaces. The major protein that transports cholesterol in the blood is high-density lipoprotein (HDL). Removal of cholesterol from the plasma membrane in other cells is thought to involve HDL together with LCAT and apolipoprotein A (Van Deenen and De Gier, 1974; Levy, 1981). Esterification of acyl chains to cholesterol by LCAT removes free, unesterified cholesterol and promotes further diffusion of cholesterol from the lipid bilayer. Apolipoprotein A, which is associated with HDL, acts as a cofactor in this LCAT reaction. Cholesteryl ester transfer proteins transfer cholesteryl ester from HDL to low-density lipoprotein (LDL), which recycles HDL for further uptake of cholesterol. The secretions of the female reproductive tract contain acceptor molecules capable of promoting cholesterol efflux from sperm (Langlais et al, 1981; Davis, 1982; Langlais and Roberts, 1985; Ehrenwald et al, 1988a,b). These acceptors include HDL, apolipoprotein A, and cholesteryl ester transfer activity in serum, follicular fluid, and female tract secretions, and LCAT in follicular fluid (Langlais and Roberts, 1985; Parks and Ehrenwald, 1990). Under the influence of estrogen, the amounts and/or activities of cholesterol acceptor proteins increase in serum, reproductive tract fluids, and follicular fluid. Peak levels occur during the periovulatory period, when these fluids have the maximum ability to capacitate sperm (Langlais and Roberts, 1985; Parks and Ehrenwald, 1990). Serum albumin, which is present in most capacitation media, can act as a cholesterol acceptor (Go and Wolf, 1985). In vitro, albumin may act as the cholesterol acceptor, but its activity is much lower than that of HDL in vivo (Langlais et al, 1988). Other serum proteins present as contaminants of serum albumin preparations may also have this activity (Ravnik and Muller, 1989).

The negative surface charge of sulfated cholesterol stabilizes its interaction with the lipid bilayer and inhibits its removal by cholesterol acceptor proteins. Arylsulfatase C is an integral membrane protein that desulfates cholesteryl sulfate. This enzyme is present in the epithelia of the female reproductive tract and in cumulus cells (Langlais and Roberts, 1985). Desulfation of cholesteryl sulfate and subsequent efflux of membrane cholesterol may play important roles in preparing the sperm membrane for the acrosome reaction (Langlais and Roberts, 1985).

Membrane stabilization by cholesterol and cholesteryl sulfate are greatest when there is a high degree of unsaturation in the fatty acid side chains of membrane phospholipids. Mammalian sperm do have particularly high levels of long-chain, unsaturated fatty acyl chains in their membrane lipids. In the sperm of most species, some 30 percent of fatty acid side chains have 22 carbons with 5 to 6 double bonds (i.e., docosahexaenoic acid and docosapentaenoic acid) (Watson, 1981). In the absence of cholesterol, such membranes have high fluidity and permeability. In addition to these highly fluid lipid components, fusogenic lipid levels also increase in sperm membranes during capacitation (Davis et al, 1979; Soupart et al, 1979). The three-dimensional shape of cholesterol suggests that it would form a stable complex with the fusogenic lipids, LPC, and free fatty acids in membranes, reducing the fusogenic activity of these lipids (Rand et al, 1975). Experimental evidence for this is that cholesterol and cholesteryl sulfate are able to inhibit hemolysis of erythrocytes treated with LPC (Kellaway and Saunders, 1969).

Methylation of membrane phospholipids may also promote fusion. Removal of methyl groups from the predominant membrane phospholipid, phosphatidylcholine (PC; also called lecithin), converts it to phosphatidylethanolamine (PE). Because of the small head group of PE, it is susceptible to forming nonbilayer phases that disrupt the lipid bilayer. Formation of these reverse hexagonal, or HexII, phase structures is believed to be involved in the initiation of membrane fusion during exocytosis in other cells. Inhibitors of phospholipid methylation prevent the acrosome reaction in hamster sperm (Meizel, 1985).

In addition to desulfation of cholesteryl sulfate, sperm capacitation may involve desulfation of seminolipid, a sulfogalactolipid found only in mammalian sperm and testis (Murray and Narasimhan, 1990). The desulfated form of this lipid is highly fusogenic, promoting HexII phase structures in membranes (Tanphaichitr et al, 1990). Desulfation of seminolipid may be involved in the buildup of fusogenic lipids in sperm membranes during capacitation. Arylsulfatase A, an enzyme associated with mammalian sperm and seminal plasma, is capable of desulfating seminolipid (Yang and Srivastava, 1974; Dudkiewicz, 1984). Sperm content of seminolipid decreases during capacitation of mouse sperm (Tanphaichitr et al, 1990).

Lipid changes during capacitation help to explain some of the nonspecificity of this process. Capacitation requires no specific stimulus and will occur in many environments, including most sites within the female reproductive tract, the reproductive tract of females of any species, and simple culture media. The only exogenous requirements for the lipid changes described in this section are a cholesterol acceptor and a temperature high enough to promote adequate cholesterol diffusion from the lipid bilayer. The presence of highly unsaturated acyl chains, high cholesterol content, and the machinery to produce fusogenic lipids preprograms the sperm plasma membrane to achieve the physical characteristics required for the acrosome reaction.

Cytoskeletal Elements

Characterization of the cytoskeletal elements of the sperm membrane has received little experimental attention, and the participation of these elements in the events of capacitation and the acrosome reaction has not been studied. Cytochemical studies have demonstrated the presence in mammalian sperm of actin and the molecules constituting intermediate filaments in other cells (Olson et al, 1987). In addition, the mammalian sperm head contains a novel cytoskeletal protein that probably acts to stabilize the equatorial region (Longo et al, 1987; Olson and Winfrey, 1988; Toshimori et al, 1991). The sperm cell is highly polarized, having multiple membrane domains in the plasma membrane that differ in their ultrastructural appearance, membrane fluidity, and functional properties. Cytoskeletal elements are believed to participate in the maintenance of such domains. Although the involvement of the sperm cytoskeleton in capacitation and the acrosome reaction remains purely speculative, the importance of such elements in maintaining membrane domains argues for a role of cytoskeletal elements in the membrane reorganization events of capacitation.

Hyperactivated Motility

Hyperactivated motility has been described for the sperm of a large number of species following capacitation in vivo and/or in vitro (Yanagimachi, 1988; Drobnis and Katz, 1991). Although this form of motility accompanies capacitation and is probably required for penetration of the zona, it is not a reliable marker for capacitation. Experimentally, hyperactivation can be separated from capacitation (Yanagimachi, 1981; Suarez and Pollard, 1990), and it can even be induced reversibly in uncapacitated sperm (Suarez et al, 1987). In fact, the motility patterns displayed by caput epididymal sperm, which are incapable of progressive motility, are similar to those of capacitated sperm (Suarez, 1988). It has been suggested that part of decapacitation by epididymal and seminal fluids is imparting linear motility upon the ejaculated sperm cell (Suarez, 1987; Suarez and Pollard, 1990).

On the cellular level, flagellar dynamics have been studied using so-called sperm models. In these preparations, the sperm membranes are removed with detergent and sperm motility is reactivated by adding ATP (Brokaw, 1986, 1987; Lindemann and Kanous, 1989; Tash and Means, 1987; Tash, 1989, 1990). Such studies have revealed that sperm motility is regulated primarily by $[cAMP]_i$ and $[Ca^{2+}]_i$. The former regulates the initiation and maintenance of motility while the latter regulates the extent of flagellar bending. Other factors, including pH_i are also important but play relatively minor roles. Vigorous motility requires cAMP (e.g., Lindemann, 1978). Thus, the increase in $[cAMP]_i$ during capacitation is consistent with the occurrence of hyperactivated motility (Lindemann and Kanous, 1989). The activity of cAMP is mediated via cAMP-dependent protein kinases and phosphorylation of proteins associated with the flagellar axoneme.

The effects of Ca^{2+} on flagellar motility are mediated through calmodulin (Lindemann and Goltz, 1988; Tash, 1990). In the presence of very low $[Ca^{2+}]_i$, sperm models swim with straight trajectories (Lindemann and Goltz, 1988). High-curvature bends, which result in asymmetric waveforms and nonlinear trajectories in intact sperm, are produced by increasing the $[Ca^{2+}]$ into the micromolar range (Lindemann et al, 1987; Lindemann and Goltz, 1988). Waves of increased $[Ca^{2+}]_i$, traveling along the flagellum, produce local, high-curvature bends that reverse as $[Ca^{2+}]_i$ returns to lower levels. The

$[Ca^{2+}]_i$ at which the curvature reverses is close to an inflection point on the Ca^{2+}–calmodulin binding curve, suggesting that Ca^{2+} interaction with calmodulin governs the direction of flagellar bending. Further increases in $[Ca^{2+}]_i$, to levels above the low millimolar range, result in cessation of flagellar activity. This model is consistent with our current understanding of $[Ca^{2+}]_i$ during capacitation. Increasing $[Ca^{2+}]_i$ results in the high-curvature flagellar bends, asymmetric beat shapes, and nonlinear trajectories of hyperactivated motility.

Kinetics of Capacitation

The time required by sperm to undergo capacitation may be an important functional endpoint, yet it can be difficult to assess. Not only is there variation among species in capacitation time, but variation among fertile males within one species and among sperm from one male (Shalgi et al, 1981; Perreault and Rogers, 1982; Yanagimachi, 1984). Sperm samples collected from the epididymis, the ejaculate, or the female tract are always heterogeneous in their physiology (Bedford, 1983). Subpopulations are present that differ in their rate of capacitation under given incubation conditions. After incubation in vitro, only a fraction of the sperm will acrosome-react in response to an inducer, while other sperm will not respond and still others will have become immotile or will have died. To some extent, this variation in sperm physiology may function to provide functionally competent sperm to the site of fertilization under a variety of physical, chemical, and temporal conditions in vivo (Bedford, 1983).

For the purposes of in vitro assays, the capacitation rate varies considerably among species; for example, in vitro capacitation requires only 1.5 to 2 h for mouse sperm, but requires 5 to 10 h for human sperm. The rate of capacitation is also sensitive to sperm preparation techniques and incubation conditions such as temperature and composition of the medium (Rogers, 1978, 1985; Bavister, 1981). In particular, the protein supplementation can profoundly affect the time required for capacitation (e.g., Rogers, 1985; Bronson and Rogers, 1988). In vivo, different regions in the female reproductive tract capacitate sperm at different rates, and the time required depends upon the stage in the estrus cycle (Parks and Ehrenwald, 1990; Yanagimachi, 1988).

CELLULAR OR MOLECULAR DEFECTS IN SPERM CAPACITATION AND THE ACROSOME REACTION

When evaluating sperm function, an important perspective is that not all motile, morphologically normal sperm are equal. A traditional semen evaluation provides important endpoints in assessing male fertility, including estimations of the total numbers of sperm produced by the male, the percentage of sperm that are motile, and the proportion of sperm having normal morphology. However, if male fertility is compromised by abnormal sperm capacitation and acrosome reaction, more sophisticated techniques will be required to detect these lesions in living, motile sperm. Sperm function can be evaluated at four levels: (1) fertility trials, (2) semen evaluation, (3) in vitro functional assays, and (4) cellular and molecular assays. A toxicologic effect occurs at the molecular level, but our ability to measure this damage is limited by our inadequate knowledge of sperm function. A better understanding of normal cellular and molecular mechanisms of capacitation and the acrosome reaction will provide insights into cellular and molecular abnormalities and their functional consequences. At this time, evaluation of sperm dysfunction resulting from toxicologic exposure of the male is limited primarily to fertility evaluation and semen evaluation. Toxicologists and sperm biologists are just beginning to apply assays of sperm function.

Functional assays of sperm capacitation and the acrosome reaction include measurements of (1) longevity and the kinetics of capacitation in vitro, (2) hyperactivation of sperm motility, (3) response to induction of the acrosome reaction, and (4) interaction with oocytes (cumulus penetration, zona binding, zona penetration, oocyte fusion). Assays of sperm function have been developed and are being applied to studies of male infertility. Such approaches will be used increasingly to study male reproductive toxicology. Recently, several of these functional assays have been refined in order to provide more cellular information (Bavister, 1990).

The pattern and vigor of sperm motility can now be determined with accuracy and precision using computer-aided sperm analysis (CASA) (Katz and Davis 1987; Toth et al, 1991; Davis et al, 1991). The trajectory of the sperm head during hyperactivation is characterized by decreased linearity (LIN) and increased amplitude of lateral head displacement (ALH). These characteristics can be detected using CASA (e.g., Mack et al, 1988; Neill and Olds-Clark, 1987; Robertson et al, 1988) and can be used to detect distinct subpopulations of sperm (Davis et al, 1991; Drobnis et al, 1992).

With the identification of specific inducers of the acrosome reaction, it has become possible to probe sperm for capacitation without using oocytes. Sperm can be treated with the inducing substance and the acrosomal response detected in a representative sample of the entire population of sperm (Lee and Storey, 1985; Suarez et al, 1986). When assessing acrosomal status of sperm, it is desirable to determine the acrosomal status of motile sperm specifically because sperm often lose their acrosomes when they die (Yanagimachi, 1981). The large acrosomes can be visualized directly in motile sperm of guinea pigs and hamsters. The smaller acrosomes of paddle-shaped sperm from species such as rabbits, dogs, and ruminants can also be visualized using light microscopy, but usually this is not possible for highly motile sperm. The sperm of humans, mice, and rats are too small for the acrosome to be visualized using light microscopy. New staining techniques (Cross and Meizel, 1989; Wolf, 1989) permit determination of acrosomal status on viable sperm that exclude supravital dye. These new techniques are an improvement over acrosomal staining techniques that failed to differentiate living from dead sperm.

Functional assays of sperm–oocyte interaction have also been improved and better standardized. The zona-free hamster oocyte (ZFHO) assay is based on the fact that the hamster oolemma has no species barrier to sperm fusion (Yanagimachi et al, 1976). This assay is used worldwide to test for sperm function in humans (World Health Organization, 1987). Fusion with ZFHO requires that a subpopulation of sperm is present that has capacitated, acrosome-reacted, and remained viable. Thus, it evaluates longevity as well as capacitation. Recent refinements of assays of sperm–zona interaction now allow separate assessment of zona binding, acrosome reaction on the zona, and zona penetration (Overstreet et al, 1980b; Gould et al, 1985; Morales et al, 1989; Overstreet and Cross, 1988; Coddington et al, 1990). Nonliving frozen or salt-stored zonae can be utilized (Yanagimachi et al, 1979; Yoshimatsu et al, 1988), and hemizonae cut from a single oocyte can be used to reduce the variability among zonae when treatments are being compared (Burkman et al, 1988, 1990). These in vitro assays can provide significant information regarding the level of sperm dysfunction.

In this section, I have outlined some of the cellular and molecular lesions that could disrupt normal capacitation and the acrosome reaction. I have suggested some methods for assessing these aspects of sperm function and given examples from studies of subfertility in men. However, even in the field of male infertility, there has been a paucity of research into the cellular and molecular mechanisms underlying sperm dysfunction. In consequence, at this time, a discussion of this type remains largely speculative.

Sperm Surface

The physical properties of the sperm surface are important to its function in vivo. Inappropriate surface characteristics could result in reduced ability of sperm to (1) migrate through the complex biopolymers that coat the epithelial surfaces of the female reproductive tract, (2) form appropriate interactions with epithelial cells and/or immunocompetent cells of the female tract, and (3) penetrate the oocyte investing coats.

One functional assay for surface characteristics (as well as motility) of human sperm is penetration into cervical mucus (World Health Organization, 1987), although this assay is used for freshly ejaculated sperm rather than for sperm during capacitation. Motile sperm from subfertile men have reduced ability to penetrate cervical mucus (Alexander, 1981; Hull et al, 1984; Keel et al, 1987). Videomicrography techniques have demonstrated that greater resistance to penetration of the mucus microstructure is experienced by abnormal sperm suggesting that their surface properties differ from those of normal sperm (Katz et al, 1990). Although there are currently no standardized assays to detect functional differences in the sperm surface during capacitation and the acrosome reaction, such assays could be devised. Hamster sperm adhere to cumulus cells and erythrocytes until they capacitate (Gwatkin et al, 1972, 1976; Hartmann, 1983). Human sperm lose the ability to penetrate cervical mucus and gain the ability to penetrate the cumulus during capacitation (Katz et al, 1988). These changes in surface properties could form the basis of assays for normal surface function during sperm capacitation.

At the cellular level, techniques are available to probe the cell surface. Lectin probes can be applied to detect modification and/or rearrangement of glycoconjugates. In one study, the sperm of subfertile men with abnormal ability to undergo the acrosome reaction had a decreased ability to bind to the GAG heparin (Lalich et al, 1989). This may or may not be a specific interaction as the authors suggest, but this result indicates differences in the surface characteristics of dysfunctional sperm. Surface proteins can also be evaluated by general labeling of the surface proteins of motile sperm, extracting the sperm proteins, and applying protein separation techniques to compare treated versus normal sperm (e.g., Oliphant and Singhas, 1979; Esbenshade and Clegg, 1980). Another approach is to measure the surface charge of sperm directly. This can be accomplished by focusing of sperm in an electric field (Vaidya et al, 1971; Rosado et al, 1973; Moore, 1979) or by probing the sperm surface with charged molecules or particles (Yanagimachi et al, 1972).

Lipid Composition

The lipid composition of the sperm plasma membrane is probably the most important factor in the process of capacitation and the acrosome reaction. In sperm membranes, a significant proportion of the fatty acid side chains of phospholipids are unusually long and unsaturated. Lipids having these characteristics promote membrane fluidity, permeability, and fusion. The presence of high cholesterol and sulfated sterol levels in the membrane before capacitation is probably required to prevent premature capacitation and acrosome reactions, while their removal during capacitation is required to promote the acrosome reaction at the appropriate time. These membrane lipid constituents, and perhaps others, may be critical for the coordination of these sperm processes in vivo.

Sperm having defective membrane lipid composition would have abnormal membrane permeability and fusibility and might display reduced longevity, premature capacitation, or inability to acrosome-react. These functional lesions might be detected by motility analysis, acrosomal challenge, and/or incubation with oocytes at various time points during incubation under capacitating conditions in vitro. Functional problems of these types have been seen in subfertile men. The ZFHO assay has been in use for a decade and has been applied extensively to assessing human sperm function. In some subfertile men, sperm are not capable of fusing with the oolemma of ZFHO or they fuse at reduced rates (Rogers, 1985). Cryopreserved sperm will fuse with oocytes after a shorter incubation time than is optimal for nonfrozen sperm, suggesting that capacitation is premature and longevity is reduced for the treated sperm (Critser et al, 1987). In other cases, treatment with the calcium ionophore A23187 overcomes deficits in ZFHO fusion (Aitken et al, 1984). Binding of sperm to the zona (Overstreet et al, 1980b; Burkman et al, 1988; Franken et al, 1990) and penetration through the zona (Overstreet et al, 1980b) are reduced for subfertile men. Less work has been directed to measuring acrosomal function in subfertile men, but there is some indication that in some cases, sperm dysfunction is associated with premature or inadequate acrosome reactions (Schill et al, 1988; Fenichel et al, 1991).

Alterations in lipid composition resulting in increased membrane permeability could affect the ability of sperm to control $[Ca^{2+}]_i$ and/or pH_i. Because membrane transport proteins have their hydrophobic components in contact with the lipid bilayer, lipid composition can also influence membrane permeability indirectly by influencing protein function. In a similar manner the activity of other membrane proteins could be affected, including receptors and signal transduction elements. Detection of decreased function of membrane proteins is discussed in the next section.

Abnormalities in lipid composition can be detected directly using biochemical methods, including altered C/P has been detected in the sperm from subfertile men (Huacuja et al, 1981; Sugkraroek et al, 1991). In addition to C/P, the phospholipid head groups, phospholipid-bound fatty acids, and the sperm-specific sulfogalactolipid could be evaluated in sperm before and during capacitation.

A special case of abnormal membrane lipids results from peroxidation (Horton and Fairhurst, 1987). Unsaturated fatty acids in cell membranes are particularly susceptible to attack by reactive oxygen species, making sperm more sensitive than most cells. Lipid peroxidation results in an increase in membrane permeability, loss of homeostasis, and eventual death (Slater, 1984; Horton and Fairhurst, 1987). Sperm from subfertile men do have increased lipid peroxidation (Aitken and Clarkson, 1987; Rao et al, 1989). Mammalian sperm, as well as most cells, both produce reactive oxygen species (Holland and Storey, 1981; Aitken et al, 1989a,b) and have protective systems to protect membranes from peroxidation (Alvarez and Storey, 1984, 1989; Alvarez et al, 1987). An increase in lipid peroxidation can result from any one (or more) of three factors: (1) an increase in the production of reactive oxygen species, (2) a decrease in the activity of protective systems, and/or (3) an increased susceptibility of lipids. Each of these factors can be measured separately in order to isolate the affected system.

Membrane Proteins

Membrane proteins important in sperm capacitation and the acrosome reaction include receptor molecules, signal transduction molecules, and transport molecules. The activity of these proteins may be altered by an abnormal lipid environment, as mentioned above, but there could also be a deficiency in the presence, exposure, and/or activity of membrane proteins. Almost any assay for sperm function in vivo could be altered by an abnormal membrane protein function, including decreased longevity, premature acrosome reaction, failure of the acrosome reaction, abnormal hyperactivation, and abnormal interaction with the zona and oolemma. Each of these functional assays has been used to detect sperm dysfunction, and differences exist in each between sperm from fertile and sperm from subfertile men.

As receptors are identified in the sperm plasma membrane, it will be possible both to quantitate receptor numbers using labeled antibody techniques and to develop assays for their receptor activity. Once the signal transduction elements involved in each receptor-mediated event are better characterized, it will be possible to evaluate the activity of these systems using specific agonists and antagonists. Clearly, these systems will have to be much better characterized before experiments can be designed to dissect out specific elements having abnormal function. For example, the $[cAMP]_i$ can be extracted from sperm for measurement, but the results would be difficult to interpret because the role of cAMP in capacitation and the acrosome reaction is not clear.

Normal function of membrane transport molecules is essential to normal sperm function. In particular, $[Ca^{2+}]_i$ and pH_i are involved in capacitation and the acrosome reaction. Increased permeability to Ca^{2+} and protons before capacitation could result in decreased longevity and premature acrosome reactions, whereas a failure to increase transport of these ions in response to the appropriate stimulus could result in failure of hyperactivation and/or the acrosome reaction. It is now possible to measure $[Ca^{2+}]_i$ and pH_i in living sperm both in bulk suspension and in individual cells (e.g., Florman et al, 1989). Better characterization of these transport molecules will allow us to use specific transport inhibitors in order to differentiate altered permeability of the lipid bilayer from abnormal permeability resulting from dysfunctional transporters.

Abnormal $[Ca^{2+}]_i$, pH_i, and/or $[cAMP]_i$ would inhibit fertilization by affecting capacitation and the acrosome reaction, but it is likely that these intracellular defects would also cause changes in sperm motility. Because the axoneme is sensitive to small changes in intracellular concentration of these factors, the details of sperm trajectories may provide information about abnormal sperm function even if the motility itself is adequate to support normal sperm function. The motility pattern is abnormal in sperm from subfertile men (Katz et al, 1981; Aitken et al, 1982a,b,c). Changes in the pattern and vigor of sperm motility have also been detected following systemic toxicologic exposure of the male. The velocity, LIN, and ALH of rat sperm changed significantly after acute or chronic treatment of the males with epichlorohydrin, while the percentage of motile sperm was not affected (Toth et al, 1991; Slott et al, 1990). In an occupational health study, it was determined that the wives of dry cleaners required longer to establish pregnancies than the wives of laundry workers (Eskenazi et al, 1991b). The semen of the dry cleaners and the laundry workers were not different in the percentage of motile sperm, but the ALH and LIN were different for the dry cleaners and were related to the level of perchloroethylene in expired air (Eskenazi et al, 1991a). These results emphasize that subtle effects on the physiology of motile sperm can result from systemic male exposure to toxicants.

Other Elements

The activity of acrosomal enzymes may also be a cause of sperm dysfunction. For example, the acrosin activity is lower in sperm from subfertile men (Mohsenian et al, 1982; Goodpasture et al, 1982; Koukoulis et al, 1989). With respect to capacitation and the acrosome reaction, timely activation, release, and/or exposure of these enzyme activities could be abnormal. For example, premature conversion of proacrosin to acrosin might result in reduced activity during zona penetration. Evaluation of these factors on relevant populations of sperm could prove to be important components of the evaluation of sperm function.

CONCLUDING REMARKS

Reduced fertility can result from inadequate function of the motile sperm in semen. Recent studies have demonstrated that subtle changes in sperm physiology can result from toxicologic exposure at levels that do not cause a difference in the percentage of motile sperm. It is likely that at least some of these sperm defects result in abnormal capacitation and acrosome reaction. Although this area is a major focus in sperm biology, the processes of sperm capacitation and induction of the acrosome reaction are complex and remain to be well characterized. Studies of these phenomena are difficult because they occur naturally in a complex series of microenvironments within the female reproductive tract and because only a small proportion of the sperm in an ejaculate will reach the oviduct following coitus. The characteristics that confer success upon this subpopulation have not been determined. Nonetheless, techniques of in vitro capacitation and the acrosome reaction have allowed more rapid characterization of these processes. The challenge to applied sperm biologists is to develop and apply assays of sperm function that will serve as effective endpoints for detecting decreased male fertility. Recent advances have been made in our ability to evaluate sperm function in vitro, including the development of CASA techniques for motility analysis, identification of inducers of the acrosome reaction that are useful probes for capacitation, refinement of acrosomal assays to discriminate between viable and dead sperm, and development of more informative assays of sperm–oocyte interaction. The results of these assays will be more easily interpreted if they are evaluated for their relationship with sperm function in vivo. Differences in sperm function detected by functional assays can then be studied further by applying cellular techniques to test hypotheses regarding the lesions underlying the detected sperm dysfunction. This process will benefit from the continuing efforts of basic sperm biologists to determine the mechanisms that support capacitation and the acrosome reaction.

REFERENCES

Aitken RJ, Clarkson JS: Cellular basis of defective sperm function and its association with the genesis of reactive oxygen species by human spermatozoa. *J Reprod Fertil* 81:459–469, 1987.

Aitken RJ, Best FSM, Richardson DW, Djahanbakhch O, Lees MM: The correlates of fertilizing capacity in normal fertile men. *Fertil Steril* 38:68–76, 1982a.

Aitken RJ, Best FSM, Richardson DW, Djahanbakhch O, Templeton AA, Lees MM: An analysis of semen quality and sperm function in cases of oligozoospermia. *Fertil Steril* 38:705–711, 1982b.

Aitken RJ, Best FSM, Richardson DW, Djahanbakhch O, Mortimer D, Templeton AA, Lees MM: An analysis of sperm function in cases of unexplained infertility: Conventional criteria, movement characteristics, and fertilizing capacity. *Fertil Steril* 38:212–221, 1982c.

Aitken RJ, Ross A, Hargreave T, Richardson D, Best F: Analysis of human sperm function following exposure to the ionophore A23187: Comparison of normospermic and oligozoospermic men. *J Androl* 5:321–329, 1984.

Aitken RJ, Clarkson JS, Fishel S: Generation of reactive oxygen species, lipid peroxidation, and human sperm function. *Biol Reprod* 41:183–197, 1989a.

Aitken RJ, Clarkson JS, Hargreave TB, Irvine DS, Wu FCW: Analysis of the relationship between defective sperm function and the generation of reactive oxygen species in cases of oligospermia. *J Androl* 10:214–220, 1989b.

Alexander NJ: Evaluation of male infertility with an in vitro cervical mucus penetration test. *Fertil Steril* 36:201–208, 1981.

Alvarez JG, Storey BT: Lipid peroxidation and the reactions of superoxide and hydrogen peroxide in mouse spermatozoa. *Biol Reprod* 30:833–841, 1984.

Alvarez JG, Storey BT: Role of glutathione peroxidase in protecting mammalian spermatozoa from loss of motility caused by spontaneous lipid peroxidation. *Gamete Res* 23:77–90, 1989.

Alvarez JG, Touchstone JC, Blasco L, Storey BT: Spontaneous lipid peroxidation and production of hydrogen peroxide and superoxide in human spermatozoa: Superoxide dismutase as major enzyme protectant against oxygen toxicity. *J Androl* 8:338–348, 1987.

Amann RP: Can the fertility potential of a seminal sample be predicted accurately? *J Androl* 10:89–98, 1989.

Ashraf M, Peterson RN, Russell LD: Activity and location of cation-dependent ATPases on the plasma membrane of boar spermatozoa. *Biochem Biophys Res Comm* 107:1273–1278, 1982.

Austin CR: Observations on the penetration of the sperm into the mammalian egg. *Aust J Sci Res* 4:581–596, 1951.

Austin CR: The "capacitation" of the mammalian sperm. *Nature* 170:326, 1952.

Austin CR, Bishop MWH: Role of the rodent acrosome and perforatorium in fertilization. *Proc R Soc Lond (Biol)* 149:241, 1958.

Babcock DF, Singh JP, Lardy HA: Purification and characterization of a calcium transport inhibitor protein from bovine seminal plasma. *Dev Biol* 69:85–93, 1979.

Barros C, Yanagimachi R: Polyspermy-preventing mechanisms in the golden hamster egg. *J Exp Zool* 180:251–266, 1972.

Bavister BD: Analysis of culture media for in vitro fertilization and criteria for success, in Mastroianni L Jr, Biggers JD (eds): *Fertilization and Embryonic Development in vitro*. New York, Plenum Press, 1981, pp 41–60.

Bavister BD: Tests of sperm fertilizing ability, in Asch RH, Balmaceda JP, Johnston I (eds): *Gamete Physiology*. Norwell, MA, Serono Symposia, 1990, pp 77–105.

Bearer EL, Friend DS: Modifications of anionic-lipid domains preceding membrane fusion in guinea pig sperm. *J Cell Biol* 92:604–615, 1982.

Bedford JM: Changes in the electrophoretic properties of rabbit spermatozoa during passage through the epididymis. *Nature* 200:1178–1180, 1963.

Bedford JM: The importance of capacitation for establishing contact between eggs and sperm in the rabbit. *J Reprod Fertil* 13:365–367, 1967.

Bedford JM: Limitation of the uterus in the development of the fertility (capacitation) of spermatozoa. *J Reprod Fertil* 8(suppl):19–26, 1969.

Bedford JM: Sperm capacitation and fertilization in mammals. *Biol Reprod* 2 (suppl): 128–158, 1970a.

Bedford JM: The saga of mammalian sperm from ejaculation to syngamy, in Gibian H, Plotz EJ (eds): *Mammalian Reproduction*. New York, Springer-Verlag, 1970b, pp 124–182.

Bedford JM: Sperm transport, capacitation and fertilization, in Balin H, Glasser S (eds): *Reproductive Biology*. Amsterdam, Excerpta Medica, 1972, pp 338–392.

Bedford JM: Significance of the need for sperm capacitation before fertilization in eutherian mammals. *Biol Reprod* 28:108–120, 1983.

Bedford JM, Cooper GW: Membrane fusion events in fertilization of vertebrate eggs, in Poste G, Nicolson GL (eds): *Membrane Surface Reviews (Membrane Fusion)* (5th ed). Amsterdam, North-Holland, 1978, pp 65–125.

Bentley JK, Garbers DL, Domino SE, Noland TD, VanDop C: Spermatozoa contain a guanine nucleotide binding protein ADP-ribosylated by pertussis toxin. *Biochem Biophys Res Commun* 138:728–734, 1986.

Berridge MJ: The molecular basis of communication within the cell. *Sci Am* 253:142–152, 1985.

Berridge MJ: Inositol lipids and calcium signalling. *Proc R Soc Lond (Biol)* 234:359–378, 1988.

Berruti G, Franchi E, Camatini M: Ca^{++} localization in boar spermatozoa by the pyroantimonate technique and X-ray microanalysis. *J Exp Zool* 237:257–262, 1986.

Blackmore PF, Beebe SJ, Danforth DR, Alexander N: Progesterone and 17-hydroxy-progesterone. Novel stimulators of calcium influx in human sperm. *J Biol Chem* 265:1376–1380, 1990.

Blandau RJ: Gamete transport—comparative aspects, in Hafez ESE, Blandau RJ (eds): *The Mammalian Oviduct*. Chicago, University of Chicago Press, 1969, pp 129–162.

Bleau G, Vandenheuvel WJA, Anderson OF, Gwatkin RBL: Desmosteryl sulfate of hamster spermatozoa, a potent inhibitor of capacitation in vitro. *J Reprod Fertil* 43:175–178, 1975.

Bleil JD, Wassarman PM: Autoradiographic visualization of the mouse egg's sperm receptor bound to sperm. *J Cell Biol* 102:1363–1371, 1986.

Bloch K: Cholesterol: Evolution of structure and function, in Vance DE, Vance JE (eds): *Biochemistry of Lipids and Membranes*. New York, Benjamin-Cummings, 1985, pp 1–24.

Boatman DE: Oviductal modulators of sperm fetilizing ability, in Bavister BD, Cummins J, Roldan ERS (eds): *Fertilization in Mammals*. Norwell, MA, Serono Symposia, 1990, pp 223–238.

Bradley MP, Forrester IT: Sperm calcium homeostasis during maturation, in Lobl TJ, Hafez ESE (eds): *Male Fertility and its Regulation*. Boston, MTP Press, 1985, pp 437–449.

Breitbart H, Darshan R, Rubinstein S: Evidence for the presence of ATPase-dependent calcium pump and ATPase activities in bull sperm head membranes. *Biochem Biophys Res Comm* 122:479–489, 1984.

Brokaw CJ: Sperm motility. *Meth Cell Biol* 27:41–56, 1986.

Brokaw CJ: Regulation of sperm flagellar motility by calcium and cAMP-dependent phosphorylation. *J Cell Biochem* 35:175–184, 1987.

Bronson RA, Rogers BJ: Pitfalls of the zona-free hamster egg penetration test: Protein source as a major variable. *Fertil Steril* 50:851–854, 1988.

Brown CR, Hartree EF: Distribution of trypsin-like proteinase in the ram spermatozoa. *J Reprod Fertil* 36:195–198, 1974.

Brown CR, Hartree EF: Effects of acrosin inhibitors on the soluble and membrane-bound forms of ram acrosin and a reappraisal of the role of the enzyme in fertilization. *Hoppe Seylers Z Physiol Chem* 357:57–65, 1976.

Burgess TL, Kelly RB: Constitutive and regulated secretion of proteins. *Annu Rev Cell Biol* 3:243–294, 1987.

Burkman LJ: Hyperactivated motility of human spermatozoa during in vitro capacitation and implications for fertility, in Gagnon C (ed): *Controls of Sperm Motility: Biological and Clinical Aspects*. Boca Raton, CRC Press, 1990, pp 303–329.

Burkman LJ, Overstreet JW, Katz DF: A possible role for potassium and pyruvate in the modulation of sperm motility in the rabbit oviductal isthmus. *J Reprod Fertil* 71:367–376, 1984.

Burkman LJ, Coddington CC, Franken DR, Kruger TF, Rosenwaks S, Hodgen GD: The hemizona assay (HZA): Development of a diagnostic test for the binding of human spermatozoa to the human hemizona pellucida to predict fertilization potential. *Fertil Steril* 49:688–697, 1988.

Burkman LJ, Coddington CC, Franken DR, Oehninger SC, Hodgen GD: The hemizona assay (HZA): Assessment of fertilizing potential by means of human sperm binding to the human zona pellucida, in Keel BA, Webster BW (eds): *Handbook of the Laboratory Diagnosis and Treatment of Infertility*. Boca Raton, CRC Press, 1990, pp 213–228.

Chang MC: Fertilizing capacity of spermatozoa deposited in Fallopian tubes. *Nature* 168:997–998, 1951.

Chang MC: A detrimental effect of seminal plasma on the fertilizing capacity of sperm. *Nature* 179:258–260, 1957.

Chapman D, Benga G: Biomembrane fluidity—Studies of model and natural biomembranes, in Chapman D (ed): *Biological Membranes*. New York, Academic Press, 1984, pp 1–56.

Cherr GN, Lambert H, Katz DF: In vitro studies of the golden hamster sperm acrosome reaction: Completion on the zona pellucida and induction by homologous soluble zonae pellucidae. *Dev Biol* 114:119–131, 1986.

Cheung WY: Calmodulin. *Sci Am* 246:48–56, 1982.

Clegg ED: Mechanisms of mammalian sperm capacitation, in Hartmann JF (ed): *Mechanism and Control of Animal Fertilization*. New York, Academic Press, 1983, pp 177–212.

Coddington CC, Fulgham DL, Alexander NJ, Johnson D, Herr JC, Hodgen GD: Sperm bound to zona pellucida in hemizona assay demonstrate acrosome reaction when stained with T-6 antibody. *Fertil Steril* 54:504–508, 1990.

Cooper GW, Overstreet JW, Katz DF: The motility of rabbit spermatozoa recovered from the female reproductive tract. *Gamete Res* 2:35–42, 1979.

Critser JK, Arneson BW, Aaker DV, Huse-Benda AR, Ball GD: Cryopreservation of human spermatozoa: II. Postthaw chronology of motility and of zona-free hamster ova penetration. *Fertil Steril* 47:980–984, 1987.

Cross NL, Meizel S: Methods for evaluating the acrosomal status of mammalian sperm. *Biol Reprod* 41:635–641, 1989.

Cross NL, Overstreet JW: Glycoconjugates of the human sperm surface: Distribution and alterations that accompany capacitation in vitro. *Gamete Res* 16:23–35, 1987.

Cross NL, Morales P, Overstreet JW, Hanson FW: Induction of acrosome reactions by the human zona pellucida. *Biol Reprod* 38:235–244, 1988.

Cummins JM, Yanagimachi R: Sperm-egg ratios and the site of the acrosome reaction during in vivo fertilization in the hamster. *Gamete Res* 5:239–256, 1982.

Cummins JM, Yanagimachi R: Development of ability to penetrate the cumulus oophorus by hamster spermatozoa capacitated in vitro in relation to the timing of the acrosome reaction. *Gamete Res* 15:187–212, 1986.

Davis BK: Inhibitory effect of synthetic phospholipid vesicles containing cholesterol on the fertilizing capacity of rabbit spermatozoa. *Proc Soc Biol Med* 152:257–261, 1976.

Davis BK: Interaction of lipids with the plasma membrane of sperm cells: I. The antifertilization action of cholesterol. *Arch Androl* 5:249–254, 1980.

Davis BK: Uterine fluid proteins bind sperm cholesterol during capacitation in the rabbit. *Experientia* 38:1063–1064, 1982.

Davis BK, Hungund BJ: Effects of modified membrane vesicles from seminal plasma on the fertilizing capacity of rabbit spermatozoa. *Biochem Biophys Res Comm* 69:1004–1010, 1976.

Davis BK, Byrne R, Hungund B: Studies on the mechanism of capacitation: II. Evidence for lipid transfer between plasma membrane of rat sperm and serum albumin during capacitation in vitro. *Biochim Biophys Acta* 558:257–266, 1979.

Davis RO, Overstreet JW, Asch RH, Ord T, Silber SJ: Movement characteristics of human epididymal sperm used for fertilization of human oocytes in vitro. *Fertil Steril* 56:1128–1135, 1991.

Dravland E, Joshi MS: Sperm-coating antigens secreted by the epididymis and seminal vesicle of the rat. *Biol Reprod* 25:649–658, 1981.

Drobnis EZ, Katz DF: Videomicroscopy of mammalian fertilization, in Wassarman PM (ed): *Elements of Mammalian Fertilization.* Boca Raton, CRC Press, 1991, pp 269–300.

Drobnis EZ, Overstreet JW: Natural history of mammalian spermatozoa in the female reproductive tract. *Oxford Rev Reprod Biol* 14:1–45, 1992.

Drobnis EZ, Yudin AI, Cherr GN, Katz DF: Kinematics of hamster sperm during penetration of the cumulus cell matrix. *Gamete Res* 21:367–383, 1988a.

Drobnis EZ, Yudin AI, Cherr GN, Katz DF: Hamster sperm penetration of the zona pellucida: Kinematic analysis and mechanical implication. *Dev Biol* 130:311–323, 1988b.

Drobnis EZ, Davis RO, Sousa SL, Robertson KR, Overstreet JW: Cold shock changes the pattern of human sperm motility. *J Androl* 13 (suppl): 7, 1992.

Dudkiewicz AB: Purification of boar acrosomal arylsulphatase A and possible role in the penetration of cumulus cells. *Biol Reprod* 30:1005–1014, 1984.

Dukelow WR, Chernoff HN, Williams WL: Properties of decapacitation factor and presence in various species. *J Reprod Fertil* 14:393–399, 1965.

Eddy EM, O'Brien DA: Biology of the gamete: Maturation, transport, and fertilization, in Working PK (ed): *Toxicology of the Male and Female Reproductive Systems. New York, Hemisphere Publishing, 1989, pp 31–100.*

Egli GE, Newton M: The transport of carbon particles in the human female reproductive tract. *Fertil Steril* 12:151–155, 1961.

Ehrenwald E, Parks JE, Foote RH: Cholesterol efflux from bovine sperm: I. Induction of the acrosome reaction with lysophosphatidylcholine after reducing sperm cholesterol. *Gamete Res* 20:145–157, 1988a.

Ehrenwald E, Parks JE, Foote RH: Cholesterol efflux from bovine sperm: II. Effects of reducing sperm cholesterol on penetration of zona-free hamster and in vitro matured bovine ova. *Gamete Res* 20:413–420, 1988b.

Endo Y, Lee MA, Kopf GS: Evidence for the role of a guanine nucleotide-binding regulatory protein in the zona pellucida-induced mouse sperm acrosome reaction. *Dev Biol* 119:210–216, 1987a.

Endo Y, Mattei P, Kopf GS, Schultz RM: Effects of a phorbol ester on mouse eggs: Dissociation of sperm receptor activity from acrosome reaction-inducing activity of the mouse zona pellucida protein, ZP3. *Dev Biol* 123:574–577, 1987b.

Endo Y, Lee MA, Kopf GS: Characterization of an islet-activating protein-sensitive site in mouse sperm that is involved in the zona pellucida-induced acrosome reaction. *Dev Biol* 129:12–24, 1988.

Eng LA, Oliphant G: Rabbit sperm reversible decapacitation by membrane stabilization with a highly purified glycoprotein from seminal plasma. *Biol Reprod* 19:1083–1094, 1978.

Eskenazi B, Wyrobek AJ, Fenster L, Katz DF, Sadler M, Lee J, Hudes M, Rempel DM: A study of the effect of perchloroethylene exposure on semen quality in dry cleaning workers. *Am J Ind Med* 20:575–591, 1991a.

Eskenazi B, Fenster L, Hudes M, Wyrobek AJ, Katz DF, Gerson J, Rempel DM: A study of the effect of perchloroethylene exposure on the reproductive outcomes of wives of dry-cleaning workers. *Am J Ind Med* 20:593–600, 1991b.

Farooqui AA: Metabolism and role of sulpholipids in mammals. *Int J Biochem* 9:709–716, 1978.

Fayrer–Hosken RA, Brackett BG, Brown J: Reversible inhibition of rabbit sperm fertilizing ability by cholesterol sulfate. *Biol Reprod* 36:878–883, 1987.

Fenichel P, Basteris B, Donzeau M, Ayraud N, Farahifar D, Hsi B-L: Dynamics of human sperm acrosome reaction: Relation with in vitro fertilization. *Fertil Steril* 55:994–999, 1991.

Fleming AD, Yanagimachi R: Effects of various lipids on the acrosome reaction and fertilizing capacity of guinea pig spermatozoa with special reference to the possible involvement of lysophospholipids in the acrosome reaction. *Gamete Res* 4:253–273, 1981.

Florman HM, Babcock DF: Progress toward understanding the molecular basis of capacitation, in Wassarman PM (ed): *Elements of Mammalian Fertilization.* Boca Raton, FL, CRC Press, 1991, pp 105–132.

Florman HM, Tombes RM, First NL, Babcock DF: An adhesion-associated agonist from the zona pellucida activates G protein-promoted elevations of internal Ca^{2+} and pH that mediate mammalian sperm acrosomal exocytosis. *Dev Biol* 135:133–146, 1989.

Franken DR, Kruger TF, Menkveld R, Oehninger S, Coddington CC, Hodgen GD: Hemizona assay and teratozoospermia: Increasing sperm insemination concentrations to enhance zona pellucida binding. *Fertil Steril* 54:497–503, 1990.

Franklin LE, Barros C, Fussell EN: The acrosomal region and the acrosome reaction in sperm of the golden hamster. *Biol Reprod* 3:180–200, 1970.

Fraser LR: *p*-Aminobenzamidine, an acrosin inhibitor, inhibits mouse sperm penetration of the zona pellucida but not the acrosome reaction. *J Reprod Fertil* 65:1–11, 1982.

Fraser LR: Mechanisms controlling mammalian fertilization, in Finn CA (ed): *Oxford Review of Reproductive Biology* (6th ed). Oxford, Clarendon Press, 1984, pp 174–225.

Fraser LR: Sperm capacitation and its modulation, in Bavister BD, Cummins J, Roldan ERS (eds): *Fertilization in Mammals.* Norwell, MA, Serono Symposia, 1990, pp 141–153.

Fraser LR, Ahuja KK: Metabolic and surface events in fertilization. *Gamete Res* 20:491–519, 1988.

Fraser LR, Monks NJ: Cyclic nucleotides and mammalian sperm capacitation. *J Reprod Fertil* (suppl) 42:9–21, 1990.

Friend DS, Orci L, Perrelet A, Yanagimachi R: Membrane particle changes attending the acrosome reactions in guinea pig spermatozoa. *J Cell Biol* 74:561–577, 1977.

Gaddum–Rosse P: Some observations on sperm transport through the uterotubal junction of the rat. *Am J Anat* 160:333–341, 1981.

Gaddum–Rosse P: Mammalian gamete interactions: What can be gained from observations on living eggs? *Am J Anat* 174:347–356, 1985.

Garbers DL: Molecular basis of fertilization. *Annu Rev Biochem* 58:719–742, 1989.

Garbers DL, Kopf GS: The regulation of spermatozoa by calcium and cyclic nucleotides, in Greengard P, Robinson GA (eds): *Advances in Cyclic Nucleotide Research* (vol. 13). New York, Raven Press, 1980, pp 251–306.

Gennis RB: *Biomembranes: Molecular structure and function,* New York, Springer-Verlag, 1989.

Gilman AG: G proteins: Transducers of receptor-generated signals. *Annu Rev Biochem* 56:615–649, 1987.

Go KJ, Wolf DP: The role of sterols in sperm capacitation. *Adv Lipid Res* 20:317–330, 1983.

Go KJ, Wolf DP: Albumin-mediated changes in sperm sterol content during capacitation. *Biol Reprod* 32:145–153, 1985.

Goodpasture JC, Zavos PN, Cohen MR, Zaneveld LJD: Relationship of human sperm acrosin and proacrosin to semen parameters: I. Comparisons between symptomatic men of infertile couples and asymptomatic men and between different split ejaculates. *J Androl* 3:151, 1982.

Gordon LM, Mobley PW: Membrane lipids, membrane fluidity, and enzyme activity, in Aloia RC, Boggs JM (eds): *Membrane Fluidity in Biology* (vol. 4: *Cellular Aspects*). New York, Academic Press, 1985, pp 1–49.

Gould JE, Overstreet JW, Hanson FW: Assessment of human sperm function after recovery from the female reproductive tract. *Biol Reprod* 31:888–894, 1984.

Gould JE, Overstreet JW, Hanson FW: Interaction of human spermatozoa with the human zona pellucida and zona-free hamster oocyte following capacitation by exposure to human cervical mucus. *Gamete Res* 12:47–54, 1985.

Green DPL: The mechanism of the acrosome reaction, in Johnson MH (ed): *Development in Mammals*. New York, North-Holland, 1978, pp 65–81.

Green DPL: Sperm thrusts and the problem of penetration. *Biol Rev* 63:79–105, 1988.

Gwatkin RBL: Effects of chemicals on fertilization, in Dixon RL (ed): *Reproductive Toxicology*. New York, Raven Press, 1985, pp 209–218.

Gwatkin RBL: Zona binding sites of the spermatozoa, in Dietl J (ed): *The Role of the Zona Pellucida in Mammalian Reproduction*. Berlin, Springer-Verlag, 1988, pp 99–105.

Gwatkin RBL, Williams DT: Receptor activity of hamster and mouse solubilized zona pellucida before and after the zona reaction. *J Reprod Fertil* 49:55–59, 1977.

Gwatkin RBL, Andersen OF, Hutchinson CF: Capacitation of hamster spermatozoa in vitro: The role of cumulus components. *J Reprod Fertil* 30:389–394, 1972.

Gwatkin RBL, Carter WH, Patterson H: Association of mammalian sperm with cumulus cells and the zona pellucida studied by scanning electron microscopy, in Johari H, Becker RP (eds): *Scanning Electron Microscopy* (2d ed). Chicago, ITT Research Institute, 1976, pp 379–384.

Harrison RA, Roldan ER: Phosphoinositides and their products in the mammalian sperm acrosome reaction. *J Reprod Fertil* (suppl) 42:51–67, 1990.

Hartmann JF: Mammalian fertilization: Gamete surface interactions in vitro, in Hartmann JF (ed): *Mechanisms and Control of Animal Fertilization*. New York, Academic Press, 1983, pp 325–364.

Holland MK, Storey BT: Oxygen metabolism of mammalian spermatozoa. Generation of hydrogen peroxide by rabbit epididymal spermatozoa. *Biochem J* 198:273–280, 1981.

Horton AA, Fairhurst S: Lipid peroxidation and mechanisms of toxicity. *CRC Crit Rev Toxicol* 18:27–79, 1987.

Hortsch M, Goodman CS: Cell and substrate adhesion molecules in *Drosophila*. *Annu Rev Cell Biol* 7:505–557, 1991.

Houslay MD, Stanley KK: *Dynamics of Biological Membranes,* New York, Wiley, 1982.

Huacuja L, Delgado NM, Calzada L, Wens A, Reyes R, Pedron N, Rosado A: Exchange of lipids between spermatozoa and seminal plasma in normal and pathological human semen. *Arch Androl* 7:343–349, 1981.

Hull MGR, McLeod FN, Joyce DN, Ray BD, McDermott A: Human in vitro fertilization in vivo sperm penetration of cervical mucus and unexplained infertility. *Lancet* 2:245–248, 1984.

Hunter RHF: Transport and storage of spermatozoa in the female tract. *Int Cong Anim Reprod Artif Insem* 9(2):227–233, 1980.

Hunter RHF: Pre-ovulatory arrest and peri-ovulatory redistribution of competent spermatozoa in the isthmus of the pig oviduct. *J Reprod Fertil* 72:203–211, 1984.

Hunter RHF: Human fertilization in vivo, with special reference to progression, storage and release of competent spermatozoa. *Hum Reprod* 2:239–332, 1987.

Hunter RHF: *The Fallopian Tubes,* New York, Springer-Verlag, 1988.

Hunter RHF, Hall JP Jr: Capacitation of boar spermatozoa: The influence of post-coital separation of the uterus and Fallopian tubes. *Anat Rec* 180:597–604, 1974.

Hunter RHF, Nichol R: Transport of spermatozoa in the sheep oviduct: Preovulatory sequestering of cells in the caudal isthmus. *J Exp Zool* 228:121–128, 1983.

Hunter RHF, Wilmut I: The rate of functional sperm transport into the oviducts of mated cows. *Anim Reprod Sci* 5:167–173, 1983.

Inoue M, Wolf DP: Fertilization-associated changes in the murine zona pellucida: A time sequence study. *Biol Reprod* 13:546–551, 1975.

Irvine DS, Aitken RJ: Measurement of intracellular calcium in human spermatozoa. *Gamete Res* 15:57–71, 1982.

Johnson LL, Katz DF, Overstreet JW: The movement characteristics of rabbit spermatozoa before and after activation. *Gamete Res* 4:275–282, 1981.

Jones HP, Lenz RW, Palevitz BA, Cormier MJ: Calmodulin localization in mammalian spermatozoa. *Proc Natl Acad Sci USA* 77:2772–2776, 1980.

Katz DF, Davis RO: Automatic analysis of human sperm motion. *J Androl* 8:170–181, 1987.

Katz DF, Drobnis EZ: Analysis and interpretation of the forces generated by spermatozoa, in Bavister BD, Cummins J, Roldan ERS (eds): *Fertilization in Mammals.* Norwell, MA, Serono Symposia, 1990, pp 125–137.

Katz DF, Drobnis EZ, Overstreet JW: Factors regulating mammalian sperm migration through the female reproductive tract and oocyte vestments. *Gamete Res* 22:443–469, 1989.

Katz DF, Yanagimachi R: Movement characteristics of hamster spermatozoa within the oviduct. *Biol Reprod* 22:759–764, 1980.

Katz DF, Overstreet JW, Hanson FW: Variations within and amongst normal men of movement characteristics of seminal spermatozoa. *J Reprod Fertil* 62:221–228, 1981.

Katz DF, Yudin AI, Drobnis EZ, Hanson FW, Cragun JR, Chang RJ: Human sperm penetration of cervical mucus and the hamster cumulus in relation to capacitation. *Am J Obstet Gynecol* 159(Abstr Suppl):92, 1988.

Katz DF, Morales P, Samuels SJ, Overstreet JW: Mechanisms of filtration of morphologically abnormal human sperm by cervical mucus. *Fertil Steril* 54:513–516, 1990.

Keel BA, Kelly RW, Webster BW, Zumbach KL, Roberts DK: Application of a bovine cervical mucus penetration test. *Arch Androl* 19:33–41, 1987.

Kellaway IW, Saunders L: The haemolytic activity of sols containing progesterone solubilized by phosphatidylcholine and lysophosphatidylcholine. *J Pharm Pharmacol* 21:1895–1945, 1969.

Kligman I, Glassner M, Storey BT, Kopf GS: Zona pellucida–mediated acrosomal exocytosis in mouse spermatozoa: Characterization of an intermediate stage prior to the completion of the acrosome reaction. *Dev Biol* 145:344–355, 1991.

Koehler JK: The mammalian sperm surface: Studies with specific labeling techniques. *Int Rev Cytol* 54:73–108, 1978.

Koehler JK: Surface alterations during the capacitation of mammalian spermatozoa. *Am J Primatol* 1:131–141, 1981.

Koehler JK, Gaddum-Rosse P: Media induced alterations of the membrane associated particles of the guinea pig sperm tail. *J Ultrastruct Res* 51:106–118, 1975.

Kopf GS: Zona pellucida-mediated signal transduction in mammalian spermatozoa. *J Reprod Fertil* (suppl) 42:33–49, 1990a.

Kopf GS: The zona pellucida-induced acrosome reaction: A model for sperm signal transduction, in Bavister BD, Cummins J, Roldan ERS (eds): *Fertilization in Mammals*. Norwell, MA, Serono Symposia, 1990b, pp 253–266.

Kopf GS, Gerton GL: The mammalian sperm acrosome and the acrosome reaction, in Wassarman PM (ed): *Elements of Mammalian Fertilization*. Boca Raton, FL, CRC Press, 1991, pp 153–203.

Kopf GS, Woolkalis MJ, Gerton GL: Evidence for a guanine nucleotide-binding regulatory protein in invertebrate and mammalian sperm. *J Biol Chem* 261:7327–7331, 1986.

Koukoulis GN, Vantman MD, Dennison L, Banks SM, Sherins RJ: Low acrosin activity in a subgroup of men with idiopathic infertility does not correlate with sperm density, percent motility, curvilinear velocity, or linearity. *Fertil Steril* 52:120–127, 1989.

Kuzan FB, Fleming AD, Seidel GE Jr: Successful fertilization in vitro of fresh intact oocytes by perivitelline (acrosome-reacted) spermatozoa of the rabbit. *Fertil Steril* 41:766–770, 1984.

Lalich RA, Vedantham S, McCormick N, Wagner C, Prins GS: Relationship between heparin binding characteristics and ability of human spermatozoa to penetrate hamster ova. *J Reprod Fertil* 86:297–302, 1989.

Lalumiere G, Bleau G, Chapdelaine A, Roberts KD: Cholesteryl sulfate and sterol sulfatase in the human reproductive tract. *Steroids* 27:247–260, 1976.

Langlais J, Roberts KD: A molecular membrane model of sperm capacitation and the acrosome reaction of mammalian spermatozoa. *Gamete Res* 12:183–224, 1985.

Langlais, J, Zollinger M, Plante L, Chapdelaine A, Bleau G, Roberts KD: Localization of cholesteryl sulfate in human spermatozoa in support of a hypothesis for the mechanism of capacitation. *Proc Natl Acad Sci USA* 78:7266–7270, 1981.

Langlais J, Kan FWK, Granger L, Raymond L, Bleau G, Roberts KD: Identification of sterol acceptors that stimulate cholesterol efflux from human spermatozoa during in vitro capacitation. *Gamete Res* 20:185–201, 1988.

Lee MA, Storey BT: Evidence from plasma membrane impermeability to small ions in acrosome intact mouse spermatozoa bound to mouse zonae pellucidae, using aminoacridine fluorescent pH probe. *Biol Reprod* 33:235–246, 1985.

Lee MA, Trucco GS, Bechtol KB, Wummer N, Kopf GS, Blasco L, Storey BT: Capacitation and acrosome reactions in human spermatozoa monitored by a chlortetracycline fluorescence assay. *Fertil Steril* 48:649–658, 1987.

Legault Y, Bleau G, Chapdelaine A, Roberts KD: The binding of sterol sulfates to hamster spermatozoa. *Steroids* 27:247–260, 1979.

Levitzki A: From epinephrine to cyclic AMP. *Science* 241:800–806, 1988.

Levy RI: Cholesterol, lipoproteins, apoproteins, and heart disease: Present status and future prospects. *Clin Chem* 27:653–662, 1981.

Leyton L, Saling P: Evidence that aggregation of mouse sperm receptors by ZP3 triggers the acrosome reaction. *J Cell Biol* 108:2163–2168, 1989.

Lindemann CB: A cAMP-induced increase in the motility of demembranated bull sperm models. *Cell* 13:9–18, 1978.

Lindemann CB, Goltz JS: Calcium regulation of flagellar curvature and swimming pattern in Triton X-100 extracted rat sperm. *Cell Motil Cytoskel* 10:420–431, 1988.

Lindemann CB, Kanous KS: Regulation of mammalian sperm motility. *Arch Androl* 23:1–22, 1989.

Lindemann CB, Goltz JS, Kanous KS: Regulation of activation state and flagellar wave form in epididymal rat sperm: Evidence for the involvement of both Ca^{2+} and cAMP. *Cell Motil Cytoskel* 8:324–332, 1987.

Liu DY, Lopata A, Pantke P, Baker HWG: Horse and marmoset monkey sperm bind to the zona pellucida of salt-stored human oocytes. *Fertil Steril* 56:764–767, 1991.

Longo FJ, Krohne G, Franke WW: Basic proteins of the perinuclear theca of mammalian spermatozoa and spermatids: A novel class of cytoskeletal elements. *J Cell Biol* 105:1105–1120, 1987.

Lui CW, Meizel S: Further evidence in support of a role for hamster sperm hydrolytic enzymes in the acrosome reaction. *J Exp Zool* 207:173–186, 1979.

Macek MB, Shur BD: Protein-carbohydrate complementarity in mammalian gamete recognition. *Gamete Res* 20:93–109, 1988.

Mack SO, Wolf DP, Tash JS: Quantitation of specific parameters of motility patterns in large numbers of human sperm by digital image processing. *Biol Reprod* 38:270–281, 1988.

Majerus PW: The metabolism of phosphoinositide-derived messenger molecules. *Science* 234:1519–1526, 1986.

Mattner PE, Braden AWH: Spermatozoa in the genital tract of the ewe: I. Rapidity of transport. *Aust J Biol Sci* 16:473–481, 1963.

Meizel S: The mammalian sperm acrosome reaction. A biochemical approach, in Johnson MH (ed): *Development in Mammals.* New York, North-Holland, 1978, pp 1–64.

Meizel S: The importance of hydrolytic enzymes to an exocytotic event, the mammalian sperm acrosome reaction. *Biol Rev* 59:125–157, 1984.

Meizel S: Molecules that initiate or help stimulate the acrosome reaction by their interaction with the mammalian sperm surface. *Am J Anat* 17:285–302, 1985.

Meizel S, Deamer DW: The pH of the hamster sperm acrosome. *J Histochem Cytochem* 26:98–105, 1978.

Meizel S, Lui CW: Evidence for the role of a trypsin-like enzyme in the hamster sperm acrosome reaction. *J Exp Zool* 195:137–144, 1976.

Meizel S, Mukerji SK: Biochemical studies of proacrosin and acrosin from hamster cauda epididymal spermatozoa. *Biol Reprod* 14:444–450, 1976.

Meizel S, Turner KO: Progesterone acts at the plasma membrane of human sperm. *Mol Cell Endocrinol* 77:R1–R5, 1991.

Meizel S, Pillai MC, Diaz-Perez E, Thomas P: Initiation of the human sperm acrosome reaction by components of human follicular fluid and cumulus secretions including steroids, in Bavister BD, Cummins J, Roldan ERS (eds): *Fertilization in Mammals.* Norwell, MA, Serono Symposia, 1990, pp 205–222.

Millette CF: Distribution and mobility of lectin binding sites on mammalian spermatozoa, in Edidin M, Johnson MH (eds): *Immunobiology of Gametes.* Cambridge, Cambridge University Press, 1977, pp 51–71.

Mohsenian M, Syner FN, Moghissi KS: A study of sperm acrosin in patients with unexplained infertility. *Fertil Steril* 37:223, 1982.

Moore HDM: The net surface charge of mammalian spermatozoa as determined by isoelectric focusing: Changes following sperm maturation, ejaculation, incubation in the female tract, and after enzyme treatment. *Int J Androl* 2:449–462, 1979.

Moore HDM, Bedford JM: The interaction of mammalian gametes in the female, in Hartmann JR (ed): *Mechanism and Control of Animal Fertilization*. New York, Academic Press, 1983, pp 453–497.

Moore PB, Dedman JR: Calmodulin, a calmodulin acceptor protein, and calcimedins: Unique antibody localizations in hamster sperm. *J Cell Biochem* 25:99–107, 1984.

Morales P, Cross NL, Overstreet JW, Hanson FW: Acrosome intact and acrosome reacted human sperm can initiate binding to the zona pellucida. *Dev Biol* 133:385–392, 1989.

Mortimer D: Selectivity of sperm transport in the female genital tract, in Cohen J, Hendry WF (eds): *Spermatozoa, Antibodies and Infertility*. Oxford, Blackwell Scientific, 1978, pp 37–53.

Mortimer D: Sperm transport in the human female reproductive tract. *Oxford Rev Reprod Biol* 5:30–61, 1983.

Morton B, Albagli L: Modification of hamster sperm adenyl cyclase by capacitation in vitro. *Biochem Biophys Res Commun* 50:695–703, 1973.

Morton DB: Acrosomal enzymes: Immunological localization of acrosin and hyaluronidase in ram spermatozoa. *J Reprod Fertil* 45:375–378, 1975.

Mrsny RJ, Meizel S: Potassium ion influx and Na^+, K^+-ATPase activity are required for the hamster sperm acrosome reaction. *J Cell Biol* 91:77–82, 1981.

Murray RK, Narasimhan R: Glycoglycerolipids of animal tissues, in Kates M (ed): *Handbook of Lipid Research* (vol 6: *Glycolipids, Phosphoglycolipids and Sulfoglycolipids*). New York, Plenum Press, 1990, pp 321–361.

Myles DG, Primakoff P: Sperm surface domains, in Springer TA (ed): *Hybridoma Technology in the Biosciences and Medicine*. New York, Plenum, 1985, pp 239–250.

Myles DG, Hyatt H, Primakoff P: Binding of both acrosome-intact and acrosome-reacted guinea pig sperm to the zona pellucida during in vitro fertilization. *Dev Biol* 121:559–567, 1987.

Myles DG, Koppel DE, Primakoff P: Sperm surface domains and fertilization, in Bavister BD, Cummins J, Roldan ERS (eds): *Fertilization in Mammals*. Norwell, MA, Serono Symposia, 1990, pp 169–177.

Neill JM, Olds-Clarke P: A computer-assisted assay for mouse sperm hyperactivation demonstrates that bicarbonate but not bovine serum albumin is required. *Gamete Res* 18:121–140, 1987.

Nicolson GL, Yanagimachi R: Mobility and the restriction of mobility of plasma membrane lectin-binding components. *Science* 184:1294–1296, 1974.

Nicolson GL, Usui N, Yanagimachi R, Yanagimachi H, Smith JR: Lectin-binding sites on the plasma membranes of rabbit spermatozoa. *J Cell Biol* 74:950–962, 1977.

Nishizuka Y: Studies and perspectives of protein kinase C. *Science* 233:305–312, 1986.

O'Rand MG: Modification of the sperm membrane during capacitation. *Ann NY Acad Sci* 383:392–404, 1982.

O'Rand MG, Fisher SJ: Localization of zona pellucida binding sites on rabbit spermatozoa and induction of the acrosome reaction by solubilized zonae. *Dev Biol* 119:551–559, 1987.

Okabe M, Takada K, Adachi T, Kohama Y, Miura T: Inconsistent reactivity of an antisperm monoclonal antibody and its relationship to sperm capacitation. *J Reprod Immunol* 9:67–70, 1986a.

Okabe M, Takada K, Adachi T, Kohama Y, Miura T, Aonuma S: Studies on sperm capacitation using monoclonal antibody—Disappearance of an antigen from the anterior part of mouse sperm head. *J Pharmacol Dyn* 9:55–60, 1986b.

Oliphant G: Removal of sperm-bound seminal plasma components as a prerequisite to induction of the acrosome reaction. *Fertil Steril* 27:28–38, 1976.

Oliphant G, Singhas CA: Iodination of rabbit sperm plasma membrane: Relationship of specific surface proteins to epididymal function and sperm capacitation. *Biol Reprod* 21:937–944, 1979.

Oliphant G, Reynolds AB, Thomas TS: Sperm surface components involved in the control of the acrosome reaction. *Am J Anat* 174:269–283, 1985.

Olson GE, Winfrey VP: Characterization of postacrosomal sheath of bovine spermatozoa. *Gamete Res* 20:329–342, 1988.

Olson GE, Winfrey VP, Flaherty SP: Cytoskeletal assemblies of mammalian spermatozoa. *Ann NY Acad Sci* 513:222–246, 1987.

Overstreet JW: Transport gametes in the reproductive tract of the female mammal, in Hartmann JR (ed): *Mechanism and Control of Animal Fertilization.* New York, Academic Press, 1983, pp 499–543.

Overstreet JW, Cross NL: The biology of human sperm–zona pellucida interaction. *Ann NY Acad Sci* 541:337–345, 1988.

Overstreet JW, VandeVoort CA: Sperm transport in the female genital tract, in Asch RH, Balmace da JP, Johnston I (eds): *Gamete Physiology.* Norwell, MA, Serono Symposia, 1990, pp 43–52.

Overstreet JW, Gould JE, Katz DF, Hanson FW: In vitro capacitation of human spermatozoa after passage through a column of cervical mucus. *Fertil Steril* 34:604–606, 1980a.

Overstreet JW, Yanagimachi R, Katz DF, Hayashi K, Hanson FW: Penetration of human spermatozoa into the human zona pellucida and the zona-free hamster egg: a study of fertile donors and infertile patients. *Fertil Steril* 33:534–542, 1980b.

Parks JE, Ehrenwald E: Cholesterol efflux from mammalian sperm and its potential role in capacitation, in Bavister BD, Cummins J, Roldan ERS (eds): *Fertilization in Mammals.* Norwell, MA, Serono Symposia, 1990, pp 155–167.

Parrish RF, Polakoski KL: Mammalian sperm proacrosin–acrosin system. *Int J Biochem* 10:391–395, 1979.

Perreault SD: Impaired gamete function: Implications for reproductive toxicology, in Working PK (ed): *Toxicology of the Male and Female Reproductive Systems*. New York, Hemisphere Publishing, 1989, pp 217–229.

Perreault SD, Rogers BJ: Capacitation pattern of human spermatozoa. *Fertil Steril* 38:258, 1982.

Perreault SD, Zirkin BR, Rogers BJ: Effect of trypsin inhibitors on acrosome reaction of guinea pig spermatozoa. *Biol Reprod* 26:343–351, 1982.

Pillai MC, Meizel S: Trypsin inhibitors prevent the progesterone-initiated increase in intracellular calcium required for the human sperm acrosome reaction. *J Exp Zool* 258:384–393, 1991.

Ralt D, Goldenberg M, Fetterolf P, Thompson D, Dor J, Mashiach S, Garbers DL, Eisenbach M: Sperm attraction to a follicular fluid factor(s) correlates with human egg fertilizability. *Proc Nat Acad Sci USA* 88:2840–2844, 1991.

Rand RP, Pangborn WA, Purdon AD, Tinker DO: Lysolecithin and cholesterol interact stoichiometrically forming bimolecular lamellar structures in the presence of excess water, lysolecithin or cholesterol. *Can J Biochem* 53:189–195, 1975.

Rao B, Soufir JC, Martin M, David G: Lipid peroxidation in human spermatozoa as related to midpiece abnormalities and motility. *Gamete Res* 24:127–134, 1989.

Rasmussen H: The cycling of calcium as an intracellular messenger. *Sci Am* 261:66–73, 1989.

Ravnik SE, Muller CH: Relationship between support of human sperm capacitation and lipid transfer activity among different albumin preparations. *J Androl* 10(Suppl): P-21, 1989.

Ribbes H, Plantavid M, Bennet PJ, Chap H, Douste-Blazy L: Phospholipase C from human sperm specific for phosphoinositides. *Biochim Biophys Acta* 919:245–254, 1987.

Robertson L, Wolf DP, Tash JS: Temporal changes in motility parameters related to acrosomal status: Identification and characterization of populations of hyperactivated human sperm. *Biol Reprod* 39:797–805, 1988.

Rogers BJ: Mammalian sperm capacitation and fertilization in vitro: A critique of methodology. *Gamete Res* 1:165–223, 1978.

Rogers BJ: The sperm penetration assay: Its usefulness reevaluated. *Fertil Steril* 43:821–840, 1985.

Rogers BJ: Examination of data from programs of in vitro fertilization in relation to sperm integrity and reproductive success, in Burger EJ, Tardiff RG, Scialli AR, Zenick H (eds): *Sperm Measures and Reproductive Success: Institute for Health Policy Analysis*. New York, Alan R Liss, 1989, pp 69–93.

Roldan ERS, Fleming AD: Is a Ca^{2+}-ATPase involved in Ca^{2+} regulation during capacitation and the acrosome reaction of guinea-pig spermatozoa? *J Reprod Fertil* 85:297–308, 1989.

Roldan ERS, Harrison RAP: Polyphosphoinositide breakdown and subsequent exocytosis in the Ca^{2+}/ionophore-induced acrosome reaction of mammalian spermatozoa. *Biochem J* 259:397–406, 1989.

Roldan ERS, Harrison RAP: Molecular mechanisms leading to exocytosis during the sperm acrosome reaction, in Bavister BD, Cummins J, Roldan ERS (eds): *Fertilization in Mammals*. Norwell, MA, Serono Symposia, 1990, pp 179–196.

Rosado A, Velázquez A, Lara-Ricalde R: Cell polarography. II. Effect of neuraminidase and follicular fluid upon the surface characteristics of human spermatozoa. *Fertil Steril* 24:349–354, 1973.

Rufo GA, Singh JP, Babcock DF, Lardy HA: Purification and characterization of a calcium transport inhibitor protein from bovine seminal plasma. *J Biol Chem* 257:4627–4632, 1982.

Rufo GA, Schoff PK, Lardy HA: Regulation of calcium content in bovine spermatozoa. *J Biol Chem* 259:2547–2552, 1984.

Saling PM: Mammalian sperm interaction with extracellular matrices of the egg. *Oxford Rev Reprod Biol* 11:339–388, 1989.

Saling PM, Storey BT: Mouse gamete interactions during fertilization in vitro. Chlortetracycline as a fluorescent probe for the mouse acrosome reaction. *J Cell Biol* 83:544–555, 1979.

Saling PM, Wolf DP, Storey BT: Calcium-dependent binding of mouse epididymal spermatozoa to the zona pellucida. *Dev Biol* 65:515–525, 1978.

Saling PM, Sowinski J, Storey BT: An ultrastructural study of epididymal mouse spermatozoa binding to zonae pellucidae in vitro: Sequential relationship to the acrosome reaction. *J Exp Zool* 209:229–238, 1979.

Saling PM, Bunch DO, Le Guen P, Leyton L: ZP3-induced acrosomal exocytosis: A new model for triggering, in Bavister BD, Cummins J, Roldan ERS (eds): *Fertilization in Mammals*. Norwell, MA, Serono Symposia, 1990, pp 239–252.

San Agustin JT, Hughes P, Lardy HA: Properties and function of caltrin, the calcium transport-inhibitor of bull seminal plasma. *FASEB J* 1:60–66, 1987.

Schill W-B, Topfer-Petersen E, Heissler E: The sperm acrosome: Functional and clinical aspects. *Hum Reprod* 3:139–145, 1988.

Schwarz MA, Koehler JK: Alterations in lectin binding to guinea pig spermatozoa accompanying in vitro capacitation and the acrosome reaction. *Biol Reprod* 21:1295–1307, 1979.

Shalgi R, Kraicer PF: Timing of sperm transport, sperm penetration and cleavage in the rat. *J Exp Zool* 204:353–360, 1978.

Shalgi R, Kaplan R, Nebel L, Kraicer PF: The male factor in fertilization of rat eggs in vitro. *J Exp Zool* 217:399–402, 1981.

Shams-Borham G, Harrison RAP: Production, characterization, and use of ionophore-induced, calcium-dependent acrosome reaction in ram spermatozoa. *Gamete Res* 4:407–432, 1981.

Shapiro BM, Cook S, Quest AF, Oberdorf J, Wothe D: Molecular mechanisms of sea-urchin sperm activation before fertilization. *J Reprod Fertil* (suppl) 42:3–8, 1990.

Shinitzky M: Membrane fluidity and cellular functions, in Shinitzky M (ed): *Physiology of Membrane Fluidity*. Boca Raton, FL, CRC Press, 1984, pp 1–51.

Slater TF: Free-radical mechanisms in tissue injury. *Biochem J* 222:1–15, 1984.

Slott VL, Suarez JD, Simmons JE, Perreault SD: Acute inhalation exposure to epichlorohydrin transiently decreases rat sperm velocity. *Fundam Appl Toxicol* 15:596–606, 1990.

Smith TT, Yanagimachi R: Capacitation status of hamster spermatozoa in the oviduct at various times after mating. *J Reprod Fert* 86:255–261, 1989.

Soupart P, Anderson ML, Albert DH, Coniglio JG, Repp JE: Accumulation, nature and possible functions of the malachite green affinity material in ejaculated human spermatozoa. *Fertil Steril* 32:450–454, 1979.

Stambaugh R: Enzymatic and morphological events in mammalian fertilization. *Gamete Res* 1:65–85, 1978.

Stein DM, Fraser LR: Cyclic nucleotide metabolism in mouse epididymal spermatozoa during capacitation in vitro. *Gamete Res* 10:283–299, 1984.

Suarez SS: Sperm transport and motility in the mouse oviduct: Observations in situ. *Biol Reprod* 36:203–210, 1987.

Suarez SS: Hamster sperm motility transformation during development of hyperactivation in vitro and epididymal maturation. *Gamete Res* 19:51–65, 1988.

Suarez SS, Pollard JW: Capacitation, the acrosome reaction, and motility in mammalian sperm, in Gagnon C (ed): *Controls of Sperm Motility: Biological and Clinical Aspects*. Boca Raton, FL, CRC Press, 1990, pp 77–90.

Suarez SS, Katz DF, Meizel S: Changes in motility that accompany the acrosome reaction in hyperactivated hamster spermatozoa. *Gamete Res* 10:253–265, 1984.

Suarez SS, Wolf DP, Meizel S: Induction of the acrosome reaction in human spermatozoa by a fraction of human follicular fluid. *Gamete Res* 14:107–121, 1986.

Suarez SS, Vincenti L, Ceglia MW: Hyperactivated motility induced in mouse sperm by calcium ionophore A23187 is reversible. *J Exp Zool* 244:331–336, 1987.

Suarez SS, Drost M, Redfern K, Gotlieb W: Sperm motility in the oviduct, in Bavister BD, Cummins J, Roldan ERS (eds): *Fertilization in Mammals*. Norwell, MA, Serono Symposia, 1990, pp 111–124.

Sugkraroek P, Kates M, Leader A, Tanphaichitr N: Levels of cholesterol and phospholipids in freshly ejaculated sperm and Percoll-gradient-pelleted sperm from fertile and unexplained infertile men. *Fertil Steril* 55:820–827, 1991.

Talbot P: Sperm penetration through oocyte investments in mammals. *Am J Anat* 174:331–346, 1985.

Tanphaichitr N, Smith J, Kates M: Levels of sulfogalactosylglycerolipid in capacitated motile and immotile mouse spermatozoa. *Biochem Cell Biol* 68:528–535, 1990.

Tash JS: Role of cAMP, calcium, and protein phosphorylation in sperm motility, in Gagnon C (ed): *Controls of Sperm Motility: Biological and Clinical Aspects.* Boca Raton, FL, CRC Press, 1987, pp 229–250.

Tash JS: Protein phosphorylation: The second messenger signal transducer of flagellar motility. *Cell Motil Cytoskel* 14:332–339, 1989.

Tash JS, Means AR: Ca^{2+} regulation of sperm axonemal motility. *Meth Enzymol* 139:808–823, 1987.

Tesarik J: Topographic relations of intramembranal particle distribution patterns in human sperm membranes. *J Ultrastruct Res* 89:42–55, 1984.

Tesarik J, Flechon JE: Distribution of sterols and anionic lipids in human sperm plasma membrane: Effects of in vitro capacitation. *J Ultrastruct Mol Struct Res* 97:227–237, 1986.

Thomas P, Meizel S: Phosphatidylinositol 4,5-bisphosphate hydrolysis in human sperm stimulated with follicular fluid or progesterone is dependent upon Ca^{2+} influx. *Biochem J* 264:539–546, 1989.

Toshimori K, Tanii I, Oura C, Eddy EM: A monoclonal antibody, MN13, that recognizes specifically a novel substance between the postacrosomal sheath and the overlying plasma membrane in the mammalian sperm head. *Mol Reprod Dev* 29:289–293, 1991.

Toth GP, Stober JA, Zenick H, Read EJ, Christ SA, Smith MK: Correlation of sperm motion parameters with fertility in rats treated subchronically with epichlorohydrin. *J Androl* 12:54–61, 1991.

Toyoda Y, Yokoyama M, Hoshi T: Studies on the fertilization of mouse eggs in vitro. *Jpn J Anim Reprod* 16:147–157, 1971.

Vaidya RA, Glass RW, Dandekar P, Johnson K: Decrease in electrophoretic mobility of rabbit spermatozoa following intra-uterine incubation. *J Reprod Fertil* 24:299–301, 1971.

Van Deenen LLM, De Gier J: Lipids of the red cell membrane, in Surgenor D, Mac N (eds): *The Red Blood Cell* (2d ed). New York, Academic Press, 1974, pp 147–211.

Vernon RB, Muller CH, Eddy EM: Further characterization of a secreted epididymal glycoprotein in mice that binds to sperm tails. *J Androl* 8:123–128, 1987.

Volglmayr JK, Sawyer FR: Surface transformation of ram spermatozoa in uterus, oviduct and cauda epididymal fluids in vitro. *J Reprod Fertil* 78:315–325, 1986.

Ward CR, Storey BT: Determination of the time course of capacitation in mouse spermatozoa using a chlortetracycline fluorescence assay. *Dev Biol* 104:287–296, 1984.

Wassarman PM: Early events in mammalian fertilization. *Annu Rev Cell Biol* 2:109, 1987.

Wassarman PM: Zona pellucida glycoproteins. *Ann Rev Biochem* 57:415–442, 1988.

Watson PF: The effects of cold shock on sperm cell membranes, in Morris GJ, Clark A (eds): *Effects of Low Temperatures on Biological Membranes.* New York, Academic Press, 1981, pp 189–218.

Weinman S, Ores-Carton C, Escaig F, Feinberg J, Puszkin S: Calmodulin immunoelectron microscopy: Redistribution during ram spermatogenesis and epididymal maturation. *J Histochem Cytochem* 34:1181–1193, 1986.

Wolf DE: Diffusion and the control of membrane regionalization. *Ann NY Acad Sci* 513:247–261, 1987.

Wolf DE, Hagopian SS, Ishijima S: Changes in sperm plasma membrane lipid diffusibility following hyperactivation during in vitro capacitation in the mouse. *J Cell Biol* 102:1372–1377, 1986.

Wolf DP: Acrosomal status quantitation in human sperm. *Am J Reprod Immunol* 20:106–113, 1989.

World Health Organization: *WHO Laboratory Manual for the Examination of Human Semen and Semen-Cervical Mucus Interaction,* Cambridge, Cambridge University Press, 1987.

Yanagimachi R: Mechanisms of fertilization in mammals, in Mastroianni L, Biggers JD (eds): *Fertilization and Embryonic Development in Vitro.* New York, Plenum Press, 1981, pp 81–182.

Yanagimachi R: Zona-free hamster eggs: Their use in assessing fertilizing capacity and examining chromosomes of human spermatozoa. *Gamete Res* 10:178–232, 1984.

Yanagimachi R: Mammalian fertilization, in Knobil E, Niell J (eds): *The Physiology of Reproduction.* New York, Raven Press, 1988, pp 135–185.

Yanagimachi R, Chang MC: Sperm ascent through the oviduct of the hamster and rabbit in relation to the time of ovulation. *J Reprod Fertil* 6:413–420, 1963.

Yanagimachi R, Mahi CA: The sperm acrosome reaction and fertilization in the guinea-pig: A study in vivo. *J Reprod Fertil* 46:49–54, 1976.

Yanagimachi R, Phillips DM: The status of acrosomal caps of hamster spermatozoa immediately before fertilization in vivo. *Gamete Res* 9:1–19, 1984.

Yanagimachi R, Noda YD, Fujimoto M, Nicolson GL: The distribution of negative surface charges on mammalian spermatozoa. *Am J Anat* 135:497–520, 1972.

Yanagimachi R, Yanagimachi H, Rogers BJ: The use of zona-free animal ova as a test system for the assessment of the fertilizing capacity of human spermatozoa. *Biol Report* 15:471–476, 1976.

Yanagimachi R, Lopapta A, Odom CB, Bronson RA, Mahi CA, Nicolson GL: Retention of biological characteristics of zona pellucida in highly concentrated salt solution: The use of salt-stored eggs for assessing the fertilizing capacity of spermatozoa. *Fertil Steril* 31:562–574, 1979.

Yang CH, Srivastava PN: Purification and properties of arylsulfatases from rabbit sperm acrosomes. *Proc Soc Exp Biol Med* 145:721–725, 1974.

Yeagle PL: Cholesterol and the cell membrane. *Biochim Biophys Acta* 822:267–287, 1985.

Yeagle P: *The Membranes of Cells,* New York, Academic Press, 1987.

Yeagle PL: Lipid regulation of cell membrane structure and function. *FASEB J* 3:1833–1842, 1989.

Yoshimatsu N, Yanagimachi R, Lopata A: Zonae pellucidae of salt-stored hamster and human eggs: Their penetrability by homologous and heterologous spermatozoa. *Gamete Res* 21:115–126, 1988.

Yudin AI, Cherr GN, Katz DF: Structure of the cumulus matrix and zona pellucida in the golden hamster: A new view of sperm interaction with oocyte-associated extracellular matrices. *Cell Tissue Res* 251:555–564, 1988a.

Yudin AI, Gottlieb W, Meizel S: Ultrastructural studies of the early events of the human sperm acrosome reaction as initiated by human follicular fluid. *Gamete Res* 20:11–24, 1988b.

Zamboni L: Fertilization in the mouse, in Moghissi KS, Hafez ESE (eds): *Biology of Mammalian Fertilization and Implantation.* Springfield, IL, CC Thomas, 1972, pp 213–262.

Zao PZR, Meizel S, Talbot P: Release of hyaluronidase and *N*-acetylhexosaminidase during in vitro incubation of hamster sperm. *J Exp Zool* 234:63–74, 1985.

Zinaman M, Drobnis EZ, Morales P, Brazil CK, Kiel M, Cross NL, Hanson FW, Overstreet JW: The physiology of sperm recovered from the human cervix: Acrosomal status and response to inducers of the acrosome reaction. *Biol Reprod* 41:790–797, 1989.

CHAPTER 6

The Menstrual Cycle

Preston C. Sacks

The menstrual cycle represents a complex interaction among the nervous, endocrine, and reproductive systems culminating in the production of a mature gamete and a favorable environment for the establishment and maintenance of pregnancy. The physiologic changes that take place during a normal menstrual cycle are outlined in Chap. 2. In terms of reproduction, the menstrual cycle has five basic goals: (1) the development of a mature oocyte—folliculogenesis, (2) release of the mature oocyte from the ovary—ovulation, (3) gamete and zygote transport, (4) development of an endometrium favorable for implantation, and (5) support of early pregnancy via the corpus luteum. A defect in any one of these elements may lead to reproductive failure. The first part of this chapter reviews the pathophysiology and causes of abnormalities in the menstrual cycle that may lead to reproductive failure.

Abnormalities in the menstrual cycle may stem from exogenous or endogenous forces that affect the reproductive axis. Exogenous factors include physical forces, lifestyle, radiation, drug effects, and chemical exposures. Endogenous factors include age, body habitus, and underlying medical illnesses (Fig. 6-1). The effects of specific factors on the menstrual cycle are reviewed in the second part of this chapter.

PATHOPHYSIOLOGY OF MENSTRUAL DYSFUNCTION

Folliculogenesis: The Oocyte

Unlike males, who maintain the ability to produce spermatogonia throughout adult life, females produce oogonia only during early fetal life. The number of oocytes in the human female reaches a peak of 6 to 7 million at 20 weeks' gestation. Subsequently, there is massive atresia, and at birth the ovaries contain approximately 2 million oocytes (Baker, 1963). Atresia continues throughout childhood, until at the time of puberty, approximately 300,000 to 400,000 oocytes remain—only 5 percent of the original number.

Prenatally, the oocyte progresses to the diplotene stage of meiosis 1, where it remains in a state of arrested development until it undergoes either recruit-

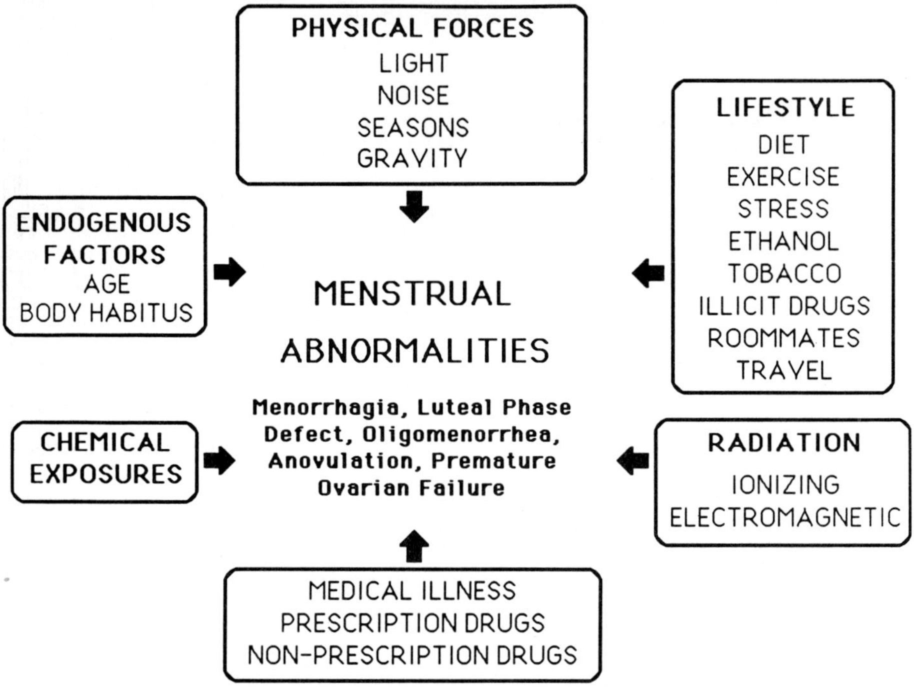

Figure 6-1. Factors influencing menstrual cycle regularity.

ment or atresia. It is important to remember that no oogonia are produced after the 20th week of gestation; therefore, any event that damages an oocyte does so in a permanent fashion (unless repair mechanisms are adequate). The damage may be an alteration in the chromosomal makeup of the oocyte (mutagenesis) or complete destruction of the oocyte (ablation). Mutagenesis does not directly affect the menstrual cycle but leads to reproductive loss via failures in fertilization and defective embryogenesis. This is discussed further in Chap. 7.

Any condition that causes the absence of functional oocytes reduces estradiol production by the ovary. The pituitary gland responds by secreting increased amounts of follicle-stimulating hormone (FSH), leading to hypergonadotropic hypogonadism (ovarian failure). This may occur via a variety of mechanisms (Table 6-1). There may be a defect in the production of primordial germ cells (ovarian dysgenesis), a defect in the response of oocytes to gonadotropins (resistant ovary syndrome) (Kim, 1974), ovarian antibodies disrupting access to gonadotropin receptors (Damewood et al, 1986), or accelerated follicular atresia. Destruction of the oocytes may also occur secondary to infectious processes, such as mumps (Morrison et al, 1975) and tuberculosis (Nogales-Ortiz et al, 1979), inborn errors of metabolism (galactosemia) (Kaufman et al, 1981), or exposures to toxic substances. Oocytes may be destroyed at any time

Table 6-1. Causes of hypergonadotropic hypogonadism

Natural menopause
Surgical castration
Ovarian dysgenesis (Turner syndrome)
Resistant ovarian syndrome (Savage syndrome)
Autoimmune disorders
Addison disease, Graves disease, Hashimoto thyroiditis, hypoparathyroidism, myasthenia gravis
Infectious processes
Mumps
Tuberculosis
Galactosemia
Chemotherapeutic agents
Busulfan, chlorambucil, cisplatinum, cyclophosphamide, etoposide, 5-fluorouracil, L-phenylalanine mustard, mechlorethamine, melphalan, nitrogen mustard, vinblastine
Chemical exposure
Tobacco
Heavy metals
Ionizing radiation

during an individual's life, and the destruction may only come to clinical attention when signs of amenorrhea, infertility, or menopause present.

The sensitivity of the oocyte to damage from a toxic exposure is dependent on the developmental stage at the time of the exposure (Baker, 1971). Oogonia undergoing mitosis are extremely sensitive to the effects of ionizing radiation, while mature graafian follicles are relatively resistant. Similarly, chemotherapeutic drugs are less likely to produce ovarian destruction if given prepubertally (Whitehead et al, 1982; Siris et al, 1976) or to patients less than 25 years old (Schilsky et al, 1981). This information may be clinically useful, as recently there has been a promising report on reducing the incidence of ovarian failure in patients undergoing chemotherapy whose ovaries have been synchronized by pretreatment with oral contraceptives (Chapman and Sutcliffe, 1981).

Examples of agents causing oocyte destruction include ionizing radiation and cancer chemotherapeutic agents, especially the alkylating agents. There is also laboratory (Takizawa et al, 1985) and epidemiologic (Jick et al, 1977) evidence to link chronic tobacco use with oocyte destruction. These specific exposures are discussed in further detail later in this chapter.

Folliculogenesis: Follicular Development

Follicular recruitment and development occur in response to appropriate stimulation of the oocyte by the pituitary gonadotropins, FSH and LH

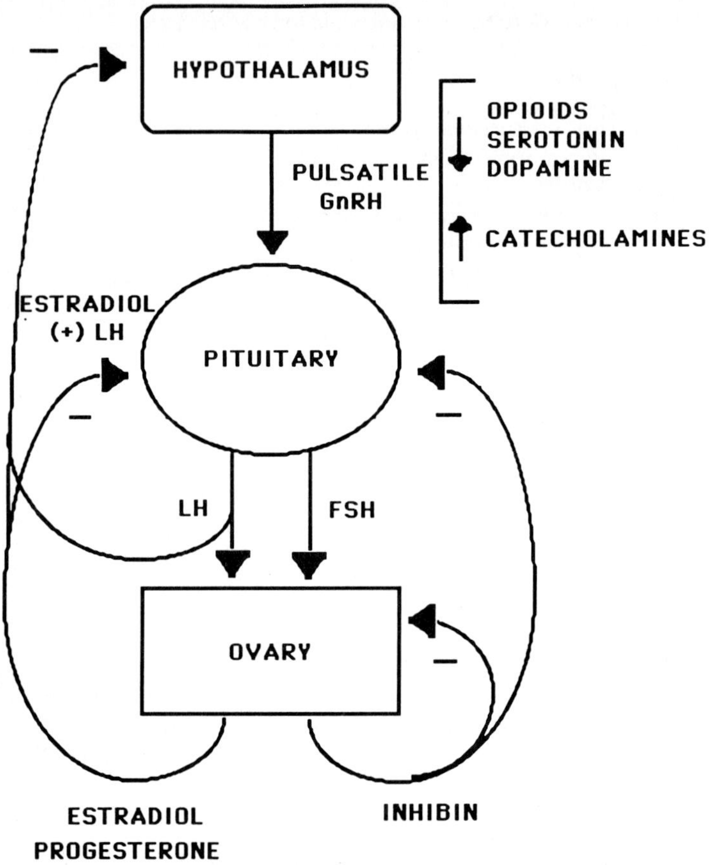

Figure 6-2. Feedback regulation of the hypothalamic–pituitary–ovarian axis.

(luteinizing hormone). Pulsatile secretion of gonadotropin-releasing hormone (GnRH) by the hypothalamus stimulates the production and pulsatile release of FSH and LH. These gonadotropins stimulate the follicle to develop and mature as outlined in Chap. 2. The hypothalamic–pituitary–ovarian (HPO) axis is extremely reliant on short and long feedback loops for regulation (Fig. 6-2). Any derangement in secretion or regulation in this system may result in abnormal follicular development, menstrual irregularities, and reproductive dysfunction.

Secretion of GnRH

Work by Knobil (1980) demonstrated that GnRH is secreted from the arcuate nucleus of the hypothalamus in pulsatile fashion with a pulse frequency in humans of 45 to 90 min. When either extreme of pulse frequency occurs (continuous GnRH secretion or absent GnRH secretion), there is reduction in secretion of FSH and LH by the pituitary, and a hypogonadotropic state ensues. Pulsatile secretion of GnRH can be affected by a number of factors. Opioids, dopamine, and serotonin decrease the pulse frequency of GnRH release, and catecholamines (particularly norepinephrine) increase the pulse frequency (Yen, 1986a).

A variety of clinical situations affect the menstrual cycle through changes in the secretion of GnRH. These clinical situations include stress, anorexia nervosa, exercise-induced amenorrhea, weight loss, malnutrition, hyperprolactinemia, and narcotic abuse. Furthermore, a number of pharmaceuticals affect GnRH secretion by altering opioid, dopamine, serotonin, and catecholamine levels.

Secretion of FSH and LH

FSH and LH are glycoprotein hormones produced, stored, and secreted by the adenohypophysis. They are secreted in pulsatile fashion in response to pulsatile GnRH released from the hypothalamus. The release of FSH and LH is modified by feedback of estradiol, progesterone, and inhibin on the pituitary gland (see Chap. 2). Alteration of the production or secretion of pituitary gonadotropins may occur by a variety of mechanisms.

Pituitary destruction secondary to infarction (Sheehan syndrome), radiation, or an infiltrating or infectious process eliminates the production of gonadotropins. Hypopituitarism may also occur transiently (lasting up to 6 months) following the withdrawal of glucocorticoids (Graber et al, 1965). There are also cases of Kallmann syndrome (hypogonadotropic hypogonadism and anosmia) in which there is an isolated deficiency in the production of gonadotropins from the pituitary leading to hypogonadotropic hypogonadism (Tagatz et al, 1970; Spitz et al, 1974).

More commonly, abnormalities in feedback regulation are the cause for deranged FSH and LH release, leading to menstrual irregularity. The most widely studied is the polycystic ovarian syndrome (PCOS). In PCOS there is an increase in ovarian androgen production, with peripheral aromatization of these androgens to estrogens. The net effect is a hyperandrogenic, hyperestrogenic milieu (DeVane et al, 1975). The tonic hyperestrogenic state leads to a persistent increase in LH secretion, which further stimulates the ovary to produce androgens. This sets up a vicious cycle of endocrine changes that alters the pulsatile release of gonadotropins and inhibits follicular development (Fig. 6-3). As a result, ovulation, production of progesterone, and conversion of the endometrium to a secretory pattern do not take place.

Several clinical conditions lead to high levels of estrogens or estrogen precursors (androgens) and produce a hormonal profile similar to that found in patients with PCOS. These conditions include obesity, adrenal hyperplasia, Cushing syndrome, and overproduction or underexcretion of estrogens or androgens.

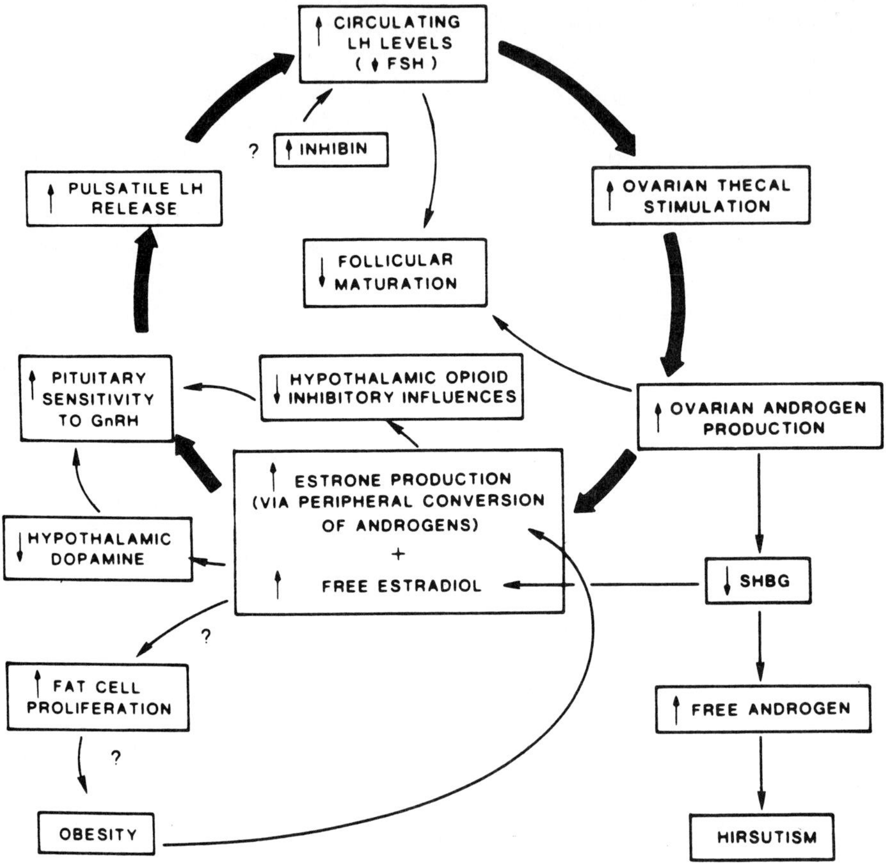

Figure 6-3. Vicious cycle of changes that occurs in patients with PCOS (SHBG = sex hormone-binding hormone). [*From Rebar RW (1983), with permission.*]

Steroid Metabolism

The ovary contains two main regions—stroma and cortex. The stroma gives rise to the theca component of the growing follicle, while the granulosa cells, which surround the maturing follicle, are found in the cortex. The theca cells secrete androgens in response to stimulation by LH and human chorionic gonadotropin (hCG). These androgens are then available for conversion to estradiol by the aromatase enzyme system present in the granulosa cells of the maturing follicle. FSH secreted by the pituitary in the early follicular phase stimulates the activity of the aromatase enzyme system and enhances the follicular production of estradiol. This machinery is the basis of the *two-cell model* of ovarian function. In essence, the maturing ovarian follicle transforms an androgenic microenvironment found in the ovarian stroma to an estrogen-predominant microenvironment necessary for follicular maturation (Ross and

Schreiber, 1986). In PCOS there is an increase in the ovarian stroma and consequently an increase in ovarian androgen production. As a result, the estrogenic microenvironment does not develop, and follicular maturation ceases.

Following production by the ovary, estradiol is secreted into the systemic circulation and binds to sex hormone-binding globulin (SHBG) and albumin. In general, 36 percent of plasma estradiol is bound to SHBG, 61 percent is carried by albumin, and only 2 to 3 percent remains free and available for receptor binding (Wu et al, 1976). An alteration in the serum concentration of SHBG (as occurs in liver and thyroid disorders) changes the amount of free estradiol available for biological activity at the receptor level.

Estradiol is the predominant and most potent estrogen in reproductive-age women. Conversion of androgens to estrone in adipose tissue increases the total estrogen concentration in the circulation, though estrone is a relatively weak estrogen when compared with estradiol. Significant levels of estrone, however, may feed back to decrease the pituitary release of FSH and increase the secretion of LH, leading to anovulation. Obesity, hyperthyroidism, Cushing syndrome, and congenital adrenal hyperplasia affect the menstrual cycle through tonic increases in estrone.

The catabolism and clearance of estrogens involves activity by the liver and kidney (Witt et al, 1990). Free (nonprotein-bound) hormone is converted by the liver into sulfates or glucosidurates, which are metabolically inactive and water-soluble. These conjugates are available for excretion via the kidney or may enter the bile and undergo enterohepatic circulation, to be excreted in the feces. Hepatic and renal diseases, as well as changes in fecal excretion (high-fiber and vegetarian diets) affect the serum concentration of estradiol.

Ovulation

The hormonal events in the periovulatory period have been extensively studied. Changes in the concentration of estradiol and progesterone (Hoff et al, 1983; Liu and Yen, 1983) lead to a midcycle surge in LH secretion. This LH surge causes resumption of meiosis and final maturation of the oocyte. Finally, a cascade of enzymatic processes leads to rupture of the ovarian capsule (Espey, 1974) and a prostaglandin-mediated release of the oocyte (Lumsden et al, 1986). Administration of high doses of nonsteroidal anti-inflammatory drugs (NSAIDs) effectively inhibits ovulation in nonprimates, monkeys, and humans by blocking prostaglandin production in the mature follicle (Priddy et al, 1990). This has been proposed as one cause of the luteinized unruptured follicle syndrome (Priddy et al, 1990).

Gamete and Zygote Transport

Gamete and zygote transport occurs by segmental peristaltic and antiperistaltic activities of the fallopian tube. In the rat and rabbit, tubal peristalsis is decreased by high levels of progesterone and restored with increasing estrogen concentrations (Thibault, 1972). Studies on the human oviduct show that pharmacologic doses of epinephrine, norepinephrine, ergonovine, and prostaglandin F increase tubal contractility, and α-adrenergic blockers and prostaglandin E cause tubal relaxation (Coutinho, 1971). Alterations in tubal motility, however, do not affect menstrual function.

Endometrial Development

There are numerous hormonal and enzymatic processes that are involved in the production of a functioning, synchronous endometrium that will provide a nurturing environment for early development of the embryo. The factors involved with endometrial development are described in detail in Chap. 7.

Progesterone and Corpus Luteal Function

The main function of the corpus luteum (CL) is the synthesis of progesterone (Stouffer, 1990; Filicori, 1990). The production of progesterone is dependent on stimulation by LH or hCG. Adequate levels of progesterone transform the endometrium to a synchronous secretory pattern and maintain pregnancy prior to the onset of placental functioning (approximately 7 weeks' gestation). In the absence of pregnancy, involution of the CL occurs in 14 to 16 days, progesterone secretion diminishes, and menstruation ensues.

Insufficient production or activity of progesterone is implicated in luteal phase deficiency (Lee, 1987), and removal of the CL prior to day 21 after the missed menses is associated with pregnancy loss (Csapo et al, 1973). Replacement of progesterone in the luteal phase may correct the luteal phase defect, and pregnancy loss secondary to luteolysis can be prevented by the administration of progesterone.

Luteal phase defects are commonly seen in patients with hyperprolactinemia, hyperandrogenism, and hypothyroidism. Inadequacy of the luteal phase is also common in the perimenarchal and perimenopausal periods and is frequently associated with athletic conditioning and changes in body weight.

Finally, luteolysis may occur following ingestion of the progesterone antagonist RU486 (Yen et al, 1987), or in association with toxic levels of heavy metals (Franks et al, 1989). This has led to the use of RU486 (Ulmann, 1987) and lead (Hall, 1905) as abortifacents.

FACTORS LEADING TO MENSTRUAL DYSFUNCTION

Menstrual Cycle as an Outcome Variable

Most studies on reproductive toxicology use birth outcome as the measurement variable. Few studies comment on fecundity, and even fewer measure changes in the menstrual cycle. Consequently, much of the information on the effects of a particular agent on the menstrual cycle comes from animal research and small epidemiologic studies. The significance of animal test data is unclear (discussed further in Chap. 8), and the assessment of menstrual cycle variation in a population is a difficult and often unreliable task (Bean et al, 1979).

Variability in the length of the menstrual cycle of "normal" women is common. Chiazze and coworkers (1968) studied 30,655 menstrual cycles from 2316 women. In a minimum of 10 study cycles, only 13 percent of the women had a range of cycle length less than 6 days. Treolar and coworkers (1967) studied menstrual cycle variation and noted a high degree of irregularity, especially in the early menarchal and premenopausal periods. With this large a degree of background variability, the validity of menstrual length as an outcome variable must be questioned. The extremes of menstrual variation (amenorrhea and oligomenorrhea) appear to be the most reliable and valid parameters to study (Shortridge, 1988); however, using these parameters will exclude subtle changes in menstruation.

Biological monitoring is the most accurate method of detecting subtle changes in the menstrual cycle, but the cost of such monitoring makes it impractical in large series. Recently, Hughes (1988) proposed an ovarian cycle monitoring protocol consisting of daily measurements of salivary or vaginal electrical resistance, midcycle urine testing for LH surge, and luteal phase salivary progesterone monitoring. He estimated the cost at $350 per cycle and felt such a protocol would "aid in the early detection of reproductive hazards . . . that might present as an ovulatory disturbance."

Today, most of the information available on the toxic effects of exposures on the reproductive system suffers from lack of adequately controlled epidemiologic studies and biological monitoring data. However, by combining the information obtained from large epidemiologic studies with laboratory data from small studies on humans, subhuman primates, and rodents, one may begin to evaluate the effects of agents on the human menstrual cycle.

Identification of Reproductive Toxicants

The Food and Drug Administration requires new drug applications to show extensive preclinical testing to assess a drug's effect on the reproductive process. Chemical substances are not as tightly regulated, and the National Institute for Occupational Safety and Health (NIOSH) Registry of Toxic Effects of Chemical Substances (RTECS) lists 79,000 chemicals, of which 16 percent are reported to demonstrate some evidence of reproductive toxicity (Wells et al, 1988). Only a handful of these chemicals have been studied in detail, and little is known about their reproductive effects in humans.

With women representing an increasingly significant proportion of the total U.S. work force, there has been a heightened interest in the reproductive effects of occupational exposures. Disorders of reproduction are listed among the 10 leading work-related diseases and injuries by the NIOSH (Centers for

Disease Control, 1985). In response to U.S. House Resolution 126 (A-80), the Council on Scientific Affairs of the American Medical Association appointed an advisory panel to study reproductive hazards in the workplace. In 1984 and 1985, the council published reports on the effects of physical forces (Council on Scientific Affairs, 1984) and toxic chemicals (Council on Scientific Affairs, 1985) on the reproductive system. These publications were intended to guide physicians in counseling patients on hazardous exposures. In these reports, however, there is a notable lack of scientific evidence for the effects on the menstrual cycle of the agents listed.

In order to improve the investigation of reproductive toxicants, the Division of Surveillance at NIOSH in 1988 proposed a new method of surveillance for reproductive toxicants (Wells et al, 1988). It proposed to systematically review the toxicology literature in search of chemicals that show evidence of reproductive toxicity. This will aid in the selection of chemicals and occupational settings for further study. It is hoped that with increased surveillance and controlled studies on exposures, the next decade will show great advances in our understanding of environmental and occupational factors that influence the menstrual cycle.

Factors Influencing the Menstrual Cycle: Physical Forces

Light

Exposure to light plays an important role in the reproductive process of lower animals. Melatonin, secreted by the pineal gland during periods of darkness, inhibits ovulation in rats (Collu et al, 1971) and has been associated with the seasonal mating habits of animals. This appears to be through the ability of high levels of melatonin to inhibit the LH surge (Fraschini et al, 1968). Studies in humans show a variation in melatonin levels throughout the menstrual cycle, with a nadir on the day of the LH surge (Wetterberg et al, 1976). This suggests that decreasing levels of melatonin at midcycle may have a permissive effect on LH release.

In humans, light is known to affect both melatonin secretion (Lewy et al, 1980) and soltriol (dihydroxyvitamin D_3) production (Holick et al, 1980). Melatonin secretion is stimulated by darkness, and soltriol production stimulated by exposure to sunlight. These effects are further modified by the variation in environmental luminance (photoperiod) found with changes in latitude and season (Wetterberg, 1990; Webb et al, 1988).

There is a common belief that the menstrual cycle is linked to changes in environmental light associated with the phases of the moon. In fact, the terms *moon, menses,* and *month* are all derived from the same root word. In the late 19th century, Cook (1894) observed the menstrual cycle of Eskimo women and noted a high degree of suppression of menstrual function during the 4-month Arctic night. Menstruation resumed its normal cyclicity, however, during periods of normal light.

In an effort to test the effects of light on the menstrual cycle, Dewan and coworkers (1978) studied the effects of photic stimulation on the menstrual cycles of women with irregular cycles. When these women were exposed to light during nights 14 to 17 of the cycle, a striking degree of menstrual regularity developed. Unfortunately, no hormonal studies were obtained in these patients. This lends support to the folk belief that increased reflectance of light from a full moon synchronizes menstruation.

Thus, there is evidence that light may affect menstrual cycle regularity in humans, possibly through changes in melatonin and soltriol levels (Stumpf et al, 1989). The significance of these effects, however, remains to be demonstrated.

Noise

Noise is the most widespread physical exposure in the workplace (Kjellberg, 1990), yet the effect of noise exposure on the menstrual cycle has not been well studied (Singh, 1972). Noise has been associated with a stimulation of the neuroadrenal axis, leading to an increase in the concentration of cortisol and catecholamines (Borg, 1981). As corticotropin-releasing factor (CRF) is known to play a modulatory role in the secretion of GnRH, it is possible that exposure to excessive noise could lead to menstrual disturbances. Studies in humans show an increase in serum LH concentration and total urinary gonadotropin excretion following short-term exposure to excessive noise during the follicular phase of the menstrual cycle (Singh, 1972; Beardwood et al, 1975; Beardwood et al, 1977). The reproductive hormonal effects of chronic noise exposure have yet to be documented.

In the only large clinical study on noise exposure, Rachootin and Olsen (1983) found a twofold increase in hormonal disturbances (amenorrhea, anovulation, luteal insufficiency, endocrine malfunctions) and infertility among female workers occupationally exposed to high levels of noise. This finding has not been duplicated.

Seasons

Variations in menstrual regularity (Israel, 1967) and conception rates (Timonen et al, 1965) are noted to occur with changes in the seasons. The variation in conception rates appears to be due to factors in the female, as a seasonal variation in births is noted in a population of women using artificial insemination by donor (Paraskevaides et al, 1988). These effects may be independent or may occur secondary to seasonal changes in environmental luminance (see the section "Light").

Biochemical studies show a seasonal variation in the secretion of LH, with a higher urinary LH bioactivity noted in the spring compared with autumn months (Carletti et al, 1964). Kivelä and coworkers (1988) studied 20 healthy female nurses in Finland. These women had a significantly higher midcycle nocturnal concentration of LH and a decrease in serum melatonin concentration during the summer months. Ronkainen and coworkers (1985) studied the steroid milieu of female runners during the spring and autumn months. There were lower luteal phase serum levels of estradiol, progesterone, and testosterone in the autumn without changes in levels of gonadotropin, prolactin, or SHBG. These studies support a seasonal effect on the reproductive axis. Whether this effect is primary or secondary to changes in environmental luminance has yet to be clarified.

Gravity

The effects of gravity (or the absence of it) are of particular interest to the National Aeronautics and Space Administration (NASA), as disturbances of the menstrual cycle by space travel would seriously affect the prospects for reproduction in space. Numerous hormonal changes occur with changes in gravity and adaption to a zero-gravity environment (Money, 1981). Weightlessness causes a redistribution of body fluids, inhibiting release of antidiuretic hormone (ADH) and causing diuresis. Spaceflight is associated with elevations in the serum concentrations of parathyroid hormone (PTH), cortisol, thyroxine, and thyroid-stimulating hormone (TSH). These endocrine responses may extinguish as adaptation to weightlessness occurs (Santy, 1984).

Unfortunately, few data are available on the effects of space travel on the menstrual cycle (Jennings and Santy, 1989). Instead, prolonged bed rest must be used as a model for weightlessness. Sandler and Winters (1978) noted no change in menstrual cycle length following 17 days of bed rest. As this study did not encompass one entire menstrual cycle, it could not have measured subtle effects. Rock and Fortney (1984) studied the effects of weightlessness on the luteal phase of the menstrual cycle. Three of the 12 patients studied had a deficiency in progesterone secretion during the luteal phase. No control population was used for comparison. Clearly more research needs to be done in this area before long-range space travel can be considered for women.

The U.S. Air Force studied the effects of positive G stress on 102 women (Gillingham et al, 1986). These women were exposed to gravitational stress exercises during menses and intermenstrually. No specific gynecologic effect was noted. One person did report the onset of menstruation and uterine cramping during gravitational straining and felt this was precipitated by the increase in gravitational force.

Factors Influencing the Menstrual Cycle: Endogenous Forces

Age

Menstrual irregularities are common at both extremes of the reproductive life. Vollman (1977) plotted the frequency distribution of menstrual cycle lengths in women from early menarchal to the climacteric phase. The degree of variation in menstrual cycle length and regularity is markedly elevated during the early menarchal and perimenopausal phases. The frequency distribution curve of median cycle length is U-shaped (Fig. 6-4). The mean cycle length at age 12 is 35.1 days, and the cycle length steadily decreases to a minimum of 27.1 days at age 43. Thereafter, the mean cycle length rises to 51.9 days at 55 years of age. Thus age is an important factor to be considered when using menstrual cycle length as an outcome variable.

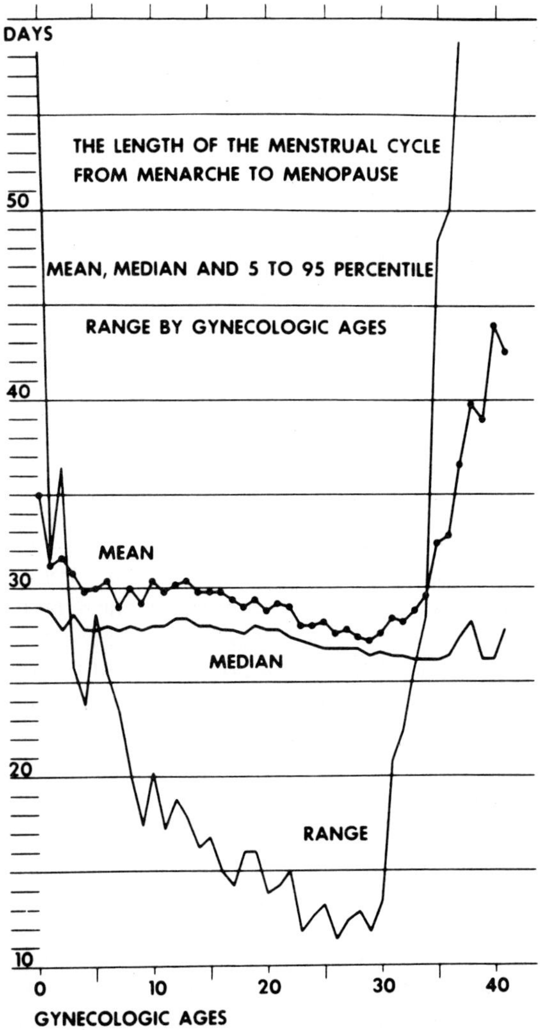

Figure 6-4. The length of the menstrual cycle from menarche to menopause. [*From Vollman (1977), with permission.*]

Hormonal studies have been carried out in early menarchal women and show a high degree (as high as 100 percent) of anovulation during the first postmenarchal year (Apter et al, 1978; Lemarchand-Béraud et al, 1982). Dynamic testing with GnRH in these patients revealed a high degree of luteal phase abnormalities. The rate of anovulation gradually decreased to 37 percent during the fifth year following maturation of the HPO axis.

The median age of menopause in the United States is 50 years (Gambrell, 1982); however, the perimenopausal period extends for several years prior to the cessation of menses. As a woman approaches menopause there is a sharp decrease in fertility rates (Stein, 1985). This decrease in fertility has been explained by elevated FSH levels and short follicular phases noted several years prior to actual menopause (Sherman and Korenman, 1975; Sherman et al, 1976; Metcalf and Livesey, 1985). This rise in gonadotropin levels is most likely a result of the reduced number of functional residual ovarian follicles present as a woman approaches menopause (Richardson et al, 1987; Alexander et al, 1990). These changes occur in relation to the age of menopause, not necessarily the age of the individual, suggesting a biological age factor (Gindoff and Jewelewicz, 1986). A biological age factor is corroborated by the data obtained during in vitro fertilization. Scott and coworkers (1989) showed that elevated basal serum levels of FSH were a better predictor of pregnancy and stimulation outcome at in vitro fertilization than age. Therefore, it may be extremely difficult to eliminate early menopausal women from any study of reproductive function based solely on age criteria.

Body Habitus: Obesity

Obesity has long been associated with menstrual irregularity and subfertility. Many authors have equated obesity with PCOS, though the terms are not synonymous. Of the seven original cases of PCOS reported by Stein and Leventhal (1935), only five patients were above ideal body weight. Obesity is found in roughly 16 to 49 percent of patients with PCOS (Goldzieher and Green, 1962). In fact, obesity may be a secondary finding in PCOS, occurring in relation to prolonged estrogen exposure (Yen, 1986b). There are, however, significant similarities between the mechanisms underlying the menstrual changes present in both obese patients and those with PCOS. Both have elevated ovarian androgen production and significant peripheral aromatization of these androgens to estrogens. Also, both conditions are independently associated with insulin resistance and hyperinsulinemia (Chang et al, 1983). It is therefore helpful to discuss obesity-related menstrual dysfunction in the context of the hormonal changes known to occur in PCOS.

Obesity is associated with an increased risk of oligoovulation secondary to changes in androgen and estrogen metabolism. There is an increase in androgen metabolism, an increase in aromatase activity, and a decrease in SHBG levels. These changes produce a hyperandrogenic–hyperestrogenic state similar to that found in PCOS (Azziz, 1989). The alteration in the androgen/

estrogen ratio leads to deranged feedback regulation and elevated levels of LH (see Fig. 6-3). Disordered follicular maturation and anovulation often ensue.

It has been postulated that the increase in ovarian androgen production is related to the elevation in the serum concentration of insulin produced by the obese state (Poretsky and Kalin, 1987). There is evidence that insulin and insulinlike growth factors have gonadotropic activity and stimulate the production of androgens by the ovary (Poretsky and Kalin, 1987). Hyperinsulinemia may also be the underlying mechanism of PCOS, whether there is obesity or not (Burghen et al, 1980). This would explain the similarity between the endocrinologic changes found in both conditions.

It is important to remember, though, that despite measurable alterations in hormonal levels, most obese patients continue to have regular menses. The extent to which obesity affects menstruation depends on the type of obesity present. Women with early-onset obesity (hyperplastic) are much more likely to have menstrual abnormalities than those with adult-onset (hypertrophic) obesity (31 versus 4 percent) (Friedman and Kim, 1985). This is most likely related to the increase in the number of stromal cells that occurs in hyperplastic, and not hypertrophic, obesity. As stromal cells are the site of aromatization (Cleland et al, 1983), there is a greater potential for estrogen conversion in patients with hyperplastic obesity.

As would be expected, weight reduction (Bates and Whitworth, 1982) in anovulatory obese individuals is associated with a decrease in serum insulin concentrations, a fall in plasma androgen concentrations, an increase in the serum concentration of SHBG, and a resumption of ovulation, menstrual regularity, and fertility (Harlass et al, 1984; Pasquali et al, 1989).

Weight Loss below Ideal Body Weight

Frisch and Revelle (1970) correlated the onset of menses with attainment of a critical body weight (47.8 kg). Subsequently Frisch and McArthur (1974) demonstrated a minimum percentage of body fat (22 percent of body weight) that must be maintained for the onset or maintenance of menstrual cycles. As acute weight loss is common in patients with anorexia nervosa, this group has been used as a paradigm to study the effects of weight loss on the reproductive axis. This may not be fully valid, as persons of normal weight with bulimia also have a high degree of menstrual irregularities (Cantopher et al, 1988). Nevertheless, the changes seen with anorexia parallel those of simple weight loss and correlate with the descriptive findings of Frisch and Revelle.

Using GnRH testing, Sherman and coworkers (1975) systematically studied patients with anorexia nervosa. The average body weight on admission to the hospital was 60 percent of ideal body weight (IBW), and the hormonal profile was similar to that found in prepubescence (low estradiol, FSH, and LH levels, with an absent response to GnRH stimulation). Following diet therapy, when these patients' weight rose to 80 percent of IBW, serum FSH levels returned to normal. LH levels remained low, and the response to GnRH, while increased, did not approach normal levels. When patients regained weight to 90 percent of IBW they had hormonal profiles equal to those of normal controls.

Abnormal LH secretion in patients whose weight falls below 95 percent of IBW has been confirmed in several studies (Bates, 1985; Beumont et al, 1976). When a weight gain program is instituted, the normal LH/FSH ratio is restored, and ovulatory menstrual cycles resume when body weight is restored to greater than 95 percent of IBW (Bates, 1985; Knuth et al, 1977; McArthur et al, 1976).

Simple weight loss related to dieting causes the same changes in hypothalamic function as anorexia nervosa does, though not to the same degree (Pirke et al, 1985; Vigersky et al, 1977; Pirke et al, 1989). The effect seen is directly proportional to the amount of weight loss and varies with the age of the dieter (younger women are more easily affected) (Schweiger et al, 1987). This further supports a common hypothalamic mechanism for menstrual disturbances in anorexia and simple weight loss.

Factors Influencing the Menstrual Cycle: Lifestyle

Ethanol Abuse

Severe drinking problems affect an estimated 4 percent of the U.S. female population (Cahalan, 1970). Chronic ethanol abuse is associated with anovulation, luteal phase defects, and amenorrhea (Ryback, 1977; Hugues et al, 1980; Välimäki et al, 1984). It is difficult, however, to attribute the menstrual problems seen with chronic exposure to alcohol solely to a toxic effect of ethanol. Alcoholic women often have severe liver disease and malnutrition—both known to affect the menstrual cycle (Mello et al, 1989). The exact mechanism of action of alcohol on the reproductive system has yet to be fully elucidated. In general, there is an alcohol-induced increase in levels of serum estradiol (Mendelson et al, 1988), which has been postulated to inhibit the secretion of FSH and LH (Mello, 1989). Whether this effect is primarily related to alcohol's effect on the hypothalamus, pituitary, or ovary is currently under investigation.

Recently there is evidence that "social drinkers" and binge drinkers may also have menstrual disturbances. Mendelson and Mello (1988) studied 12 social drinkers (average between 2.5 and 5 drinks per day) and 5 heavy drinkers (average more than 5 drinks per day) by measuring serum LH, estradiol, and progesterone levels during a 35-day stay in a research setting. All these patients reported normal menstrual cycles prior to entering the study. During the study period, 5 of the 12 (42 percent) social drinkers had menstrual disturbances (anovulatory cycle, delayed ovulation, and hyperprolactinemia) and 3 of the 5 heavy drinkers had hyperprolactinemia.

To summarize, chronic alcohol abuse and heavy social drinking have profound effects on menstrual cycle regularity, though the mechanism underlying these effects remains unclear.

Special Diets

Vegetarians have a higher-than-expected incidence of menstrual abnormalities, and the reproductive effects of a vegetarian diet have been studied. Vegetarians have alterations in estrogen excretion patterns and lower serum estradiol levels (Goldin, 1982). This is not related to the absence of dietary meat, but to an excess in dietary fiber. Vegetarians tend to have a fiber-rich diet, which increases the fecal excretion of estrogen. There is a subsequent decrease in the serum concentration of estradiol (Aldercreutz et al, 1986), a decrease in midcycle LH levels, reduced luteal phase progesterone secretion, and a high incidence of anovulation (Pirke et al, 1986). Other special diets that provide adequate nutrition have not been shown to have marked effects on the female reproductive system (Hagerty et al, 1988).

Nutrition

During World War I, many authors commented on the high incidence of amenorrhea and felt this was related to nutritional deficiency. They coined the term *war amenorrhea* (Engels et al, 1964). This nutritional theory was supported by a study in Holland during the "hunger winter" of 1944–1945 that reported a 50 percent incidence of amenorrhea that resolved with the return of food (Smith, 1947). It appears as though near starvation is required before the menstrual cycle is significantly altered.

Illicit Drugs

In the 1960s several investigators reported a high frequency of menstrual abnormalities in heroin addicts presenting to methadone clinics (Gaulden et al, 1964; Stoffer, 1968). Most of these patients were polysubstance abusers; however, the menstrual disturbances were most closely correlated with heroin abuse. Santen and coworkers (1975) studied 76 heroin addicts attending a methadone clinic. Ninety-five percent of these women reported normal menses prior to heroin addiction, while only 30 percent were regular while on heroin. In-depth hormonal studies on a subset of this study population disclosed an absence of cyclic gonadotropin release and low basal gonadotropin levels consistent with hypothalamic amenorrhea. These effects appear to be reversible, as menstruation returns following the discontinuance of heroin abuse (Gaulden et al, 1964).

Cocaine has become a very popular drug of abuse. It is estimated that one-third of college-age students have used cocaine more than 10 times (Canterbury et al, 1991). Recent studies in the rat show that serum cocaine levels similar to those found in human cocaine abusers disrupt estrous cyclicity by altering the hypothalamic amine activity necessary for pulsatile hypothalamic GnRH release (King et al, 1990). This may be related to an increase in the concentration of prolactin. Cocores and coworkers (1986) studied four chronic cocaine users and noted sexual dysfunction and hyperprolactinemia (serum levels of 60.3 to 75.4 ng/dl) in three of these women. These patients all responded to bromocriptine therapy.

Cocaine blocks dopamine reuptake and increases the concentration of dopamine available to the receptor. Chronic administration, however, may deplete dopamine reserves, decreasing synaptic dopamine levels and leading to increased secretion of prolactin from the pituitary (Dackis and Gold, 1985). Hyperprolactinemia leads to menstrual irregularities (Caufriez, 1985). With the increasing prevalence of chronic cocaine abuse, cocaine may become an important cause of prolactin-related menstrual dysfunction.

Marijuana and its active ingredient, Δ^9-tetrahydrocannabinol (THC) have been extensively studied. Sauer and coworkers (1983) showed that crude marijuana extract, but not THC, competes with estradiol for specific binding sites. In vivo studies, however, failed to demonstrate estrogenic or anti-estrogenic effects. Nir and coworkers (1973) blocked the midcycle LH surge and suppressed ovulation in rats treated with high dosages of THC. On morphologic analysis, the entrapped oocytes were noted to have matured, suggesting either a marijuana-induced effect on the mechanics of ovulation or a very time-specific effect on LH secretion. Leung and coworkers (1985) demonstrated that THC blocked progesterone release in in vitro cultures of rat luteal cells. Similar data on marijuana-induced suppression of LH secretion and ovulation exist in the rabbit (Asch et al, 1979) and subhuman primate (Smith et al, 1979). In the rhesus monkey, the mechanism of LH suppression appears to be through alterations in the hypothalamic control of gonadotropin secretion (Smith et al, 1979). There is evidence in a wide variety of animal models for direct and indirect effects of marijuana on the female reproductive system.

Placebo-controlled studies in humans have shown that marijuana smoking decreases the pulse frequency of LH secretion in postmenopausal women (Mendelson et al, 1985) and suppresses serum LH levels during the luteal phase in reproductive-age women (Mendelson et al, 1986). These effects were noted within 60 to 120 min following marijuana smoking. Despite acute suppression of luteal phase LH levels, there was no change in the serum concentrations of estradiol and progesterone, raising the question of the biological significance of these alterations. Furthermore, these hormonal changes might be complicated by the development of tolerance to the effects of marijuana, as occurs in primates (Smith et al, 1983) and therefore may have little significance in chronic users.

In a case–control study, Bauman and coworkers (1979) reported shorter menstrual cycles and abnormal luteal phases in women who reported using marijuana at least four times a week. This study has been criticized for the use of self-reporting to document drug use and menstrual cycle characteristics

(Mendelson et al, 1986). There are currently no prospective clinical data that address the effects of marijuana on fertility in the human female.

Travel

With the ability to undergo rapid changes in altitude and geographic location, women have anecdotally reported changes in menstrual function in association with jet travel. In order to document these effects, Cameron (1969) studied the menstrual cycles of flight attendants for SwissAir. He noted an initial increase in menstrual irregularity when stewardesses began jet flying. These findings were independently confirmed by Iglesias and coworkers (1980), who documented a high incidence of menstrual changes (39 percent) in stewardesses of Mexicana Airline. This appears to be specifically related to jet travel. Farrell and Allen (1973) studied U.S. Air Force female flight nurses and noted that flight nurses who flew exclusively in jet aircraft had a higher incidence of menstrual irregularity (hypomenorrhea, menorrhagia, dysmenorrhea) than those who flew in propeller craft. Whether the changes in the menstrual cycle are related to job stress, changes in time zones ("jet lag"), or changes in altitude and pressurization is uncertain.

These changes in menstrual cycle regularity appear to be reversible. In a follow-up study, Cameron (1969) found that with discontinuation of flying, most women reverted to their preflight menstrual regularity. There was also no notable effect on the fertility of these flight attendants.

Stress

Stress is difficult to isolate as an independent factor in menstrual dysfunction, because it is often associated with changes in nutrition, body weight, and lifestyle factors. Studies specifically focusing on the effects of psychologic stress report dramatic changes in menstrual function when women are placed in very stressful situations (i.e., war, incarceration, starvation).

Sydenham (1946) and Whitacre and Barrera (1944) studied prisoners in internment camps during World War II. They both documented a high degree of amenorrhea. Bioassays performed in a select group of patients showed a hypoestrogenic milieu. Both these authors commented that the onset of amenorrhea occurred before any possible change in nutrition and was a result of intense psychologic stress.

Speroff (1984) comments on a remarkable study from 1940 (Stieve, 1940) that showed that 100 percent of women indicted for severe crimes and subsequently executed developed hypogonadism and amenorrhea immediately after sentencing. Autopsies confirmed the presence of hypogonadism. Adamopoulos and coworkers (1982) studied the pituitary function of young women who developed amenorrhea following imprisonment in a reformatory institution. Dynamic testing of the HPO axis suggested increased dopaminergic activity as the primary cause of amenorrhea. This supports the theory of increased levels of β-endorphins as the mechanism underlying stress-related menstrual disorders.

Stress causes release of CRF, which stimulates the release of both corticotropin and opioid peptides from the pituitary (Vale et al, 1981). Increased levels of opioids in the central nervous system, particularly β-endorphin, inhibit GnRH release and stimulate the release of prolactin (Grossman, 1987). Stress also activates the adrenal axis, increasing the concentration of glucocorticoids and catecholamines (both modulators of GnRH pulsatility) (Axelrod and Reisine, 1984). As a result of these changes, GnRH pulse frequency is altered, and menstrual irregularities ranging from luteal phase defects to oligoovulation to hypogonadotropic hypogonadism may ensue (Gindoff and Ferin, 1987).

Pseudocyesis represents a special menstrual disorder related to emotional stress. Pseudocyesis is a clinical syndrome in which a nonpsychotic woman develops the signs and symptoms of pregnancy, including amenorrhea, galactorrhea, increased abdominal girth, and cervical softening. There are a variety of neuroendocrinologic abnormalities that may be present including hyperprolactinemia, low levels of FSH, normal or elevated levels of LH, and elevated serum concentrations of progesterone from a persistent CL (O'Grady and Rosenthal, 1989). Unfortunately, there is no consistent neuroendocrinologic finding in these patients (Starkman et al, 1985), and attempts at demonstrating increased opioid activity as the principal defect have been unsuccessful (DeVane et al, 1985). Nevertheless, stress is felt to be the underlying mechanism for the changes present in this puzzling condition.

Tobacco Use

The health hazards of tobacco use are so clear that warnings appear on every package of cigarettes sold in the United States and 51 other countries around the world (Council on Scientific Affairs, 1990). Much publicity has been given to the association between tobacco use and pregnancy complications. By comparison, the effects of tobacco on menstrual cycle regularity and fertility are not well publicized.

There are more than 2500 compounds contained in tobacco and tobacco smoke (Stedman, 1968), each of which may be a hazard to reproduction. Two components have received the most attention—nicotine and benzo(a)pyrene.

Injection of benzo(a)pyrene into rat ovaries and intraperitoneal injection of benzo(a)pyrene into pregnant mice cause destruction of oocytes (Mattison, 1982; Takizawa et al, 1985). This laboratory evidence has been used to explain the decreased age of menopause noted among smokers (Jick et al, 1977). Destruction of oocytes in a dose-related fashion by tobacco exposure is postulated to lead to earlier ovarian failure (Mattison, 1982).

Aside from an earlier menopause, tobacco use has also been associated with an increase frequency of menstrual abnormalities (Tokuhata, 1968; Brown et al, 1988). Mattison (1982) reviewed the effects of nicotine on the HPO axis in rodents. In rodents treated with nicotine there is a delayed and blunted secretion of prolactin and a decrease in the midcycle LH surge. As a result, these nicotine-exposed rodents showed a high incidence of anovulation.

In summary, there is experimental and epidemiologic evidence in humans and rodents showing that exposure to constituents in tobacco smoke leads to oocyte destruction and alterations in the secretion of prolactin and LH. This may explain the higher incidence of menstrual abnormalities and earlier age of menopause noted among tobacco smokers.

Roommates

Menstrual synchrony has been noted to occur in women living in close situations. McClintock (1971) studied women who attended a predominantly female university and noted a high incidence of menstrual synchrony among roommates and closest friends. Graham and McGrew (1980) studied female undergraduates living together on a coeducational campus and confirmed a high degree of menstrual synchrony. This was not related to social interaction with males, but correlated with the amount of time the particular females spent together.

Pheromones, known to influence the estrous cycles of lower mammals, have long been suspected as the cause for menstrual synchrony in humans. In an attempt to prove this, Preti and coworkers (1986) performed a randomized double-blind study on the effects of the odor of axillary secretions on the female menstrual cycle. They studied 19 subjects by placing female donor axillary secretions under the nose of the study subjects. Within a period of three cycles, a high degree of synchrony with the donor cycle was obtained. As the authors point out, this is the same time frame an ovarian follicle requires for maturation (85 days).

Whether menstrual synchrony occurs in women working in relatively close contact in the workplace remains to be demonstrated. McClintock (1971) refers to seven female lifeguards whose menstrual cycles were scattered at the beginning of the summer, but who developed remarkable menstrual synchrony after 3 months spent working together. Studies of the menstrual cycles of recently hired female workers are necessary to show if a true association exists.

Physical Conditioning

The effects of strenuous exercise on the menstrual cycle have been extensively reviewed (Shangold, 1985; Loucks, 1990). Studies have focused on elite athletes [long-distance runners (Dale et al, 1979) and ballet dancers (Frisch et al, 1980; Abraham et al, 1982)] and show a short luteal phase (Shangold et al, 1979) with a substantial incidence of amenorrhea. Most of these studies have noted a strong correlation between weight and body fat composition and the incidence of menstrual irregularities.

There is also evidence to suggest that the energy drain associated with rigorous training is an independent factor influencing menstrual function. This is evident when dancers are injured or take vacations from training. In these cases, menstrual regularity resumes despite no appreciable change in body weight or composition (Warren, 1980; Abraham et al, 1982).

Several hormonal changes occur during exercise training, which may explain the high incidence of menstrual dysfunction. There is a decrease in progesterone production (Prior, 1990) and lower midluteal plasma progesterone concentrations (Shangold et al, 1979). In women runners, there is an exercise-induced release of prolactin (Boyden et al, 1982; Shangold et al, 1981), a decrease in serum estradiol levels, and an increase in dehydroepiandosterone sulfate and androstenedione concentrations. The fact that these changes in serum concentrations occur without a change in SHBG suggests that a production deficiency exists (Baker, 1981). The net effect is a hypoestrogenic, anovulatory situation, which places these athletes at high risk for osteoporosis (Shangold et al, 1990).

It is known that running causes an increase in β-endorphins leading to "runner's high." Carr and coworkers (1981) documented an exercise-induced increase in plasma levels of β-endorphin. Physical conditioning augments this exercise-induced release, explaining why trained athletes appear to be at higher risk for hormonal disturbances than their recreational counterparts. Repeated activation of the CRF–corticotopin–POMC system (as occurs with physical conditioning) leads to an increase in the level of β-endorphins. This explains the hormonal profile noted in trained athletes (hypoestrogenism, exaggerated prolactin release, anovulation) (McArthur, 1985).

Research attention has recently focused on untrained, recreational athletes. Induction of strenuous exercise in untrained women led to disruption of menstrual regularity, CL dysfunction, and luteal phase defects (Beitins et al, 1991). The effect was greatest in those women with weight loss, though also notable in those patients who maintained their body weight (Bullen et al, 1985). Broocks and coworkers (1990) noted depressed estradiol and lower luteal phase progesterone levels in recreational athletes (women who ran between 20 and 30 km per week). Of note, these women reported "regular" menstrual cycles, and disturbances in ovarian function were only identified by hormonal monitoring. Similar findings were noted in the study by Prior and coworkers (1990), underscoring the importance of careful gynecologic evaluation of even recreational athletes.

Factors Influencing the Menstrual Cycle: Radiation

Radiation: Nonionizing

Extremely low frequency radiation (ELFR) from electric and magnetic fields has received much attention for its association with the development of certain tumors (Wertheimer and Leeper, 1979; Wertheimer and Leeper, 1982).

The most widely studied ELFR exposure in reproductive women is that related to video display terminals (VDTs). VDTs are a source of nonionizing radiation. Chronic exposure to VDTs has been reportedly associated with an increase in the rate of spontaneous abortions (Goldhaber et al, 1988): however, large case–control and cohort studies fail to show an association (Bryant and Love, 1989; Blackwell and Chang, 1988; Schnorr et al, 1991). Currently, there is no evidence that nonionizing radiation has any effect on the female reproductive tract (Wilkening and Sutton, 1990).

Radiation: Ionizing

Ionizing radiation has profound effects on all living tissue, including the female reproductive tract. The dose necessary to cause alterations in ovarian function varies with the age of the individual (Jacox, 1939) and the developmental stage of the oocyte (Baker, 1966) at the time of exposure, the total dosage of radiation, and the directness of the exposure (Stillman et al, 1981). This is most evident in reviewing studies from the 1930s to 1950s in which small doses of x-ray (105 rad total) were used to induce ovulation and menstrual regularity in women with secondary amenorrhea and sterility (Kotz and Herrmann, 1962). Israel (1952) reported regular cyclic menses for 1 year following treatment in 71 percent of his patients. This is felt to result from improved intraovarian metabolism secondary to either the hyperemia induced with radiation or the destruction of intraovarian inhibitory influences (Kotz and Herrmann, 1962; Israel, 1952; Baker, 1971). Even at this low dosage, however, premature menopause occurred in roughly 5 percent of the cases (Kotz and Herrmann, 1962).

In contrast to these studies, a total dosage to the ovaries of 500 rad has been considered significant enough to produce sterility (Jacox, 1939). In younger individuals, however, the dosage necessary to produce sterility may be much greater. In patients treated with oophoropexy and radiotherapy for Hodgkin disease, a minumum dosage of 263 to 660 rad produced permanent amenorrhea in 59 percent of the cases (Ray et al, 1970). Detailed analysis failed to reveal any significant factor (including age and total dosage) that discriminated between the patients who became amenorrheic and those with continuing ovarian function. Furthermore, the absence of menses for even 2 years following radiation does not necessarily correlate with sterility, as pregnancies have been reported several years following radiation-induced amenorrhea (Baker et al, 1972). Clearly there are factors other than age and dosage involved in the radiosensitivity of the ovary.

All studies of the effects of radiation on the female genital tract deal with single, relatively high dose exposures. There is an absence of information on the effects of intermittent or continuous low-dose exposure to radiation that occurs with medical, occupational, or environmental exposures. Currently, there are studies looking at the health effects of low-dose radiation exposure in terms of cancer risk and mortality (Wing et al, 1991; Marwick, 1990a, b).

Analysis of the data base obtained may shed some light on the menstrual effects of low-dose radiation.

Radiation to nonpelvic organs may also cause changes in menstruation. A dosage of 5000 rad to the hypothalamic–pituitary region is effective in producing hypopituitarism and hypogonadism (Samaan et al, 1975). Radioactive iodine administered for thyroid ablation is associated with temporary hypergonadotropic amenorrhea in 30 percent of the cases. This amenorrhea typically occurs between 2 and 6 months following administration, is inversely related to the age of the patient, and rarely lasts longer than 12 months (Raymond et al, 1989).

Factors Influencing the Menstrual Cycle: Medical Illness

The menstrual cycle effects of a particular medical illness may be direct or indirect. Disorders may directly affect the hormonal milieu of the reproductive axis or indirectly affect menstrual regularity through high levels of emotional stress and nutritional deficiencies. A list of medical illnesses that frequently affect the menstrual cycle appears in Table 6-2.

Endocrinopathies

Acromegaly is characterized by an oversecretion of growth hormone (GH), usually from a pituitary tumor. Often these tumors are mixed, producing both GH and prolactin in approximately 45 percent of the cases (Lamberts et al, 1979). Hypogonadism may occur secondary to hyperprolactinemia or the mass effect of the tumor suppressing gonadotropin secretion. As a result, menstrual disturbances have been reported in 32 to 87 percent of patients with acromegaly (Jadresic et al, 1982). Surgery is the primary mode of therapy for GH-secreting tumors, while in cases of mixed tumors, bromocriptine therapy is effective in reducing serum GH levels and effecting clinical improvement in 70 to 92 percent of the patients (Vance et al, 1984).

Congenital adrenal hyperplasia (CAH) is most commonly due to a defect in the 21-hydroxylase enzyme complex. As a result, there is a deficiency in the production of cortisol, with a buildup of cortisol precursors and adrenal androgens. The decrease in serum cortisol levels prompts an increase in the secretion of corticotropin, which further stimulates the adrenal gland to produce cortisol precursors and adrenal androgens. A significant amount of peripheral conversion of adrenal androgens to estrogens occurs, and the net effect is a hyperandrogenic–hyperestrogenic milieu similar to that found in PCOS.

There may be a complete defect in the 21-hydroxylase enzyme (classic CAH), or a very mild deficiency resulting in late-onset congenital adrenal hyperplasia (LOCAH). This milder variety (LOCAH) often presents with

Table 6-2. Medical conditions frequently associated with menstrual cycle irregularity

Medical condition	Proposed mechanisms of menstrual dysfunction
Primary endocrinopathies	
Acromegaly	Hypopituitarism from "mass effect" Associated hyperprolactinemia
Adrenal hyperplasia	Increased production of adrenal androgens Peripheral conversion of androgens to estrogens
Cushing syndrome	Increased production of androgens that are converted to estrogens in peripheral tissue Elevated CRF
Diabetes mellitus	Alteration in pituitary–gonadal feedback mechanisms
Prolactinoma	Inhibition of GnRH secretion
Thyroid disorders	
Hypothyroidism	Elevated TRH, causing stimulation of prolactin secretion
Hyperthyroidism	Elevated LH, estradiol, SHBG, androgens Increased conversion of androgens to estrone
Autoimmune diseases Addison disease Graves disease Hashimoto thyroiditis Hypoparathyroidism Myasthenia gravis	Production of antibodies to FSH receptor)
Renal failure	Hyperprolactinemia secondary to decreased clearance
Hepatic disorders	Increased serum estrone due to decreased conjugation and clearance
Blood dyscrasias	Suppression of GnRH pulsatility from chronic illness Infiltration of pituitary and/or ovary by hemochromatosis from transfusions
Neurologic diseases	
CNS lesions Tumors Head trauma Infections	Alteration in hypothalamic–pituitary axis
Empty sella syndrome	Hyperprolactinemia "Mass effect"
Epilepsy	Alteration of hypothalamic dopamine levels or GnRH pulsatility from seizure activity Increase in seizure activity from anovulation
Eating disorders	Loss of GnRH pulsatility from either weight loss or psychologic factors

only mild hirsutism and oligomenorrhea (Azziz and Zacur, 1989) and is confirmed by measurement of the adrenal precursor 17-hydroxyprogesterone following corticotropin stimulation (Brodie and Wentz, 1987). Menstruation normalizes with glucocorticoid therapy in greater than 60 percent of the patients with classic CAH (Mulaikal et al, 1987) and virtually all patients with LOCAH (Birnbaum and Rose, 1984).

Prolactinomas are the most common type of pituitary tumor and are extremely prevalent in the general population. In a series of 120 unselected autopsies, 27 percent of the pituitary glands were found to have microadenomas, of which 41 percent stained positive for prolactin (Burrow et al, 1981). Menstrual abnormalities are extremely common in association with prolactin-secreting microadenomas (up to 90 percent) (Franks et al, 1977). Bromocriptine is effective in bringing about the return of menses in 57 to 97 percent of patients with a prolactin-secreting microadenoma (Vance et al, 1984).

Cushing syndrome is categorized primarily by hypercortisolemia; however, increased adrenal androgen production and obesity are important facets of the disease. The adrenal androgens undergo aromatization to estrogens in the abundant adipose tissue. This tonic increase in serum estrogen levels leads to anovulation. In a 30-year review of patients with Cushing syndrome followed at the University College Hospital in London, menstrual disturbances were noted in 84 percent of the patients (Ross and Linch, 1982). Menses and fertility often return with successful treatment (Ross and Linch, 1982).

Unaided by insulin, greater than 90 percent of the girls in Joslin's series experienced amenorrhea following the onset of diabetes mellitus (Joslin et al, 1925). Insulin therapy brought about menses in a significant number, though not all, of these patients. Despite adequate glucose control, menstrual irregularities remain common in patients with diabetes (Bergqvist, 1954). Hormonal evaluation of these patients reveals a derangement in pituitary–gonadal feedback regulation (Djursing et al, 1982; Distiller et al, 1975), though a specific mechanism has yet to be established.

Untreated hyperthyroidism and hypothyroidism are both associated with menstrual irregularities. Goldsmith and coworkers (1952) studied the menstrual pattern of patients with thyroid disease and noted a 95 percent incidence of amenorrhea or oligomenorrhea in patients with hyperthyroidism. Patients with hypothyroidism reported amenorrhea or oligomenorrhea in 80 percent of the cases. All these menstrual abnormalities improved rapidly within months of initiating therapy.

Hyperthyroidism is associated with increased production of SHBG and a decrease in the clearance of estrogen and testosterone (Ruder et al, 1971). There is an increase in peripheral conversion of testosterone to estrone (Southren et al, 1974) and an alteration in the metabolism of estrogen favoring an increase in the conversion to 2-methoxyestrone (Fishman et al, 1962). Ultimately an increase in the tonic level of serum estrogen alters pituitary feedback regulation. Baseline gonadotropin levels are frequently elevated and there is an exaggerated response of LH to stimulation by GnRH (Tanaka et al, 1981). Anovulation and menstrual irregularities often ensue.

In hypothyroidism, there is a decrease in the production of SHBG and an increase in the conversion of testosterone to androstenedione (Gordon et al, 1969). This leads to an increase in the free (biologically active) serum androgen concentration. Hyperprolactinemia often accompanies hypothyroidism, presumably as a result of stimulation of the lactotrophs by elevated levels of TRH or diminishing hypothalamic dopamine turnover (Thomas and Reid, 1987). Thyroxine replacement normalizes serum androgen and prolactin levels and reinstates menstrual regularity (Boroditsky and Faiman, 1973).

Autoimmune Disorders

Autoimmune diseases are associated with premature ovarian failure. In a review of 115 cases of premature ovarian failure, immune disturbances were present in 17.4 percent of these patients (Rebar and Connolly, 1990). Autoimmune illnesses that have been commonly associated with premature ovarian failure include Addison disease (Elder et al, 1981; Vazquez and Kenney, 1973). Hashimoto thyroiditis (Vazquez and Kenney, 1973), hypoparathyroidism (Kleerekoper et al, 1974; Vazquez and Kenney, 1973), Graves disease (Rabinowe et al, 1989), and myasthenia gravis (Escobar et al, 1982). Autoimmune disorders are often polyglandular (Rabinow et al, 1989), and all autoimmune disorders have the potential to be associated with premature ovarian failure. The mechanism of ovarian dysfunction has been shown to be one of inhibition of FSH receptor binding by circulating ovarian antibodies (Chiauzzi et al, 1981), and there are a few reports of successful therapy and even pregnancy following corticosteroid administration (Cowchock et al, 1988).

Galactosemia

Galactosemia, an inborn error in the pathway of galactose metabolism, is characterized by a deficiency of the enzyme galactose-1-phosphate uridyltransferase. As a result there is tissue buildup of galactose and its metabolites. There is a high incidence of hypergonadotropic hypogonadism (60 percent) and diminished ovarian function due to destruction of the ovarian follicles by toxic metabolites (Kaufman et al, 1981, 1987).

Chronic Renal Failure

Chronic renal failure causes profound alterations in the menstrual cycle of greater than 90 percent of patients (Rice, 1973; Perez et al, 1978). Most commonly patients have hypermenorrhea and ultimately amenorrhea. Intensive endocrine studies in these patients reveal low serum concentrations of estrogen, normal basal FSH levels, and moderately elevated levels of LH (Strickler et al, 1974). Hyperprolactinemia is present in 75 to 91 percent of patients, and the mechanism is felt to be one of decreased clearance rather than overproduction of prolactin (Lim et al, 1979; Gómez et al, 1980). Hemodialysis is ineffective at reducing serum prolactin concentrations and reversing the menstrual irregularities (Gómez et al, 1980; Perez et al, 1978; Rice, 1973). Renal transplantation corrects the hyperprolactinemia (Lim et al, 1979) and is associated with resumption of menses in 60 percent of the cases (Bierman and Nolan, 1977). Recently, there has been a report on the use of recombinant erythropoietin in patients with chronic renal failure. Serum prolactin levels decreased significantly, and 5 of 9 patients treated resumed menstruation (Schaefer et al, 1989). It is not clear whether this is related to an increase in renal perfusion or secondary to the improvement in overall well-being.

Liver Disease

Amenorrhea occurs in 25 to 65 percent of patients with chronic liver disease (Green and Rubin, 1959). Välimäki and coworkers (1984) studied the hormonal status of nine patients with amenorrhea related to alcoholic liver disease. Women with alcoholic liver disease had lower plasma estradiol and higher plasma estrone levels when compared with premenopausal and postmenopausal controls. There was no compensatory elevation of gonadotropin levels suggesting primary hypothalamic dysfunction. Also, urinary estrone levels were low in the face of increased plasma concentration of estrone most likely due to reduced clearance associated with hepatic dysfunction.

Blood Dyscrasias

Sickle cell anemia and thalassemia have been associated with a high incidence of menstrual irregularities. Patients with thalassemia commonly have pubertal failure (Kletzky et al, 1979), and secondary amenorrhea has been reported to occur in 13 to 33 percent of patients who achieve sexual maturation (DeSanctis et al, 1988; Borgna-Pignatti et al, 1985). Dynamic endocrine testing in these cases shows a poor or absent response in both the pituitary and ovary, most likely secondary to iron deposition in these tissues in association with frequent transfusions (DeSanctis et al, 1988). This is not surprising, as histologic studies in patients with hemochromatosis show that iron is preferentially localized in the gonadotropin-secreting cells of the pituitary (Bergeron and Kovacs, 1978).

Samuels-Reid and Scott (1985) studied the menstrual cycle of patients seen in the Sickle Cell Disease Clinic at the Howard University. Patients with sickle cell anemia have shorter menstrual cycles than normal controls, though the mechanism underlying this finding has not been well studied. As patients with sickle cell anemia have profound nutritional deficiencies (Prasad et al, 1976), this likely contributes to their menstrual dysfunction.

Neurologic Illnesses

Reproductive endocrine disorders are unusually common in patients with temporal lobe epilepsy (TLE). Herzog and coworkers (1986) studied 50 patients with TLE. Amenorrhea occurred in 16 percent of these patients, and 20 percent reported oligomenorrhea. Hormonal studies revealed PCOS (20 percent) and hypogonadotropic hypogonadism (12 percent) in numbers significantly greater than those found in the general population. Abnormalities were noted equally among patients on antiseizure medications and those not on medication and appear to be related to an abnormal LH pulse pattern (Bilo et al, 1991).

A recent study of 20 patients with primary generalized epilepsy (PGE) revealed one case of primary amenorrhea and five cases (20 percent) of oligomenorrhea (Bilo et al, 1988). Detailed hormonal analysis in the patients with oligomenorrhea revealed PCOS in three and hypogonadal hypogonadism in two. These numbers are similar to those reported in patients with TLE.

Proposed mechanisms include (1) a disruption in the normal limbic modulation of hypothalamic GnRH regulation of pituitary function by epileptic discharges (Herzog et al, 1986; Bilo et al, 1988), (2) alterations in dopamine metabolism by seizure activity (Herzog et al, 1986; Bilo et al, 1988), or (3) a common developmental defect in neurotransmitters affecting both the neural and reproductive axes independently (Bilo et al, 1988). Alternatively, menstrual disorders may predispose patients to epileptic activity. The facts that estrogens can induce seizure activity (Logothetis et al, 1959), and that progesterone reduces the frequency of seizures in certain women (Herzog, 1986) lend credence to this hypothesis.

Intracranial tumors are associated with hypothalamic dysfunction and hypopituitarism (Jenkins et al, 1976). This may be due either to compression of the hypothalamus or pituitary by an enlarging intracranial mass or secondary to radiotherapy used in the treatment of intracranial tumors (Samaan et al, 1975). The empty sella syndrome may also cause menstrual irregularities as a result of its association with hyperprolactinemia (Bryner and Greenblatt, 1977).

Factors Influencing the Menstrual Cycle: Pharmacologic Agents

Pharmacologic agents used in the treatment of illness may have profound effects on the menstrual cycle. These effects may be mediated through (1) binding to steroid receptors and exertion of estrogenic or antiestrogenic effects, (2) stimulation of an increase in serum prolactin levels, or (3) a direct toxic effect on the ovary (Table 6-3).

Table 6-3. Pharmaceuticals that disrupt the female reproductive system

Proposed mechanism of menstrual dysfunction	Examples
Elevation of serum prolactin levels	α-Methyldopa Calcium channel blockers Cimetidine D-Penicillamine Narcotics Oral contraceptives Phenothiazines Reserpine Tricyclic antidepressants
Intrinsic estrogenic activity	Coumarin Digitalis Flavonoids Marijuana Oral contraceptives Estrogen-containing cosmetics
Direct ovarian toxicity	Busulfan Chlorambucil Cisplatinum Cyclophosphamide Etoposide 5-Fluorouracil (5-FU) L-Phenylalanine mustard (L-PAM) Mechlorethamine Melphalan Nitrogen mustard Vinblastine

Drugs with Intrinsic Estrogenic Activity

Pharmacologic agents may bind directly to the estrogen receptor and exhibit intrinsic estrogenic or antiestrogenic activity. This alters the feedback regulation of the gonadotropins and leads to ovulatory disturbances. The alterations in the HPO axis are similar to those found in PCOS.

Examples of drugs that show binding to the estrogen receptor include coumarin (Wani et al, 1975), digitalis (Rifka et al, 1976), flavonoids (Cheng et al, 1954; Wani et al, 1975), marijuana (Solomon et al, 1977; Rifka et al, 1976), and oral contraceptives. Estrogens are also present in cosmetics, where they are easily absorbed across the skin and mucous membranes. Absorption has led to a high incidence of menstrual disturbances in exposed workers at oral contraceptive manufacturing plants (Harrington et al, 1978) and precocious thelarche in premenarchal users of estrogen-containing cosmetics (Medical Letter, 1985).

Drugs Causing Hyperprolactinemia

Many agents act via modifications of prolactin release. These may increase serum prolactin concentrations via inhibition of dopamine synthesis or release, blockade of the histamine receptor or calcium channel, or through as-yet-undefined mechanisms (Katz and Adashi, 1990). Pharmaceuticals that decrease dopamine levels and reduce hypothalamic inhibition of prolactin release include reserpine, phenothiazines, amoxapine, sulpiride, metoclopramide, opiates, cocaine, and α-methyldopa (Dackis and Gold, 1985; Turkington, 1972; Cooper et al, 1981; Mancini et al, 1976; McCallum et al, 1976; Oseko et al, 1988). Cimetidine (Carlson and Ippolity, 1977), a competitive blocker of the H_2-histamine receptor, and verapamil (Gluskin et al, 1981), a calcium channel blocker, also cause elevated serum levels of prolactin via receptor-mediated control of prolactin release. There have been some reports of hyperprolactinemia in association with D-penicillamine (Kahl et al, 1985). Finally, oral contraceptives and some tricyclic antidepressants (Fowlie and Burton, 1987) appear to have a direct effect on the secretion of prolactin.

Drugs Affecting Steroid Metabolism

Drugs may affect hepatic enzymes or the production of carrier proteins by the liver and thus influence the clearance of steroid hormones. These are most commonly drugs affecting the cytochrome p450 enzyme complex in the liver and include ampicillin, barbiturates, carbamazepine, phenytoin, primidone, rifampin, and tetracycline (Gilman et al, 1980). The significance of these alterations appears minor, as menstrual irregularities are rarely reported in association with the use of these medications.

Agents Directly Toxic to the Ovary

Finally, agents may be directly toxic to the ovary and destroy oocytes. Most commonly this occurs with antineoplastic agents, especially the alkylating agents. Histologically, there is stromal fibrosis and destruction of the primordial follicles (Schilsky et al, 1980). Table 6-3 contains a list of chemotherapeutic agents with documented toxicity for the human ovary (Feldman, 1989; Chapman, 1983; Gradishar and Schilsky, 1989; Wallace et al, 1989). The effects seen vary with the age of the patient at the time of treatment (Schilsky et al, 1980; Whitehead et al, 1982; Siris et al, 1976). Furthermore, patients may have a fluctuating course of ovarian failure following chemotherapy. Resumption of menses and pregnancy have been reported in patients following years of ovarian failure (Alper et al, 1986).

Factors Influencing the Menstrual Cycle: Chemical Toxicants

Chemicals differ markedly from pharmaceuticals in their regulation and the degree of testing necessary prior to licensing for widespread use in the United States. Pharmaceuticals are regulated by the Food and Drug Administration (FDA), and prior to approval of drugs to be used by persons of reproductive potential, the FDA requires reproductive toxicity testing in one or two animal species and labeling of drugs with potential reproductive hazards (Raheja et al, 1988). By contrast, chemicals may be licensed without toxicology testing and do not require labeling of potential reproductive hazards. This occurs despite the fact that 16 percent of the 79,000 chemicals listed in the Registry of Toxic Effects of Chemical Substances (RTECS) have some evidence of reproductive toxicity (Wells et al, 1988). Furthermore, a much greater proportion of the population comes in contact with specific chemicals (i.e., solvents and dyes) than any particular pharmaceutic agent.

Mechanism of Action of Chemical Toxicants

The mechanism of action of chemical reproductive toxicants was recently reviewed (Mattison, 1983; Mattison et al, 1990). Chemical agents affecting the menstrual cycle may act directly on the endocrine system or indirectly via a toxic metabolite or alteration in the physiologic control mechanisms of the reproductive process. Finally, there may be a combination of mechanisms of action for one chemical.

Chemicals may act directly owing to a structural similarity to an endogenous compound (estrogen). These chemicals bind to the estrogen receptor and show estrogenic or antiestrogenic activity. This category includes xenobiotics that bind to steroidal hormonal receptors at high doses, for example, DDT (Welch et al, 1969), chlordecone (Kepone) (Cannon et al, 1978; Hammond et al, 1979), and other organochlorine pesticides (Kupfer, 1975), polychlorinated biphenyls (PCBs) (Korach et al, 1988), and polybrominated biphenyls (PBBs). There are numerous edible plants with recognized estrogenically active compounds (Table 6-4). Dietary intake of plant estrogens has

Table 6-4. Edible plants with recognized estrogenically active compounds

Estrogens	Isoflavones	Coumestans	Resorcylic acid lactones	Others
Licorice	Soybean	Alfalfa	Oats	Fennel
French bean	Chick-pea	Soybean sprouts	Barley	Carrot
Date palm	Cherry	Cowpea	Rye	Anise
Pomegranate		Green beans	Sesame	Hops
Apple		Red beans	Wheat	
		Split peas	Peas	

Source: Setchell (1985).

yet to be associated with menstrual irregularities in humans. In sheep, however, grazing in pastures rich in a clover that contains a weak estrogen leads to infertility known as clover disease (Bennetts et al, 1946; Setchell, 1985).

Compounds may be toxic based on direct chemical reactivity with elements of the reproductive tract. Examples of chemical reactivity include the heavy metals cadmium (Kar et al, 1959), lead (Rom, 1976; Vermande-Van Eck, and Meigs, 1960), mercury (Popescu, 1978), boron (Weir and Fisher, 1972), and organometallic agents.

Chemical reproductive toxicants may also act indirectly, requiring metabolic activation to a toxic metabolite. This category includes polycyclic aromatic hydrocarbons. Toxicants may also produce changes in the physiologic control mechanisms involved in the regulation of the endocrine system. Changes in the control mechanisms may be brought about by modification of enzymes involved in steroid secretion or clearance. Examples of this include insecticides and the halogenated hydrocarbons, PBBs, PCBs, and DDT (Mattison, 1983).

Barlow and Sullivan (1982) reviewed the animal and human data linking industrial chemicals with derangements in the reproductive hormonal axis. Table 6-5 contains a list of chemicals for which there is evidence of alterations in menstrual regularity. For a more detailed discussion of the data linking these chemicals to specific alterations, the reader is referred to the original references (Barlow and Sullivan, 1982; Zhou et al, 1988; US Congress, 1985).

Occupational Exposures

It is estimated that there are 29 million women of reproductive age in the United States work force (Bernhardt, 1990). In spite of this large number, little information is available on the effects of occupational exposures on reproductive functon (Baier, 1989). There are several occupations that have been associated with an increase in menstrual abnormalities. These reports, however, are subject to selection bias based on the difference in demographics between female workers and a nonexposed control population (Joffe, 1985, 1988). Female workers tend to leave their places of employment when they conceive or give birth, selecting out those workers with decreased fertility, possibly related to menstrual abnormalities, to remain in the workplace. This may lead to a false-positive association between an occupation and menstrual irregularities (Joffe, 1985). Furthermore, distinguishing the work-relatedness of menstrual disorders is complicated by the multiplicity of exposures encountered in the workplace, job stress, and individual differences in the genetic makeup, medical condition, and lifestyle of the individual worker (Bregman et al, 1989).

Exposure to industrial chemicals is expected in workers exposed to chemicals in manufacturing plants. Exposure may also occur in a “pristine” office environment. Recently, there has been an increasing focus on the indoor air quality of office buildings, and the “tight building syndrome” or “sick build-

Table 6-5. Chemicals for which there are animal or human data associating exposure with alterations in the female reproductive hormonal axis

Aniline dyes
Benzene
Benzo(*a*)pyrene
Boron
Cadmium
Carbon monoxide
Carbon disulfide
Carbon tetrachloride
Chlordecone (Kepone)
Chloroprene
1,2-Dibromo-3-chloropropane (DBCP)
DDT
Dichloromethane
Dioxin
Ethylene oxide
Formaldehyde
Lead
Mercury
Nitrogen dioxide
Polybrominated biphenyls (PBBs)
Polychlorinated biphenyls (PCBs)
Selenium
Styrene
Toluene
Trichloroethylene
Xylene

Source: Adapted from Barlow and Sullivan (1982) and U.S. Congress, Office of Technology Assessment (1985).

ing syndrome" (SBS) is now recognized as a clinical entity (Hicks, 1984; Burge et al, 1987). Symptoms include nasal and eye irritation, headache, lethargy, respiratory difficulties, and skin irritation. SBS is thought to be primarily related to air-conditioning and poor ventilation (Finnegan et al, 1984) allowing the buildup of indoor air pollutants (Spengler and Sexton, 1983). The World Health Organization estimates that 30 percent of new or remodeled office buildings show signs of SBS. No study to date has addressed menstrual irregularities in persons affected with SBS, though when comparing the list of indoor air pollutants detected in investigation of SBS (Table 6-6) and the list of chemicals that affect the reproductive hormonal axis (Table 6-5) one is struck by the similarities. It is hoped that further studies will establish or fail to establish an association.

Table 6-6. Common indoor air pollutants and contaminants detected in investigations of sick building syndrome

Butylcyclohexane	Formaldehyde
n-Butylacetate	Limonene
Carbon monoxide	Menthone
Carbon dioxide	1-Methylnaphthalene
Carvone	Naphthalene
C7–C11 alkanes	Ozone
Cyclohexane	1,1,2,2-Perchlorethylene
Cyclohexanol	Propylbenzene
Dust	Toluene
Ethylbenzene	1,1,1-Trichlorethane
Ethylsilane	Xylene

Source: Lyles, Greve, Bauer (1991).

SUMMARY

The initiation and maintenance of regular menstruation depends on a delicate interplay between the reproductive system and numerous exogenous and endogenous factors. The first section of this chapter reviewed the pathophysiology underlying menstrual irregularity. The female reproductive axis is extremely reliant on feedback regulation, and many conditions disrupt menstrual cycle regularity through alterations in feedback regulation. The mechanism underlying these alterations was briefly reviewed and common examples highlighted.

The second section reviewed specific factors that influence menstrual cycle regularity and therefore impact on fertility potential. This includes age, body habitus, various physical forces, lifestyle factors, medical illnesses, drugs, and chemicals. With an understanding of this information, one can effectively evaluate the relative impact a particular condition or exposure may have on menstrual cycle regularity and subsequent fertility.

Unfortunately, most of the information available on the effect of a specific exposure is from animal studies or small epidemiologic investigations. It is hoped that in the future, large, prospective, controlled studies using biochemical markers will further elucidate the underlying mechanisms and relative impact that various medical illnesses, lifestyle factors, and occupational and environmental exposures have on menstrual cycle regularity and fertility.

REFERENCES

Abraham SF, Beumont PJV, Fraser IS, LLewellyn-Jones D: Body weight, exercise and menstrual status among ballet dancers in training. *Br J Obstet Gynecol* 89:507–10, 1982.

Adamopoulos DA, Lerapetritakis G, Kapolla N: Pituitary function in young women in a reformatory institution. *Br J Obstet Gynecol* 89:511–515, 1982.

Adlercreutz H, Fotsis T, Bannwart C, Hämäläinen E, Bloigu S, Ollus A: Urinary estrogen profile determination in young Finnish vegetarian and omnivorous women. *J Steroid Biochem* 24:289–296, 1986.

Alexander SE, Aksel S, Hazelton JM, Yeoman RR, Gilmore SM: The effect of aging on hypothalamic function in oophorectomized women. *Am J Obstet Gynecol* 162:446–449, 1990.

Alper MM, Jolly EE, Garner PR: Pregnancies after premature ovarian failure. *Obstet Gynecol* (suppl) 67:59S–62S, 1986.

Apter D, Viinikka L, Vihko R: Hormonal pattern of adolescent cycles. *J Clin Endocrinol Metab* 47:944, 1978.

Asch RH, Fernandez EO, Smith CG, Pauerstein CJ: Precoital single doses of Δ9-tetrahydrocannabinol block ovulation in the rabbit. *Fertil Steril* 31:331–334, 1979.

Axelrod J, Reisine TD: Stress hormones: Their interaction and regulation. *Science* 224:452–459, 1984.

Azziz R: Reproductive endocrinologic alterations in female asymptomatic obesity. *Fertil Steril* 52:703–725, 1989.

Azziz R, Zacur HA: 21-Hydroxylase deficiency in female hyperandrogenism: Screening and diagnosis. *J Clin Endocrinol Metab* 69:577–584, 1989.

Baier EJ: Endangered species. *Occup Health Safety* 58:20–23, 1989.

Baker ER, Mathur RS, Kirk RF, Williamson HO: Female runners and secondary amenorrhea: Relation with age, parity, mileage and plasma hormonal and sex hormone binding globulin concentrations. *Fertil Steril* 35:247–248, 1981.

Baker JW, Peckham MJ, Morgan RL, Smithers DW: Preservation of ovarian function in patients requiring radiotherapy for para-aortic and pelvic Hodgkin's disease. *Lancet* 1:1307–1308, 1972.

Baker TG: A quantitative and cytological study of germ cells in human ovaries. *Proc R Soc Lord (Biol)* 158:417–433, 1963.

Baker TG: The sensitivity of oocytes in post natal rhesus monkeys to x-irradiation. *J Reprod Fertil* 12:183, 1966.

Baker TG: Radiosensitivity of mammalian oocytes with particular reference to the human female. *Am J Obstet Gynecol* 110:746–761, 1971.

Barlow SM, Sullivan FM: *Reproductive Hazards of Industrial Chemicals: An Evaluation of Animal and Human Data.* New York, Academic Press, 1982.

Bates GW: Body weight control practice as a cause of infertility. *Clin Obstet Gynecol* 28:632–644, 1985.

Bates GW, Whitworth NS: Effect of body weight reduction on plasma androgens in obese infertile women. *Fertil Steril* 38:406, 1982.

Bauman JE, Kolodny RC, Dornbush RL, Webster SK: Efectos endocrinos del uso cronico de ala mariguana en mujeres, in Julio DF (ed): *Caudernos Cientificos CEMESAM (Centro Mexicano de Estudios en Salud Mental)* (vol 10). Mexico, CEMESAM, 1979, pp 85–97.

Bean JA, Leeper JD, Wallace RB, Sherman BM, Jagger H: Variations in the reporting of menstrual histories. *Am J Epidemiol* 109:181–185, 1979.

Beardwood CJ, Kellaway LA, Querido D: Changes in the reproductive system of the rat caused by audiostimulation. *S Afr Med J* 52:409, 1977.

Beardwood CJ, Mundell CA, Utian WH: Gonadotropin excretion in response to audiostimulation of human subjects. Am J Obstet Gynecol 121:682–687, 1975.

Beitins IZ, McArthur JW, Turnbull BA, Skrinar GS, Bullen BA: Exercise induces two types of human luteal dysfunction: Confirmation by urinary free progesterone. *J Clin Endocrinol Metab* 72:1350–1358, 1991.

Bennetts HW, Underwood EJ, Sheir FL: A specific breeding problem of sheep on subterranean clover pastures in Western Australia. *Aust Vet J* 22:2–12, 1946.

Bergeron C, Kovacs K: Pituitary siderosis: A histologic immunocytologic and ultra-structural study. *Am J Pathol* 93:295–306, 1978.

Bergqvist N: The gonadal function in female diabetics. *Acta Endocrinol* (suppl) 19:3–20, 1954.

Bernhardt JH: Potential workplace hazards to reproductive health: Information for primary prevention. *J Obstet Gynecol Neonat Nurs* 19:53–62, 1990.

Beumont PJV, George GCW, Pimstone BL, Vinik AI: Body weight and the pituitary response to hypothalamic releasing hormones in patients with anorexia nervosa. *J Clin Endocrinol Metab* 43:487–496, 1976.

Bierman M, Nolan GH: Menstrual function and renal transplantation. *Obstet Gynecol* 49:186–189, 1977.

Bilo L, Meo R, Nappi C et al: Reproductive endocrine disorders in women with primary generalized epilepsy. *Epilepsia* 29:612–619, 1988.

Bilo L, Meo R, Valentino R et al: Abnormal pattern of luteinizing hormone pulsatility in women with epilepsy. *Fertil Steril* 55:705–711, 1991.

Birnbaum MD, Rose LI: Late onset adrenocortical hydroxylase deficiencies associated with menstrual dysfunction. *Obstet Gynecol* 63:445, 1984.

Blackwell R, Chang A: Video display terminals and pregnancy: A review. *Br J Obstet Gynecol* 95:446–453, 1988.

Borg E: Physiological and pathogenic effects of sound. *Acta Otolaryngol* (suppl) 381:68, 1981.

Borgna-Pignatti C, DeStefano P, Zonta L et al: Growth and sexual maturation in thalassemia major. *J Pediatr* 106:150–155, 1985.

Boroditsky RS, Faiman C: Galactorrhea–amenorrhea due to primary hypothyroidism. *Am J Obstet Gynecol* 116:661–665, 1973.

Boyden TW, Pamenter RW, Grosso D, Stanforth P, Rotkis R, Wilmore JH: Prolactin responses, menstrual cycles, and body composition of women runners. *J Clin Endocrinol Metab* 54:711–714, 1982.

Bregman DJ, Anderson KE, Buffler P, Salg J: Surveillance for work-related adverse reproductive outcomes. *Am J Public Health* (suppl) 79:53–57, 1989.

Brodie BL, Wentz AC: Late onset congenital adrenal hyperplasia: A gynecologist's perspective. *Fertil Steril* 48:175–188, 1987.

Broocks A, Pirke KM, Schweiger U et al: Cyclic ovarian function in recreational athletes. *J Appl Physiol* 68:2083–2086, 1990.

Brown S, Vessey M, Stratton I: The influence of method of contraception and cigarette smoking on menstrual patterns. *Br J Obstet Gynecol* 95:905–910, 1988.

Bryant HE, Love EJ: Video display terminal use and spontaneous abortion risk. *Int J Epidemiol* 18:132–138, 1989.

Bryner JR, Greenblatt RB: Primary empty sella syndrome with elevated serum prolactin. *Obstet Gynecol* 50:375–380, 1977.

Bullen BA, Skrinar GS, Beitins IZ, von Mering G, Turnbull BA, McArthur JW: Induction of menstrual disorders by strenuous exercise in untrained women. *N Engl J Med* 312:1349–1353, 1985.

Burge S, Hedge A, Wilson S, Bass JH, Robertson A: Sick building syndrome: A study of 4373 office workers. *Ann Occup Hyg* 31:493–504, 1987.

Burghen GA, Givens JR, Kitabchi AE: Correlation of hyperandrogenism with hyperinsulinism in polycystic ovarian disease. *J Clin Endocrinol Metab* 50:113, 1980.

Burrow GN, Wortzman G, Rewcastle NB et al: Microadenomas of the pituitary and abnormal sella tomograms in an unselected autopsy series. *N Engl J Med* 304:156, 1981.

Cahalan D: *Problem Drinkers.* San Francisco, Jossey-Bass, 1970, p 137.

Cameron RG: Effect of flying on the menstrual function of air hostesses. *Aero Med* 40:1020–1023, 1969.

Cannon SB, Veazey JM Jr, Jackson RS et al: Epidemic Kepone poisoning in chemical workers. *Am J Epidemiol* 107:529, 1978.

Canterbury RJ, Gressard CF, Vieweg WVR et al: Alcohol and cocaine use among first-year college students. *South Med J* 84:13–21, 1991.

Cantopher T, Evans C, Lacey JH, Pearce JM: Menstrual and ovulatory disturbance in bulimia. *Br Med J* 297:836–837, 1988.

Carletti B, Kehyayan E, Fraschini F: Remarkable seasonal variations of urinary gonadotropin excretion in young girls. *Experientia* 20:383, 1964.

Carlson HE, Ippolity AR: Cimetidine, an H_2-antihistamine stimulates prolactin secretion in man. *J Clin Endocrinol Metab* 45:367, 1977.

Carr DB, Bullen BA, Skrinar GS: Physical conditioning facilitates the exercise-induced secretion of beta-endorphin and beta-lipotropin in women. *N Engl J Med* 305:560–563, 1981.

Caufriez A: Menstrual disorders associated with hyperprolactinemia. *Hormone Res* 22:209–214, 1985.

Centers for Disease Control: Leading work-related diseases and injuries—United States. *MMWR* 34:219, 1985.

Chang RJ, Nakamura RM, Judd HL, Kaplan SA: Insulin resistance in nonobese patients with polycystic ovarian disease. *J Clin Endocrinol Metabol* 57:356–359, 1983.

Chapman RM: Gonadal injury resulting from chemotherapy. *Am J Ind Med* 4:149–161, 1983.

Chapman RM, Sutcliffe SB: Protection of ovarian function by oral contraceptives in women receiving chemotherapy for Hodgkin's disease. *Blood* 58:849–851, 1981.

Cheng E, Yoder L, Story CD, Burroughs W: Estrogenic activity of some isoflavone derivatives. *Science* 120:575–576, 1954.

Chiauzzi V, Cigorraga S, Escobar RE: Inhibition of follicular stimulating hormone receptor binding by circulating immunoglobulins. *J Clin Endocrinol Metab* 54:1221, 1981.

Chiazze L, Brayer FT, Macisco JJ, Parker MP, Duffy BJ: The length and variability of the human menstrual cycle. *JAMA* 203:377–380, 1968.

Cleland WH, Mendelson CR, Simpson ER: Aromatase activity of membrane fractions of human adipose tissue stromal cells and adipocytes. *Endocrinology* 113:2155, 1983.

Cocores JA, Dackis CA, Gold MS: Sexual dysfunction secondary to cocaine abuse in two patients. *J Clin Psychiatry* 47:384–385, 1986.

Collu R, Franchini F, Martini L: Blockade of ovulation by melatonin. *Experientia* 27:844–845, 1971.

Cook FA: Medical observations among the Esquimaux. *NY J Gynecol Obstet* 4:282, 1894.

Cooper DS, Geleberg AJ, Wojcik JC et al: The effect of amoxapine and imipramine on serum prolactin levels. *Arch Intern Med* 141: 1023, 1981.

Council on Scientific Affairs, American Medical Association: Effects of physical forces on the reproductive cycle. *JAMA* 251:247–250, 1984.

Council on Scientific Affairs, American Medical Association: Effects of toxic chemical on the reproductive system. *JAMA* 253:3431, 1985.

Council on Scientific Affairs: The worldwide smoking epidemic: Tobacco trade, use, and control. *JAMA* 263:3312–3318, 1990.

Coutinho EM: Physiologic and pharmacologic studies of the human oviduct. *Fertil Steril* 22:807–815, 1971.

Cowchock FS, McCabe JL, Montgomery BB: Pregnancy after corticosteroid administration in premature ovarian failure (polyglandular endocrinopathy syndrome). *Am J Obstet Gynecol* 158:118–119, 1988.

Csapo AT, Pulkkinen MO, Weist WG: Effects of luteectomy and progesterone replacement therapy in early pregnancy patients. *Am J Obstet Gynecol* 115:759, 1973.

Dackis CA, Gold MS: New concepts in cocaine addiction: The dopamine depletion hypothesis. *Neurosci Biobehav Rev* 9:1–9, 1985.

Dale E, Gerlach DH, Wilhite AL: Menstrual dysfunction in distance runners. *Obstet Gynecol* 54:47–53, 1979.

Damewood MD, Zacur HA, Hoffman GJ, Rock JA: Circulating antiovarian antibodies in premature ovarian failure. *Obstet Gynecol* 68:850–854, 1986.

DeSanctis F, Vullo C, Datz M, Wonke B, Hoffbrand AV, Bagni B: Hypothalamic–pituitary–gonadal axis in thalassemic patients with secondary amenorrhea. *Obstet Gynecol* 72:643–647, 1988.

DeVane GW, Czekala NM, Judd HL, Yen SSC: Circulating gonadotropins, estrogens, and androgens in polycystic ovarian disease. *Am J Obstet Gynecol* 121:496–500, 1975.

DeVane GW, Zera MI, Buhi WC, Kalra PS: Opioid peptides in pseudocyesis. *Obstet Gynecol* 65:183, 1985.

Dewan EM, Menkin MF, Rock J: Effect of photic stimulation on the human menstrual cycle. *Photochem Photobiol* 27:581, 1978.

Distiller LA, Sagel J, Morley JE, Joffe BI, Seftel HC: Pituitary responsiveness to luteinizing hormone–releasing hormone in insulin-dependent diabetes mellitus. *Diabetes* 24:378–380, 1975.

Djursing H, Nyholm HC, Hagen C, Carstensen L, Pedersen LM: Clinical and hormonal characteristics in women with anovulation and insulin-treated diabetes mellitus. *Am J Obstet Gynecol* 143:876–882, 1982.

Elder M, Maclaren N, Riley W: Gonadal autoantibodies in patients with hypogonadism and/or Addison's disease. *J Clin Endocrinol Metab* 52:1137, 1981.

Engels WD, Pattee CJ, Wittkower ED: Emotional settings of functional amenorrhea. *Psychosom Med* 26:682–700, 1964.

Escobar ME, Cigarroga SB, Chiauzzi VA et al: Development of the gonadotropic resistant ovary syndrome in myasthenia gravis: Suggestion of a similar autoimmune mechanism. *Acta Endocrinol* 99:431, 1982.

Espey LL: Ovarian proteolytic enzymes and ovulation. *Biol Reprol* 10:216–235, 1974.

Farrell BL, Allen MF: Physiologic/psychologic changes reported by USAF female flight nurses during flying duties. *Nurs Res* 22:31–36, 1973.

Feldman JE: Ovarian failure and cancer treatment: Incidence and interventions for the premenopausal woman. *Oncol Nurs Forum* 16:651–657, 1989.

Filicori M: Maintenance of the corpus luteum of the menstrual cycle: Hypothalamo–pituitary–ovarian axis. *Semin Reprod Endocrinol* 8:115–121, 1990.

Finnegan MJ, Pickering CAC, Burge PS: The sick building syndrome: Prevalence studies. *Br Med J* 289: 1573–1575, 1984.

Fishman J, Hellman L, Zumoff B, Gallagher TF: Influence of thyroid hormone on estrogen metabolism in man. *J Clin Endocrinol Metab* 22:389–392, 1962.

Fowlie S, Burton J: Hyperprolactinemia and nonpuerperal lactation associated with clomipramine. *Scott Med J* 32:52–53, 1987.

Franks PA, Laughlin NK, Dierschke DJ, Bowman RE, Meller PA: Effects of lead on luteal function in rhesus monkeys. *Biol Reprod* 41:1055–1062, 1989.

Franks S, Nabarro JDN, Jacobs HS: Prevalence and presentation of hyperprolactinaemia in patients with "functionless" pituitary tumours. *Lancet* 1:778–780, 1977.

Fraschini F, Mess B, Martini L: Pineal gland, melatonin and the control of luteinizing hormone secretion. *Endocrinology* 82:919–924, 1968.

Friedman CL, Kim MH: Obesity and its effect on reproductive function. *Clin Obstet Gynecol* 28:645, 1985.

Frisch RE, Wyshak G, Vincent L: Delayed menarche and amenorrhea in ballet dancers. *N Engl J Med* 303:17–19, 1980.

Frisch RE, McArthur JW: Menstrual cycles: Fatness as a determinant of minimum weight for height necessary for their maintenance or onset. *Science* 185:949–951, 1974.

Frisch RE, Revelle R: Height and weight at menarche and a hypothesis of critical body weights and adolescent events. *Science* 169:397–399, 1970.

Gambrell RD: The menopause: Benefits and risks of estrogen-progestogen replacement therapy. *Fertil Steril* 37:457–474, 1982.

Gaulden EC, Littlefield DC, Putoff OE, Seivert AL: Menstrual abnormalities associated with heroin addiction. *Am J Obstet Gynecol* 90:155, 1964.

Gillingham KK, Schade CM, Jackson WG, Gilstrap LC: Women's G tolerance. *Aviat Space Environ Med* 57:745–753, 1986.

Gilman AG, Goodman LS, Gilman A: *The Pharmacologic Basis of Therapeutics* (6th ed). New York, Macmillan, 1980.

Gindoff PR, Ferin M: Brain opioid peptides and menstrual cyclicity. *Semin Reprod Endocrinol* 5:125–133, 1987.

Gindoff PR, Jewelewicz R: Reproductive potential in the older woman. *Fertil Steril* 46:989–1001, 1986.

Gluskin LK, Strasberg B, Shah JH: Verapamil-induced hyperprolactinemia and galactorrhea. *Ann Intern Med* 95:66, 1981.

Goldhaber MK, Polen MR, Hiatt RA: The risk of miscarriage and birth defects among women who use visual display terminals during pregnancy. *Am J Ind Med* 13:695–706, 1988.

Goldin BR: Estrogen excretion pattern and plasma levels in vegetarian and omnivorous women. *N Engl J Med* 307:1542, 1982.

Goldsmith RE, Sturgis SH, Lerman J, Stanbury JB: The menstrual pattern in thyroid disease. *J Clin Endocrinol Metab* 12:846–855, 1952.

Goldzieher JW, Green JA: The polycystic ovary: I. Clinical and histologic features. *J Clin Endocrinol Metab* 22:325–338, 1962.

Gómez F, de la Cueva R, Wauters J-P, Lemarchand-Béraud T: Endocrine abnormalities in patients undergoing long-term hemodialysis: The role of prolactin. *Am J Med* 68:522–530, 1980.

Gordon GG, Louthren AL, Tochimoto S, Rand JJ, Olivo J: Effect of hyperthyroidism and hypothyroidism on the metabolism of testosterone and androstenedione in man. *J Clin Endocrinol Metab* 29:164–170, 1969.

Graber AL, Ney RL, Nicholson WE, Island DP, Liddle GW: Natural history of pituitary–adrenal recovery following long-term suppression with corticosteroids. *J Clin Endocrinol Metab* 25:11–16, 1965.

Gradishar WJ, Schilsky RL: Ovarian function following radiation and chemotherapy for cancer. *Semin Oncol* 16:425–436, 1989.

Graham CA, McGrew WC: Menstrual synchrony in female undergraduates living on a co-educational campus. *Psychoneuroendocrinology* 5:245–252, 1980.

Green P, Rubin L: Amenorrhea as a manifestation of chronic liver disease. *Am J Obstet Gynecol* 78:141–146, 1959.

Grossman A: Opioid peptides and reproductive function. *Semin Reprod Endocrinol* 5:115–124, 1987.

Hagerty MA, Howie BJ, Tan S, Shultz TD: Effect of low- and high-fat intakes on the hormonal milieu of premenopausal women. *Am J Clin Nutr* 47:653–659, 1988.

Hall: The increasing use of lead as an abortifacent. *Br Med J* 1:584–586, 1905.

Hammond B, Katzenellenbogen BS, Krauthammer, McConnell J: Estrogenic activity of the insecticide chlordecone (Kepone) and interaction with uterine estrogen receptors. *Proc Natl Acad Sci USA* 76:6641–6645, 1979.

Harlass FE, Plymate SR, Fariss BL, Belts RP: Weight loss is associated with correction of gonadotropin and sex steroid abnormalities in the obese anovulatory female. *Fertil Steril* 42:649, 1984.

Harrington JM, Stein FG, Rivera RO, deMorales AV: The occupational hazards of formulating oral contraceptives—a survey of plant employees. *Arch Environ Health* 33:12–15, 1978.

Herzog AG: Intermittent progesterone therapy and frequency of complex partial seizures in women with menstrual disorders. *Neurology* 36:1607–1610, 1986.

Herzog AG, Seibel MM, Schomer DL, Vaitukaitis JL, Geschwind N: Reproductive endocrine disorders in women with partial seizures of temporal lobe origin. *Arch Neurol* 43:341–346, 1986.

Hicks JB: Tight building syndrome: When work makes you sick. *Occup Health Safety* 53:51–56, 1984.

Hoff JD, Quigley ME, Yen SSC: Hormonal dynamics at midcycle: A reevaluation. *J Clin Endocrinol Metab* 57:792–796, 1983.

Holick MF, MacLaughlin JA, Clark MB et al: Photosynthesis of previtamin D3 in human skin and the physiologic consequences. *Science* 210:203, 1980.

Hughes CL: Monitoring of ovulation in the assessment of reproductive hazards in the workplace. *Reprod Toxicol* 2:163–169, 1988.

Hugues JN, Cofte T, Perret G et al: Hypothalamo–pituitary ovarian function in 31 women with chronic alcoholism. *Clin Endocrinol* 12:543–551, 1980.

Iglesias R, Terrés A, Chavarria A: Disorders of the menstrual cycle in airline stewardesses. *Aviat Space Environ Med* 51:518–520, 1980.

Israel SL: The empiric usage of low-dosage irradiation in amenorrhea. *Am J Obstet Gynecol* 64:971–987, 1952.

Israel SL: *Menstrual Disorders and Sterility* (5th ed). New York, Harper & Row, 1967, p 72.

Jacox H: Recovery following human ovarian irradiation. *Radiology* 32:538–545, 1939.

Jadresic A, Banks LM, Child DF, Diamant L, Doyle FH, Fraser TR, Joplin GF: The acromegaly syndrome. *Q J Med* 202:189–204, 1982.

Jenkins JS, Gilbert CJ, Ang V: Hypothalamic-pituitary function in patients with craniopharyngiomas. *J Clin Endocrinol Metab* 43:394–399, 1976.

Jennings RT, Santy PA: Reproduction in space: Part ii. Concerns for human reproduction. *Obstet Gynecol Surv* 45:7–17, 1989.

Jick H, Porter J, Morrison AS: Relation between smoking and age of natural menopause. *Lancet* 1:1354–1355, 1977.

Joffe M: Advantages of a standard method for reseearch on reproductive effects of occupation. *J Epidemiol Comm Health* 42:209–212, 1988.

Joffe M: Biases in research on reproduction and women's work. *Int J Epidemiol* 14:118–123, 1985.

Joslin EP, Root HF, White P: The growth, development and prognosis of diabetic children. *JAMA* 85:420–422, 1925.

Kahl LE, Medsger TA, Klein I: Massive breast enlargement in a patient receiving D-penicillamine for systemic sclerosis. *J Rheumatol* 12:990–991, 1985.

Kar AB, Das RP, Karkun JN: Ovarian changes in prepubertal rats after treatment with cadmium chloride. *Acta Biol Med Germ* 3:272, 1959.

Katz E, Adashi EY: Hyperprolactinemic disorders. *Clin Obstet Gynecol* 33:622–639, 1990.

Kaufman FR, Donnell GN, Lobo RA: Ovarian androgen secretion in patients with galactosemia and premature ovarian failure. *Fertil Steril* 47:1033–1034, 1987.

Kaufman FR, Kogut MD, Donnell GN, Goebelsmann U, March C, Koch R: Hypergonadotropic hypogonadism in female patients with galactosemia. *N Engl J Med* 304:994–998, 1981.

Kim MH: "Gonadotropin-resistant ovaries" syndrome in association with secondary amenorrhea. *Am J Obstet Gynecol* 120:257–263, 1974.

King TS, Schenken RS, Kang IS, Javors MA, Riehl RM: Cocaine disrupts estrous cyclicity and alters the reproductive neuroendocrine axis in the rat. *Neuroendocrinology* 51:15–22, 1990.

Kivelä A, Kauppila A, Ylöstalo P, Vakkuri O, Leppäluoto J: Seasonal, menstrual and circadian secretions of melatonin, gonadotropins and prolactin in women. *Acta Physiol Scand* 132:321–327, 1988.

Kjellberg A: Subjective, behavioral and psychological effects of noise. *Scand J Work Environ Health* (suppl 1) 16:29–38, 1990.

Kleerekoper M, Basten A, Penney R, Posen S: Idiopathic hypoparathyroidism with primary ovarian failure. *Arch Intern Med* 134:944, 1974.

Kletzky OA, Costin G, Marrs RP et al: Gonadotropin insufficiency in patients with thalassemia major. *Clin Endocrinol* 48:901, 1979.

Knobil E: The neuroendocrine control of the menstrual cycle. *Recent Prog Horm Res* 36:53–88, 1980.

Knuth UA, Hull MGR, Jacobs HS: Amenorrhea and loss of weight. *Br J Obstet Gynaecol* 84:801–807, 1977.

Korach KS, Sarver P, Chae K, McLachlan JA, McKinney JD: Estrogen receptor-binding activity of polychlorinated hydroxybiphenyls: Conformationally restricted structural probes. *Mol Pharmacol* 33:120–126, 1988.

Kotz HL, Herrmann W: Low-dosage irradiation in the induction of human ovulation. *Obstet Gynecol Surv* 17:309–328, 1962.

Kupfer D: Effects of pesticides and related compounds on steroid metabolism and function. *Crit Rev Toxicol* 4:83, 1975.

Lamberts SWJ, Klijn JGM, Kwa GH, Birkenhäger: The dynamics of growth hormone and prolactin secretion in acromegalic patients with "mixed" pituitary tumours. *Acta Endocrinol* 90:198–210, 1979.

Lee CS: Luteal phase defects. *Obstet Gynecol Surv* 42:267–274, 1987.

Lemarchand-Béraud T, Zufferey M, Reymond M, Rey I: Maturation of the hypothalamo–pituitary–ovarian axis in adolescent girls. *J Clin Endocrinol Metabol* 54:241–246, 1982.

Leung PCK, Flesher M, Moon YS, Minegishi T: Effects of cannabinoids on progesterone release in cultures of rat luteal cells. *Life Sci* 37:1691–1696, 1985.

Lewy AJ, Wëhr TA, Goodwin FK et al: Light suppresses melatonin secretion in humans. *Science* 210:1267–1269, 1980.

Lim VS, Kathpalia SC, Frohman LA: Hyperprolactinemia and impaired pituitary response to suppression and stimulation in chronic renal failure: Reversal after transplantation. *J Clin Endocrinol Metab* 48:101–107, 1979.

Liu JH, Yen SSC: Induction of midcycle gonadotropin surge by ovarian steroids in women: A critical evaluation. *J Clin Endocrinol Metab* 57:797–802, 1983.

Logothetis J, Harner R, Morrell F, Torres F: The role of estrogens in the catamenial exacerbation of epilepsy. *Neurology* 9:352–360, 1959.

Loucks AB: Effects of exercise training on the menstrual cycle: Existence and mechanisms. *Med Sci Sports Exerc* 22:275–280, 1990.

Lumsden MA, Kelly RW, Templeton AA, Van Look PFA, Swanston IA, Baird DT: Changes in the concentration of prostaglandins in preovulatory human follicles after administration of hCG. *J Reprod Fertil* 77:119–124, 1986.

Lyles WB, Greve KW, Bauer RM: Sick building syndrome. *South Med J* 84:65–71, 1991.

McArthur JW: Endorphins and exercise in females: Possible connection with reproductive dysfunction. *Med Sci Sports Exerc* 17:82–88, 1985.

McArthur JW, O'Loughlin KM, Beitins IZ, Johnson L, Hourihan J, Alonso C: Endocrine studies during the refeeding of young women with nutritional amenorrhea and infertility. *Mayo Clin Proc* 51:607–616, 1976.

McCallum RW, Sowers JR, Hershman JM, Sturdevant RA: Metoclopramide stimulates prolactin secretion in man. *J Clin Endocrinol Metab* 42:1148, 1976.

McClintock MK: Menstrual synchrony and suppression. *Nature* 229:244–245, 1971.

Mancini AM, Guitelman A, Vargas CA, Debejuk L, Aparicio NI: Effect of sulpiride on serum prolactin levels in humans. *J Clin Endocrinol Metab* 42:181, 1976.

Marwick C: More studies pending of low-dose radiation. *JAMA* 264:944, 1990a.

Marwick C: Low-dose radiation: Latest data renew questions of "safe" level. *JAMA* 264:553–557, 1990b.

Mattison DR: The effects of smoking on fertility from gametogenesis to implantation. *Environ Res* 28:410–433, 1982.

Mattison DR: The mechanisms of action of reproductive toxins. *Am J Ind Med* 4:65–79, 1983.

Mattison DR, Plowchalk DR, Meadows MJ, Al-Juburi AZ, Gandy J, Malek A: Reproductive toxicity: Male and female reproductive systems as targets for chemical injury. *Med Clin North Am* 74:391–411, 1990.

Medical Letter: Estrogens in cosmetics. *Med Lett* 27:54–55, 1985.

Mello NK, Mendelson JH, Teoh SK: Neuroendocrine consequences of alcohol abuse in women. *Ann NY Acad Sci* 562:211–240, 1989.

Mendelson JH, Mello NK: Chronic alcohol effects on anterior pituitary and ovarian hormones in healthy women. *J Pharmacol Exp Ther* 245:407–412, 1988.

Mendelson JH, Lukas SE, Mello NK, Amass L, Ellingboe J, Skupny A: Acute alcohol effects on plasma estradiol levels in women. *Psychopharmacology* 94:464–467, 1988.

Mendelson JH, Mello NK, Ellingboe J et al: Marihuana smoking suppresses luteinizing hormone in women. *J Pharmacol Exp Ther* 237:862–866, 1986.

Mendelson JH, Cristafaro P, Ellingboe J, Benedikt R, Mello N: Acute effects of marijuana on luteinizing hormone in menopausal women. *Pharmacol Biochem Behav* 23:765–768, 1985.

Metcalf MG, Livesey JH: Gonadotrophin excretion in fertile women: Effect of age and the onset of the menopausal transition. *J Endrocrinol* 105:357–362, 1985.

Money KE: Biological effects of space travel. *Can Aeronaut Space J* 27:195–201, 1981.

Morrison JC, Givens JR, Wiser WL, Fish SA: Mumps oophoritis: A cause of premature menopause. *Fertil Steril* 26:655–659, 1975.

Mulaikal RM, Migeon CJ, Rock JA: Fertility rates in female patients with congenital adrenal hyperplasia due to 21-hydroxylase deficiency. *N Engl J Med* 316:178–182, 1987.

Nir I, Ayalon D, Tsafriri A, Cordova T, Lindner HR: Suppression of the cyclic surge of luteinizing hormone secretion and of ovulation in the rat by Δ9-tetrahydrocannabinol. *Nature* 243:470–471, 1973.

Nogales-Ortiz F, Tarancón I, Nogales FF: The pathology of female genital tuberculosis. *Obstet Gynecol* 53:422–428, 1979.

O'Grady JP, Rosenthal M: Pseudocyesis: A modern perspective on an old disorder. *Obstet Gynecol Surv* 44:500–511, 1989.

Oseko F, Morikawa K, Motohashi T, Aso T: Effects of chronic sulpiride-induced hyperprolactinemia on menstrual cycles of normal women. *Obstet Gynecol* 72:267–271, 1988.

Paraskevaides EC, Pennington GW, Naik S: Seasonal distribution in conceptions achieved by artificial insemination by donor. *Br Med J* 297:1309–1310, 1988.

Pasquali R, Antenucci D, Casimirri F et al: Clinical and hormonal characteristics of obese amenorrheic hyperandrogenic women before and after weight loss. *J Clin Endocrinol Metabol* 68:173–179, 1989.

Perez RJ, Lipner H, Abdulla N, Cicotto S, Abrams M: Menstrual dysfunction of patients undergoing chronic hemodialysis. *Obstet Gynecol* 51:552–555, 1978.

Pirke KM, Schweiger U, Lemmel W, Krieg JC, Berger M: The influence of dieting on the menstrual cycle of healthy young women. *J Clin Endocrinol Metab* 60:1174–1179, 1985.

Pirke KM, Schweiger U, Laessle R, Dickhaut B, Schweiger M, Waechtler M: Dieting influences the menstrual cycle: Vegetarian versus nonvegetarian diet. *Fertil Steril* 46:1083–1088, 1986.

Pirke KM, Schweiger U, Strowitzki T et al: Dieting causes menstrual irregularities in normal weight young women through impairment of episodic luteinizing hormone secretion. *Fertil Steril* 51:263–268, 1989.

Popescu HI: Poisoning with alkylmercury compounds. *Br Med J* 1:1347, 1978.

Poretsky L, Kalin MF: The gonadotropic function of insulin. *Endocr Rev* 8:132–141, 1987.

Prasad AS, Ortega J, Brewer GJ, Oberleas D, Schoomaker EB: Trace elements in sickle cell disease. *JAMA* 235:2396–2398, 1976.

Preti G, Cutler WB, Garcia CR, Huggins GR, Lawley HJ: Human axillary secretions influence women's menstrual cycles: The role of donor extract of females. *Horm Behav* 20:474–482, 1986.

Priddy AR, Killick SR, Elstein M et al: The effect of prostaglandin synthetase inhibitors on human preovulatory follicular fluid prostaglandin, thromboxane, and leukotriene concentrations. *J Clin Endocrinol Metabol* 71:235–242, 1990.

Prior JC, Vigna YM, Schechter MT, Burgess AE: Spinal bone loss and ovulatory disturbances. *N Engl J Med* 323:1221–1277, 1990.

Rabinowe SL, Ravnikar VA, Dib SA, George KL, Dluhy RG: Premature menopause: Monoclonal antibody defined T lymphocyte abnormalities and antiovarian antibodies. *Fertil Steril* 51:450–454, 1989.

Rachootin P, Olsen J: The risk of infertility and delayed conception associated with exposures in the Danish workplace. *J Occup Med* 25:394–402, 1983.

Raheja KL, Jordan A, Fourcroy JL: Food and Drug Administration guidelines for reproductive toxicity testing. *Reprod Toxicol* 2:291–293, 1988.

Ray GR, Trueblood HW, Enright LP, Kaplan HS, Nelsen TS: Oophoropexy: A means of preserving ovarian function following pelvic megavoltage radiotherapy for Hodgkin's disease. *Radiology* 96:175–180, 1970.

Raymond JP, Izembart M, Marliac V, Dagousset F, Merceron RE, Vulpillat M, Vallée G: Temporary ovarian failure in thyroid cancer patients after thyroid remnant ablation with radioactive iodine. *J Clin Endocrinol Metab* 69:186–190, 1989.

Rebar RW: The reproductive age: Chronic anovulation, in Serra GB (ed): *The Ovary.* New York: Raven Press, 1983, p 217.

Rebar RW, Connolly HV: Clinical features of young women with hypergonadotropic amenorrhea. *Fertil Steril* 53:804–810, 1990.

Rice GG: Hypermenorrhea in the young hemodialysis patient. *Am J Obstet Gynecol* 116:539–543, 1973.

Richardson SJ, Senikas V, Nelson JF: Follicular depletion during the menopausal transition: Evidence for accelerated loss and ultimate exhaustion. *J Clin Endocrinol Metab* 65:1231–1237, 1987.

Rifka SM, Pita JC, Loriaux DL: Mechanism of interaction of digitalis with estradiol binding sites in rat uteri. *Endocrinology* 99:1091–1096, 1976.

Rock JA, Fortney SM: Medical and surgical consideration for women in spaceflight. *Obstet Gynecol Surv* (suppl) 39:525–535, 1984.

Rom WN: Effects of lead on the female and reproduction: A review. *Mt Sinai J Med (NY)* 43:542–550, 1976.

Ronkainen H, Pakarinen A, Kirkinen P, Kauppila A: Physical exercise–induced changes and season-associated differences in the pituitary–ovarian function of runners and joggers. *J Clin Endocrinol Metab* 60:416–422, 1985.

Ross EJ, Linch DC: Cushing's syndrome–killing disease: Discriminatory value of signs and symptoms aiding early diagnosis. *Lancet* 2:646–649, 1982.

Ross GT, Schreiber JR: The ovary, in Yen SSC, Jaffe RB (eds): *Reproductive Endocrinology* (2d ed). Philadelphia: WB Saunders, 1986, pp 115–139.

Ruder H, Corvol P, Mahoudeau JA, Ross GT, Lipsett MB: Effects of induced hyperthyroidism on steroid metabolism in man. *J Clin Endocrinol Metab* 33:382, 1971.

Ryback RS: Chronic alcohol consumption and menstruation. *JAMA* 238:2143, 1977.

Samaan NA, Bakdash MM, Caderao JB, Cangir A, Jesse RH, Ballantyne AJ: Hypopituitarism after external irradiation: Evidence for both hypothalamic and pituitary origin. *Ann Intern Med* 83:771–777, 1975.

Samuels-Reid J, Scott RB: Characteristics of menstruation in sickle cell disease. *Fertil Steril* 43:139–141, 1985.

Sandler H, Winters DL: *Physiological Responses of Women to Simulated Weightlessness. A Review of the Significant Findings of the First Female Bedrest Study.* NASA SP-430. NASA Scientific and Technical Information Office, 1978.

Santen RJ, Sofsky J, Bilic N, Lippert R: Mechanism of action of narcotics in the production of menstrual dysfunction in women. *Fertil Steril* 26:538–548, 1975.

Santy PA: Women in space: A medical perspective. *J Am Med Wom Assoc* 39:13–17, 1984.

Sauer MA, Rifka SM, Hawks RL, Cutler GB, Loriaux DL: Marijuana: Interaction with the estrogen receptor. *J Pharmacol Exp Ther* 224:404–407, 1983.

Schaefer RM, Wernze H, Geiger H, Heidland A: Improved sexual function in hemodialysis patients on recombinant erythropoietin: A possible role for prolactin. *Clin Nephrol* 31:1–5, 1989.

Schilsky RL, Lewis BJ, Sherins RJ, Young RC: Gonadal dysfunction in patients receiving chemotherapy for cancer. *Ann Intern Med* 93:109–114, 1980.

Schilsky RL, Sherins RJ, Hubbard SM, Wesley MN, Young RC, DeVita VT: Long-term follow-up of ovarian function in women treated with MOPP chemotherapy for Hodgkin's disease. *Am J Med* 71:552–556, 1981.

Schnorr TM, Grajewski BA, Hornung RW, Thun MJ, Egeland GM, Murray WE, Conover DL, Halperin WE: Video display terminals and the risk of spontaneous abortion. *N Engl J Med* 324:727–733, 1991.

Schweiger U, Laessle R, Pfister H et al: Diet-induced menstrual irregularities: Effects of age and weight loss. *Fertil Steril* 48:746–751, 1987.

Scott RT, Toner JP, Muasher SJ, Oehninger S, Robinson S, Rosenwaks Z: Follicle-stimulating hormone levels on cycle day 3 are predictive of in vitro fertilization outcome. *Fertil Steril* 51:651–654, 1989.

Setchell KDR: Naturally occurring non-steroidal estrogens of dietary origin, in McLachlan JA (ed): *Estrogens in the Environment II: Influences on Development*. New York: Elsevier, 1985.

Shangold MM: Athletic amenorrhea. *Clin Obstet Gynecol* 28:664–669, 1985.

Shangold M, Freeman R, Thysen B, Gatz M: The relationship between long-distance running, plasma progesterone, and luteal phase length. *Fertil Steril* 31:130, 1979.

Shangold MM, Gatz ML, Thysen B: Acute effects of exercise on plasma concentrations of prolactin and testosterone in recreational women runners. *Fertil Steril* 35:699–702, 1981.

Shangold MM, Rebar RW, Wentz AC, Schiff I: Evaluation and management of menstrual dysfunction in athletes. *JAMA* 263:1665–1669, 1990.

Sherman BM, Korenman SG: Hormonal characteristics of the human menstrual cycle throughout reproductive life. *J Clin Invest* 55:699–706, 1975.

Sherman BM, Halmi KA, Zamudio R: LH and FSH response to gonadotropin-releasing hormone in anorexia nervosa: Effect of nutritional rehabilitation. *J Clin Endocrinol Metab* 41:135, 1975.

Sherman BM, West JH, Korenman SG: The menopausal transition: Analysis of LH, FSH, estradiol, and progesterone concentrations during menstrual cycles of older women. *J Clin Endocrinol Metab* 42:629–636, 1976.

Shortridge LA: Assessment of menstrual variability in a working population. *Reprod Toxicol* 2:171–176, 1988.

Singh KB: Effects of sound on the female reproductive system. *Am J Obstet Gynecol* 112:981–991, 1972.

Siris ES, Leventhal BG, Vaitukaitis JL: Effects of childhood leukemia and chemotherapy on puberty and reproductive function in girls. *N Engl J Med* 294:1143–1146, 1976.

Smith CA: The effect of wartime starvation in Holland upon pregnancy and its product. *Am J Obstet Gynecol* 53:599–608, 1947.

Smith CG, Almirez RB, Berenberg J: Tolerance develops to the disruptive effects of delta 9 tetrahydrocannabinol on the primate menstrual cycle. *Science* 219:1453–1455, 1983.

Smith CG, Besch NF, Smith RG, Besch PK: Effects of tetrahydrocannabinol on the hypothalamic–pituitary axis in the ovariectomized rhesus monkey. *Fertil Steril* 31:335, 1979.

Solomon J, Cocchia MA, DiMartino R: Effect of delta-9-tetrahydrocannabinol on uterine and vaginal cytology of ovariectomized rats. *Science* 195:875–877, 1977.

Southren AL, Olivo J, Gordon GG, Vittek J, Brener J, Rafii F: The conversion of androgens to estrogens in hyperthyroidism. *J Clin Endocrinol Metab* 38:207–214, 1974.

Spengler JD, Sexton K: Indoor air pollution: A public health perspective. *Science* 221:9–17, 1983.

Speroff L: The effect of exercise on the menstrual cycle. *Postgrad Obstet Gynecol* 4:1–5, 1984.

Spitz IM, Diamant Y, Rosen E et al: Isolated gonadotropin deficiency: A heterogenous syndrome. *N Engl J Med* 290:10–15, 1974.

Starkman MN, Marshall JC, Ferla JL, Kelch RP: Pseudocyesis: Psychologic and neuroendocrine interrelationships. *Psychosom Med* 47:46, 1985.

Stedman RL: The chemical composition of tobacco and tobacco smoke. *Chem Rev* 68:153–207, 1968.

Stein IF, Leventhal ML: Amenorrhea associated with bilateral polycystic ovaries. *Am J Obstet Gynecol* 29:181–191, 1935.

Stein ZA: A woman's age: Childbearing and child rearing. *Am J Epidemiol* 121:327–342, 1985.

Stieve VH: Nervös bedingte Veränderungen an den Geschlechtsorganen. *Dtsch Med Wochenschr* 34:925–928, 1940.

Stillman RJ, Schinfeld JS, Schiff I et al: Ovarian failure in long-term survivors of childhood malignancy. *Am J Obstet Gynecol* 139:62–66, 1981.

Stoffer SS: A gynecologic study of drug addicts. *Am J Obstet Gynecol* 101:779, 1968.

Stouffer RL: Corpus luteum function and dysfunction. *Clin Obstet Gynecol* 33:668–689, 1990.

Strickler RC, Woolever CA, Johnson M, Goldstein M, DeVeber G: Serum gonadotropin patterns in patients with chronic renal failure on hemodialysis. *Gynecol Invest* 5:185–198, 1974.

Stumpf WE, Denny ME: Vitamin D (soltriol), light, and reproduction. *Am J Obstet Gynecol* 161:1375–1384, 1989.

Sydenham A: Amenorrhea at Stanley Camp, Hong Kong, during internment. *Br Med J* 2:159, 1946.

Tagatz G, Fialkow PJ, Smith D, Spadoni L: Hypogonadotropic hypogonadism associated with anosmia in the female. *N Engl J Med* 283:1326–1329, 1970.

Takizawa K, Yagi H, Jerina DM, Mattison DR: Experimental ovarian toxicity following intraovarian injection of benzo(a)pyrene or its metabolites in mice and rats, in Dixon RL (ed): *Reproductive Toxicology*. New York, Raven Press, 1985.

Tanaka T, Tamai H, Kuma K, Matsuzuka F, Hidaka H: Gonadotropin response to luteinizing hormone releasing hormone in hyperthyroid patients with menstrual disturbances. *Metabolism* 30:323–326, 1981.

Thibault C: Physiology and physiopathology of the fallopian tube. *Int J Fertil* 17:1–13, 1972.

Thomas R, Reid RL: Thyroid disease and reproductive dysfunction: A review. *Obstet Gynecol* 70:789–798, 1987.

Timonen S, Lokki O, Wichmann K, Vara P: Seasonal changes in obstetrical phenomena. *Acta Obstet Gynecol Scand* 44:507–533, 1965.

Tokuhata GM: Smoking in relation to infertility and fetal loss. *Arch Environ Health* 17:353–359, 1968.

Treloar AE, Boynton RE, Behn BG, Brown BW: Variation of the human menstrual cycle through reproductive life. *Int J Fertil* 12:77, 1967.

Turkington RW: Prolactin secretion in patients treated with various drugs: Phenothiazines, tricyclic antidepressants, reserpine and methyldopa. *Arch Intern Med* 130:349, 1972.

Ulmann A: Uses of RU486 for contragestation: An update. *Contraception* (suppl) 36:27–31, 1987.

U.S. Congress, Office of Technology Assessment: *Reproductive Health Hazards in the Workplace*. US Government Printing Office, 1985.

Vale W, Spiess J, Rivier C, Rivier J: Characterization of a 41-residue ovine hypothalamic peptide that stimulates secretion of corticotropin and β-endorphin. *Science* 213:1394–1397, 1981.

Välimäki M, Pelkonen R, Salaspuro M et al: Sex hormones in amenorrheic women with alcoholic liver disease. *J Clin Endocrinol Metab* 59:133–138, 1984.

Vance ML, Evans WS, Thorner MO: Bromocriptine. *Ann Intern Med* 100:78–91, 1984.

Vazquez AM, Kenney FM: Ovarian failure and antiovarian antibodies in association with hypoparathyroidism moniliasis and Addison's and Hashimoto's disease. *Obstet Gynecol* 41:414, 1973.

Vermande-Van Eck GJ, Meigs JW: Changes in the ovary of the rhesus monkey after chronic lead intoxication. *Fertil Steril* 11:223, 1960.

Vigersky RA, Anderson AE, Thompson RH et al: Hypothalamic dysfunction in secondary amenorrhea associated with simple weight loss. *N Engl J Med* 297:1141, 1977.

Vollman RF: *The Menstrual Cycle*. Philadelphia, WB Saunders, 1977 pp 19–72.

Wallace WHB, Shalet SM, Crowne ED, Morris-Jones PH, Gattamaneni HR, Price DA: Gonadal dysfunction due to cis-platinum. *Med Pediatr Oncol* 17:409–413, 1989.

Wani MC, Rector DH, Christensen HD, Kimmel GL, Cook CE: Flavenoids: 8. Synthesis and antifertility and estrogen receptor binding activities of coumarins and Δ3-isoflavenes. *J Med Chem* 18:982–985, 1975.

Warren MP: The effects of exercise on pubertal progression and reproductive function in girls. *J Clin Endocrinol Metab* 51:1150–1157, 1980.

Webb AR, Kline L, Holick MF: Influence of season and latitude on the cutaneous synthesis of vitamin D3: Exposure to winter sunlight in Boston and Edmonton will not promote vitamin D3 synthesis in human skin. *J Clin Endocrinol Metab* 67:373–378, 1988.

Weir RJ, Fisher RS: Toxicologic studies on borax and boric acid. *Toxicol Appl Pharmacol* 23:351–364, 1972.

Welch RM, Levin W, Conney AH: Estrogenic action of DDT and its analogs. *Toxicol Appl Pharmacol* 14:358, 1969.

Wells VE, Schnorr TM, Halperin: NIOSH selection of chemicals and study publications: Setting priorities for reproductive research. *Reprod Toxicol* 2:289–290, 1988.

Wertheimer N, Leeper E: Adult cancer related to electrical wires near the home. *Int J Epidemiol* 11:345–355, 1982.

Wertheimer N, Leeper E: Electrical wiring configurations and childhood cancer. *Am J Epidemiol* 109:273–284, 1979.

Wetterberg L: Lighting: Nonvisual effects. *Scand J Work Environ Health* (suppl 1) 16:26–28, 1990.

Wetterberg L, Arendt J, Paunier L et al: Human serum melatonin changes during the menstrual cycle. *J Clin Endocrinol Metab* 42:185–188, 1976.

Whitacre FE, Barrera B: War amenorrhea: A clinical and laboratory study. *JAMA* 124:399–403, 1944.

Whitehead E, Shalet SM, Jones PHM et al: Gonadal function after combination chemotherapy for Hodgkin's disease in childhood. *Arch Dis Child* 57:287–291, 1982.

Wilkening GM, Sutton CH: Health effects of nonionizing radiation. *Med Clin North Am* 74:489–507, 1990.

Wing S, Shy C, Wood JL et al: Mortality among workers at Oak Ridge National Laboratory. *JAMA* 265:1397–1402, 1991.

Witt BR, Thorneycroft IH: Reproductive steroid hormones: Generation, degradation, reception and action. *Clin Obstet Gynecol* 33:563–575, 1990.

Wu C-H, Motohashi HA, Abdel-Rahman A, Flickinger GL, Mikhail G: Free and protein-bound plasma estradiol-17β during the menstrual cycle. *J Clin Endocrinol Metab* 43:436, 1976.

Yen SSC: Neuroendocrine control of hypophyseal function: Physiological and clinical implications in reproductive endocrinology, in Yen SSC, Jaffe R (eds): *Reproductive Endocrinology* (2d ed). Philadelphia, WB Saunders, 1986a, pp 441–499.

Yen SSC: Chronic anovulation caused by peripheral endocrine disorders, in Yen SSC, Jaffe RB (eds): *Reproductive Endocrinology* (2d ed). Philadelphia, WB Saunders, 1986b, p 459.

Yen SSC, Garzo G, Liu J: Luteal contraception. *Contraception* (suppl) 36:13–25, 1987.

Zhou SY, Liang YX, Chen ZQ, Wang YL: Effects of occupational exposure to low-level carbon disulfide (CS2) on menstruation and pregnancy. *Ind Health* 26:203–214, 1988.

CHAPTER 7

Implantation

James A. Simon and Ted L. Anderson

In humans, implantation of the blastocyst begins 4 to 5 days after fertilization. At this point the zygote has undergone cleavage into a blastocyst, which consists of the embryoblast, an inner cell mass that gives rise to the embryo; the blastocyst cavity; and the trophoblast, an outer layer of cells enclosing the embryoblast that will later form the embryonic part of the placenta. Attachment of the blastocyst to the endometrial lining occurs following the disappearance of the zona pellucida (hatching). The syncytiotrophoblastic cells of the blastocyst first invade the superficial epithelial cells and the underlying stroma cells, which constitute the first two of the endometrium's three layers. At the same time the hypoblast begins to form on the innermost surface of the embryoblast. At the end of the first week following fertilization, the blastocyst is superficially implanted in the endometrium.

The second week after fertilization sees the completion of implantation. During this time the syncytiotrophoblast further invades the endometrial stroma, which contains capillaries and glands, and the stromal cells around the implantation site become filled with glycogen and lipids and become polyhedral. Some of these cells, now known as decidual cells, degenerate in the region of the penetrating syncytiotrophoblast to provide a rich source of nutrition for the embryo. The liberation of cellular components and breakdown products of the stroma may signal existing blood vessels and facilitate the growth of the nascent placental vasculature. The embryoblast then grows into a bilaminar embryonic disc composed of epiblast and hypoblast layers. This embryonic disc gives rise to all three germ layers of the embryo. Other structures that form during the second week are the amniotic cavity, amnion, yolk sac, connecting stalk, and chorion. The sequence of biochemical and physical interactions between the embryo and the uterus that results in implantation is a complex process that not infrequently fails. Implantation requires the embryo to develop appropriately, the endometrium to undergo radical cellular changes, and embryonic growth and endometrial decidualization to occur in proper synchrony. Implantation involves communication between embryo and uterus and between uterine epithelial and stromal tissues. The specific mechanisms leading to implantation failure are for the most part not yet understood, but potential for failure is probably inherent in every aspect of the process. This chapter examines many factors that may be involved in implantation failure including disorders of follicular development; luteal phase defects; detrimental effects of toxicologic, pharmacologic, and environmental factors; problems in embryo development and uterine receptivity; and angiogenesis failure.

Because implantation in humans cannot be easily studied, much of our knowledge of the mechanisms of implantation is based on animal studies. Much similarity has been found in implantation mechanisms among mammals, but a wide range of species variation also exists. The extrapolation of data from other animals to humans must be done with some caution.

DISORDERS OF ENDOMETRIAL PREPARATION

Disorders of Follicular Development

Hormones and Follicular Development

The hormonal environment plays a crucial role in the failure or success of implantation. The hypothalamus, pituitary, and ovary play essential roles in producing hormones that promote the growth and maturation of the egg and the endometrial response, first to estrogen and then to progesterone. The critical luteinizing hormone (LH) surge at ovulation not only changes the follicle into the corpus luteum (CL), resulting in the production of steroid hormones, but also signals the egg to undergo the second meiotic division, reducing the number of chromosomes from 46 to 23 and enabling the egg to be fertilized. The fertilized ovum must move through the fallopian tube, the motility of which is controlled by hormones, and must arrive at the uterus when the uterus is receptive.

Disease processes can disrupt normal follicular development, which is under the aegis of gonadotropins. Follicular growth occurs as a result of a change in the negative feedback cycle of estrogen and progesterone. This change results in production by the anterior pituitary of increased amounts of follicle-stimulating hormone (FSH), which is crucial for growth of the follicles and steroidogenesis. With follicular growth, the production of estradiol within the follicle maintains the follicle's receptiveness to FSH, allowing this environment, previously controlled by androgens, to become controlled by estrogen. As estradiol levels rise, the follicular cells produce more FSH receptors. This allows the follicles that have more FSH receptors to continue their growth, even with falling FSH levels. One or two of these dominant follicles are selected and can progress to maturation and ovulation. The combination of FSH and estradiol directs the formation of LH receptors, which must be in place before luteinization and ovulation occur. The catalyst for ovulation is a surge of LH production, prompted by a high estrogen level (estrogen positive feedback).

The crucial production of estrogen is by the vascular theca interna of the follicle and groups of stromal secretory cells, collectively called the interstitial gland of the ovary, but the majority is secreted by the follicular cells that form the membrane granulosa of the follicle (Moore, 1988).

Hormonal Defects Leading to Implantation Failure

Failure at different junctures within the ovulatory cycle and/or defects in the ovaries themselves can be due to abnormal hormonal conditions, which ultimately prohibit the normal progression of follicular growth and release and may impair fertilization and implantation. One of the most common disease processes that interferes with normal follicular development is androgen excess. Two common causes are polycystic ovarian syndrome (PCO) and adrenal hyperplasia.

Androgen Excess

An excess of androgens drastically changes the hormonal milieu and prevents normal follicular maturation. Normal concentrations of androgens result in the maturation and ovulation of one (or two) follicles per cycle with other follicles undergoing atresia. However, an excess of androgens can inhibit a normal menstrual cycle (Speroff et al, 1989). Androgen excess also may cause chronic anovulation by interfering with the necessary action of estradiol on gonadotropin receptors. When hyperandrogenism occurs, particularly when ovarian in origin, dysrhythmia of FSH and LH secretion, increased LH/FSH ratio, and failure of cyclic LH release, resulting in anovulation, are also associated with the disease process. A combination of ovarian disease and excessive androgen secretion by the adrenal gland can also produce excess androgens, while a pure adrenal source of excess androgens is less common (Vorys, 1988).

The ovary and the adrenal gland secrete three major C_{19} steroid androgens: (1) androstenedione, (2) testosterone, and (3) dehydroepiandrosterone (DHEA). Both the ovary in the female and the testis in the male produce androgens, a function that is LH-dependent. The appropriate follicular secretion of estrogen takes place only when FSH-dependent follicular maturation occurs (Yen and Jaffe, 1986). If the LH/FSH ratio is increased, usually due to acyclic peripheral estrogen feedback, chronic anovulation with increased production of ovarian androgens results.

Among the conditions involving androgen excess and increased extraglandular production of estrogen (see next section), PCO may be the most common one that causes follicular and maturation abnormalities and menstrual cycle dysfunction. However, several other conditions producing these problems also exist and may mimic PCO.

Hyperadrenalism, or Cushing syndrome/disease, involves excess secretion of cortisol due to either excess secretion of pituitary corticotropin (Cushing disease) or an autonomously functioning adenoma or adenocarcinoma (Cushing syndrome). Adrenal adenomas or carcinomas occur rarely, but they typically produce large amounts of 17-ketosteroids along with cortisol. Several types of adrenal and ovarian tumors are responsible for producing androgen autonomously, causing persistent anovulation. These tumors include hilus cell tumors, arrhenoblastomas (Sertoli–Leydig cell tumors), benign cystic teratomas, luteinized thecomas, and adrenal rest tumors of the ovary (Yen and Jaffe, 1986). The most common hormone-producing neoplasms are estrogen-producing tumors or granulosa–theca cell tumors, which account for between 15 and 20 percent of all solid ovarian tumors. These tumors normally produce estrogen, but in addition, androgen is frequently produced. The production of extra estrogen and androgen, along with the peripheral conversion of androgen to estrogen, results in chronic anovulation. Ovarian hyperthecosis, hyperprolactinemia, and hyperthyroidism or hypothyroidism also may cause or be associated with androgen excess.

Polycystic Ovarian Syndrome

PCO is not a disease in itself but is evidence of the ovary's continually staying in an abnormal state, one of chronic anovulation. The average daily production of estrogens and androgens is increased by LH stimulation and is dependent on this stimulation. In PCO patients, high circulating levels of testosterone, androstenedione, DHEA, dehydroepiandrosterone sulfate (DHAS), 17-hydroxyprogesterone (17-OHP), and estrone exist. Testosterone, androstenedione, and DHEA are produced by the ovary, while DHAS is secreted almost solely by the adrenal gland. Additionally, with PCO, patients have higher levels of LH and lower levels of FSH than do women with normal cycles. Increased LH secretion is related to increased free estradiol levels (Speroff et al, 1989), although there is no increase in estradiol secretion. Free estradiol levels increase because of the decline in sex hormone binding globulin (SHBG), caused by rising levels of testosterone or due to obesity or both. Low FSH levels are related to the negative feedback on FSH of rising levels of estrogens, free estradiol and estrone, from the peripheral conversion of androstenedione. The higher levels of inhibin secreted by polycystic ovaries help suppress FSH levels and alter GnRH production (Speroff et al, 1989).

As a result of low FSH levels, follicles are unable to mature and to progress to ovulation. The lower-than-normal FSH levels found in PCO allow growth to begin but not to proceed normally. In a normal cycle, estradiol levels, which are abnormally low in PCO, help increase the number of follicular FSH receptors, ensuring that the follicle will respond to the normal LH surge in the cycle needed for ovulation. However, a low number of FSH receptors commonly occurs in PCO, contributing to anovulation. Additionally, the excess androgens in polycystic ovaries also inhibit the needed growth of the follicles locally and cause early atresia. The abnormal cycle evidenced in polycystic ovaries does not represent some derangement of reactions by the hypothalamus, the pituitary, and the ovary, but is an appropriate response to hormone levels that are continually abnormal.

Adrenal Hyperplasia

Many types of congenital adrenal hyperplasia (CAH) exist, with errors in the hydroxylation of the precursors of cortisol as the source of the disorder. Rising production of corticotropin (ACTH), which is a compensation for the low levels of cortisol (negative feedback), induces the hyperplasia of the adrenal gland.

The most common type of CAH is 21-hydroxylase deficiency. In this disorder, which can range from mild to severe, androgens are produced in excess, which causes decreased secretion of the pituitary gonadotropins. This results in the ovaries not being stimulated, and hence menstrual periods and secondary sexual development will not occur.

Excessive or Deficient Estrogen Production during the Follicular Phase

For women of reproductive age, the majority of estrogen comes from ovarian production of estradiol. The mentrual cycle is largely controlled by estradiol as levels of this hormone must appropriately fluctuate with continual morphologic changes. Subnormal estradiol production by the follicle or interference with the action of estradiol can result in failure of a follicle to grow and to ovulate.

Abnormal cycling may occur when estradiol levels are too high, that is, do not fall low enough, to permit FSH to stimulate growth and maturation of the follicle, or when estradiol levels are too low, which may not be sufficient to induce the ovulatory surge of LH. Adrenal or ovarian tumors can produce continuous, excessive secretion of estrogen, with amenorrhea or anovulation resulting. Hyperthyroidism and hypothyroidism can interfere with estrogen clearance and metabolism, also causing anovulation. Excessive extraglandular production of estrogen may result in persistently high levels of estrogen in CAH, PCO, and some cases of Cushing syndrome (Yen and Jaffe, 1986).

In addition to ovarian estradiol, additional estrogen is formed when peripheral androstenedione is converted to estrone and testosterone is converted to estradiol. When anovulation occurs, more estrone is derived from conversion of androstenedione than is typically produced in a normal cycle. The higher levels of estrone contribute to the low FSH levels that occur in anovulation because of higher estrogen and free estradiol levels. One common occurrence is anovulation resulting from obesity as adipose tissue converts androstenedione to estrone.

Excess estrogen may not always result in anovulation; instead it may provoke a luteal phase defect. A study of follicular atresia induced by estrogen in rhesus monkeys found that atresia of a follicle selected for ovulation was controlled by the strength and duration of the stimulus (Dierschke et al, 1985). The study found that although atresia is typically an all-or-none phenomenon, partial atresia induced by estrogen during development of the follicle may also occur. While not preventing ovulation the effects of such partial atresia may remain and be expressed as a luteal phase defect.

Studies have shown that the high estradiol concentrations that occur from in vitro fertilization (IVF) and embryo transfer procedures may inhibit implantation, and higher progesterone levels may be needed to counteract this action (O'Neill et al, 1985).

Luteal Phase Defects

Deficiencies in the luteal or secretory phase may be a prime cause of implantation failure. The luteal phase, which lasts about 2 weeks, coincides with the formation and growth of the CL. The CL develops shortly after ovulation when the walls of the graafian follicle and the theca folliculi collapse into folds and develop into a glandular structure that primarily secretes progesterone (Moore, 1988). The process of luteinization is initiated by the midcycle surge of LH. Following the LH surge, the CL must continue to be stimulated by small amounts of LH to maintain adequate progesterone production. Human chorionic gonadotropin (hCG) can also interact with LH receptors to induce CL production of progesterone, but LH is believed to be the single most important factor in regulating the CL's synthesis and secretion of progesterone (Brody and Wentz, 1989). These data suggest that insufficient secretion of LH or hCG may cause the CL to fail in its critical role in supporting the early embryo.

Studies have shown that the CL is essential for successful implantation to occur, with progesterone almost uniformly implicated as the primary luteal product responsible for maintaining pregnancy. In their seminal work, Csapo and coworkers (1973) demonstrated that women luteectomized near the seventh week of pregnancy subsequently failed to maintain a normal pregnancy while luteectomized patients treated with progesterone sustained normal pregnancies. Along with progesterone, estrogen also appears to play a vital role in implantation in great apes and humans. Other luteal secretory products such as prostaglandins, relaxin, and oxytocin may also have roles to play, and if proper levels of these products are not secreted implantation theoretically could be inhibited.

Insufficiency of the CL, first reported more than 40 years ago (Jones, 1949), results in inadequate secretory transformation of the endometrium. This may be due to a failure of the CL to produce sufficient levels of progesterone, a shortened period of progesterone production by the CL, or an inadequate endometrial tissue response to progesterone. Luteal phase defects have been implicated in both infertility and habitual abortion. Published estimates suggest that 3 to 20 percent of infertile women and 25 to 60 percent of women who experience repeated spontaneous abortions have luteal phase defects (Balasch and Vanrell, 1987). Luteal phase defects are also associated with other clinical disease states including hyperprolactinemia and endometriosis.

The causes of luteal phase deficiency have not been fully determined. However, it is most likely that a variety of causes exists and that, in fact, luteal phase deficiency is caused by a number of different syndromes. For example, it has been found that the ratio of FSH to LH (Strott et al, 1970; Cook et al, 1983) and absolute FSH levels (Sherman and Korenman, 1974; Cook et al, 1983) are lower in the follicular phase of women with luteal phase defects. In addition, selectively lowering FSH levels during the follicular phase of normal women has been found to lead to luteal phase defects (Sheehan et al, 1982). As a result of such findings it has been proposed that the lack of FSH predisposes the dominant follicle to poor luteinization and the formation of an abnormal CL that cannot produce adequate levels of progesterone. It may be that inadequate FSH leads to diminished levels of intrafollicular estradiol, which lead, in turn, to inadequate formation of LH receptors.

Estrogens work in association with progesterone to control transport of the conceptus through the fallopian tube and to prepare the endometrium for the arrival of the conceptus, typically at the blastocyst stage. The endometrium, like other steroid-responsive tissues, contains steroid receptors for both estradiol and progesterone, and the synthesis of progesterone receptors is clearly associated with increased estrogen levels (Buster and Simon, 1989). Studies have shown that for many primates estrogen may be as crucial to successful implantation as progesterone (Ravindranath and Moudgal, 1989). It has been speculated that the threshold of the estrogen requirement is low and may be satisfied early in the postovulatory period (Ravindranath and Moudgal, 1989). This concept is given weight by the very low levels of estradiol found in the luteal phase of the macaque monkey menstrual cycle, in marked distinction to humans.

Inadequate progesterone receptors can adversely affect endometrial response in implantation. The ability of the endometrium to respond to progesterone is dependent on induction of adequate progesterone receptors by estradiol during the follicular phase. It appears that in rare cases adequate receptors may be lacking. As a result, the endometrium does not evolve normally despite the secretion of normal amounts of steroids by the CL. Thus, the endometrium does not advance toward the full secretory structure that allows implantation and normal embryonic development. A deficiency in the number of high-affinity progesterone–endometrial cytosol binding sites has been implicated as a cause of infertility (Keller et al, 1979). At least one study has found a 25 percent reduction in endometrial cytosol progesterone receptor binding sites in infertile women with documented luteal phase defects (Spirtos et al, 1985). This binding site deficiency is characterized by a failure of endometrial stroma to undergo the full decidual transformation normally seen in the late luteal phase. This abnormality is not corrected by administration of exogenous progesterone.

Insufficient estrogen receptors may also play a role in luteal phase defects. Studies in the rhesus monkey (Kreitmann-Gimbal et al, 1981; Ghosh and Sengupta, 1988) have shown that during a fertile cycle there is an increase in estradiol receptors around the peri-implantation period. Some authors have suggested that during the preimplantation stage increased endometrial nuclear estrogen and progesterone receptor content, which reflects a higher degree of occupancy, may be required for endometrial differentiation allowing implantation (Ghosh and Sengupta, 1988). The role for other CL secretory products is uncertain, but preliminary observations implicate some of them in implantation success or failure.

It is now clear that prostaglandins, which are produced by the endometrium and the blastocyst along with the CL, have an important role in endometrial decidualization and vascular permeability changes in a number of nonprimate species including rats, mice, and rabbits. Studies have shown that administration of an inhibitor of prostaglandin synthesis (indomethacin) can inhibit endometrial vascular permeability and implantation. Research has also demonstrated that prostaglandins are elevated in the areas of increased endometrial vascular permeability (Kennedy, 1977).

Relaxin is a known secretory product of the human CL, and it has long been established that relaxin suppresses uterine motility. Excessive uterine motility may lead to expulsion of the unimplanted embryo, cause embryo–uterine asynchrony by hastening embryo transport to the uterus, or may dislodge a partially implanted embryo. Relaxin has been shown to promote uterine growth and distensibility (Vasilenko and Mead, 1987), which would facilitate implantation and early embryo growth. Several studies in nonhuman primates and at least one study in women document a measurable rise in levels of plasma relaxin associated with the normal luteal phase (Sherwood and Downing, 1983; Stewart et al, 1990). Relaxin secretion is accelerated near the time that hCG is first detected in conception cycles, and hCG has been shown to induce relaxin secretion in nonpregnant women if administered late in the luteal phase (Quagliarello et al, 1981).

Disorders of CL Rescue

CL Rescue

Any luteolytic compound can lead to the demise of the CL at the end of the ovulatory cycle. However, if the ovum is fertilized, the CL expands dramatically, becoming one of the most vascular and metabolically active tissues in the body, and luteolysis is prevented. The CL is "rescued" by chorionic gonadotropin, first released by the preimplantation blastocyst and later by placental cells (Ravindranath and Moudgal, 1989). Other factors released by the blastocyst may also play a role in triggering the maternal recognition of pregnancy that results in CL rescue. Additional embryonic factors that have been postulated to be involved in CL rescue include histamine, prostaglandins, platelet-activating factor (PAF), certain steroids, and trophoblast protein-1 (Ravindranath and Moudgal, 1989; Roberts, 1990).

Failure of CL Rescue

In humans it appears that CL rescue occurs at the last possible moment. Studies have shown that in both humans and rhesus monkeys, the CL is poised to self-destruct and progesterone production is already declining at the time hCG first appears in maternal plasma (Roberts, 1990). As a result, any delay in secretion of hCG is likely to lead to failure of CL rescue.

A physiologic rise in the concentrations of urinary estrone-3-glucuronide and pregnanediol-3-glucuronide after implantation effectively differentiated successful early pregnancies from failed pregnancies. Particularly in early pregnancy losses suggestive of late implantation, a drop in urinary estrone-3-glucuronide and pregnanediol-3-glucuronide levels indicated a failure of CL rescue, leading the authors to speculate that a late hCG signal was unable to effectively rescue the CL (Baird et al, 1991). A delay in the hCG signal may have been caused by late fertilization, slow tubal transport, slow development of the conceptus after fertilization, or problems with the early stages of the implantation process. In ectopic pregnancies, impaired hCG production by the improperly planted trophoblast appears to be the primary reason for the inability of the CL to function at its optimal level (Kratzer and Taylor, 1990).

In humans, it is well known that after the initial rescue, both luteotropins hCG and LH are necessary for maintaining the life span of the CL. CL function has been found to be significantly correlated with the rate of change of the concentration of hCG, rather than with its absolute concentration (Kratzer and Taylor, 1990).

The mechanism of luteolysis and its role in early pregnancy loss through demise of the CL are not well understood. Studies are equivocal about the exact mechanisms of luteolysis, and conflicting data exist between the numerous studies done in ruminants, rats, and nonhuman primates, and the fewer studies done in humans.

The influence of many hormones and fatty acid derivatives (i.e., prostanoids) as luteolysins that contribute to the demise of the CL is controversial and uncertain, particularly when primate or human data and animal studies are compared. In primates and humans, it appears that the mechanism controlling luteolysis involves luteal production and local action of estrogen, oxytocin, and prostaglandin $F_{2\alpha}$ ($PGF_{2\alpha}$) (Auletta and Flint, 1988).

$PGF_{2\alpha}$ appears to be a luteolysin in ruminants and possibly in primates. In ruminants, $PGF_{2\alpha}$ is produced by the endometrium under the influence of luteal oxytocin and perhaps ovarian estradiol. In primates, a luteal or ovarian source may be responsible for producing $PGF_{2\alpha}$ and could be accountable for regression at the end of a nonfertile cycle, possibly through either a reduction of progesterone production by an antigonadotropic action or a redistribution or shunting of blood away from the CL (Auletta and Flint, 1988). In sheep, the mechanism of action by which ovine trophoblastic protein-1 (oTP-1) has been shown to rescue the CL from luteolysis is through prevention of release of $PGF_{2\alpha}$ from endometrial epithelial cells (Yoshinaga, 1989a).

The effects of oxytocin on luteolysis have been demonstrated in some ruminant studies, but its role in nonhuman primates and women is not clear. While in ruminants, oxytocin appears to act systemically, in primates the action may be local, as it is certainly not systemic. Thus, while oxytocin is luteolytic in ruminants, the role of this peptide hormone in primates remains unconfirmed (Auletta and Flint, 1988).

The level of 17-OHP may be an important factor or marker in maintaining the function of the CL, and deficient 17-OHP levels may contribute to some pregnancy losses. This steroid is secreted by the CL, not the placenta. When comparing 17-OHP levels as a sign of CL function in a study of aborters versus nonaborters, it appears that low 17-OHP levels along with a drop in serum progesterone levels may be associated with spontaneous abortion (Check et al, 1990). β-hCG levels remained above 10,000 mIU in all patients in the study, documenting adequate trophic CL stimulation. Whether or not 17-OHP levels alone or with some function of progesterone are better early indicators of pregnancy failure than hCG levels are remains to be determined. The lack of association in this study between hCG levels and 17-OHP or progesterone levels suggests that some other factor of trophoblastic origin may be important in trophic stimulus of the CL and that this factor is missing in early pregnancy failure.

Prolactin is not considered a vital luteotropin in nonhuman primates or women. However, its role as a potential luteolytic agent is well recognized. Minor elevations of prolactin levels at midcycle, even without clinical galactorrhea or apparently abnormal luteal function, result in infertility. A study of 48 women with long-standing idiopathic infertility and normal ovulatory cycles found that 45 patients have transitory midcycle low-level hyperprolactinemia detected through radioimmunoassays of prolactin levels taken on 8 specified days of the menstrual cycle (Ben-David and Schenker, 1983). Because 18 of 45 patients conceived within 3 months of treatment using the dopamine agonist 2-bromo-α-ergocryptine (bromocryptine), the authors concluded that minimal hyperprolactinemia for short time periods may affect fertilization, implantation, embryogenesis, or CL function in pregnancy.

As the extent and degree of hyperprolactinemia increases, luteal function deteriorates, and then ovulatory cyclicity ends (Auletta and Flint, 1988). Hyperprolactinemia has been shown to impede follicular maturation and steroidogenesis, ovulation, the luteinization process, and CL function (Kauppila et al, 1982).

In nonhuman primates and women, the loss of LH receptors (LHr) appears not to be a required step in spontaneous luteolysis as progesterone decreases are not necessarily prefaced by or associated with decreased LHr, especially in the early stages of luteal disintegration (McNeilly et al, 1987; Auletta et al, 1986). These data suggest that LHr may not be a component of CL disorders.

DISORDERS RESULTING IN EARLY FIRST TRIMESTER LOSSES

During the implantation period, which extends to day 17 after conception, many substances, both organic and inorganic, may disrupt the implantation process, thereby terminating the pregnancy. While most research has shown the teratogenic effects of chemicals on fetuses and embryos, some studies have indicated the possible lethal consequences of chemicals during the crucial implantation process. Data about this latter area are scarce because many failed pregnancies go undetected, leaving both patient and physician unaware of the deleterious effects of some chemicals.

Toxicologic and Pharmacologic Causes

Diethylstilbestrol

The direct effects of fetal exposure to diethylstilbestrol (DES), once prescribed to prevent miscarriages, have been well documented. Females exposed in utero suffered carcinogenic and teratogenic effects resulting in genital and gynecologic anomalies. Some DES-exposed females have been known to later have vaginal clear-cell adenocarcinoma. Males suffered teratogenic effects involving sperm abnormalities, undescended testes, hypospadias, and epididymal cysts.

The effects of DES on fertility and pregnancy outcome among women and men exposed in utero are less certain. However, the preponderance of evidence suggests that women exposed in utero to DES have an increased risk of unfavorable pregnancy outcome including increased risk of spontaneous abortion and ectopic pregnancy (Stillman, 1982; Herbst et al, 1989). A recent study of IVF in 46 DES-exposed women concluded that implantation and pregnancy outcome are impaired in these women (Karande et al, 1990). Previous studies have pointed to uterine and tubal abnormalities as a cause of poorer pregnancy outcome in DES-exposed women (Stillman, 1982; Herbst et al, 1989). This was supported by the findings of Karande and coauthors, whose study showed a trend for a worse pregnancy prognosis in women with uterine constrictions and those with a combination of a T-shaped uterus and constrictions.

It is estimated that some 50 million people have been exposed to DES because it was used to fatten beef and poultry, but the effects of this exposure are unknown (Kurzel and Cetrulo, 1985).

Tricyclic Antidepressants, Phenothiazines, and Metoclopramide

Little research has been done on the ability of tricyclic antidepressants to disrupt implantation or cause early embryonic loss. However, tricyclic antidepressants are associated with hyperprolactinemia and may be implicated as a possible cause of CL defects. In addition, tricyclic antidepressants have been shown to act as teratogens in animal studies. For women who took tricyclic antidepressants in the first trimester, a study showed no evidence of gross congenital abnormalities in 15,000 births (Kuenssberg and Knox, 1972). Experts have concluded that adverse effects of antidepressant drug use during pregnancy and lactation have not been clearly established (Schrag and Dixon, 1985).

Studies of phenothiazines have shown conflicting results. A prospective study of 12,764 women and their children who were exposed to phenothiazines during their first trimester found a statistically significant increase in major congenital anomalies when phenothiazine derivatives with a three-carbon aliphatic side chain—chlorpromazine, methotrimeprazine, trimeprazine tartrate, and oxomemazine—were used (Rumeau-Rouquette et al, 1977). This effect was not observed with the use of promethazine hydrochloride or the piperidine and piperazine antipsychotic drugs. A study of 19,952 women who received phenothiazines in the first trimester showed no significant increases in the incidence of severe congenital anomalies and perinatal deaths for those receiving the drugs versus those who served as controls (Miklovich and van den Berg, 1976). When the pregnancy results of 100 women who had received haloperidol were studied retrospectively, no notable effects on fecundity or gestational duration or several other factors were found compared with controls (Van Waes and Van der Velde, 1969).

Metoclopramide's effect on raising prolactin levels has been shown by studies in which a significantly higher prolactin response was observed in some women who were given metoclopramide during a metoclopramide provocation test (Reinthaller et al, 1990). Metoclopramide-induced hyperprolactinemia in women was shown to impair maturation of ovarian follicles and to disrupt CL function (Kauppila et al, 1982).

Heavy Metals

According to biological studies, women may be more susceptible than men to the toxic effects of lead, although both males and females can suffer deleterious effects. In several animal and human models, lead exerts a strong negative effect on pregnancy and fetal development. Several studies of humans showed an increase in chromosomal aberrations for workers occupationally exposed to lead (Rom, 1976). Lead has been shown to inhibit implantation in mice. Female mice fed a diet containing lead (as the acetate) experienced a significant decrease in the ability to maintain a pregnancy compared with a control group (Jacquet et al, 1975). The cause was a failure of the blastocyst to implant on day 7. Inorganic lead also was observed to affect implantation in mice in another study, but mechanisms for the failed pregnancies were not explored (Odenbro and Kihlstrom, 1977).

Another heavy metal, arsenic, caused infertility, fetotoxicity, fetolethality, low-birth-weight offspring, and congenital malformations in mice and rats (Schrag and Dixon, 1985).

Zinc deficiency may increase the risk of spontaneous abortions. In a study of rats fed a zinc-deficient diet, 99 percent of the implantation sites either showed a resorbed conceptus or gave rise to malformed fetuses, while only 2 to 4 percent of the implantation sites in females receiving adequate amounts of zinc were affected (Hurley, 1981). Marginal zinc deprivation of rhesus monkeys was correlated with delivery complications, low birth weight, and trace metal status (Golub et al, 1984). The zinc deprivation (4 ppm beginning at conception) did not lower maternal plasma zinc levels during embryogenesis and did not increase malformation rates. A study of first trimester serum zinc concentrations in women found that the results of maternal zinc deficiency were not firmly established, but the potential for increased risk of congenital malformations and/or spontaneous abortions certainly exists (Breskin et al, 1983).

Cocaine

Few studies have examined the effects of cocaine on early first trimester losses. A study that examined cocaine use by women during their first trimester of pregnancy found an increased rate of spontaneous abortion among users (Chasnoff et al, 1985). This may occur because of decreased blood flow to the fetus from vasoconstriction of the placenta and increased contractions of the uterus from increased norepinephrine concentrations (Smith and Smith, 1990).

The effects of cocaine on the reproductive hormones, studied in nonhuman primates, showed that changes occurred in levels of gonadotropins, prolactin, and sex steroids (Scher et al, 1982). In this study, sexually mature female rhesus monkeys were given a 2.0-mg/kg dose of cocaine by an intravenous bolus injection. The results showed an immediate increase in LH levels (average 38 percent increase at 30 min). Prolactin fell to nondetectable levels within 1 to 2 h and remained depressed for up to 24 h. The authors found these results to be consistent with the known neurochemical actions of cocaine on adrenergic and dopaminergic activity in the brain.

Paternal exposure to cocaine has been linked to abnormal development of offspring in rodents (Abel et al, 1989). The mechanism of cocaine's effects in this setting is not known, nor is this association clearly demonstrated in humans. Human sperm have recently been shown to contain significant numbers of high-affinity cocaine binding sites. Sperm carrying high concentrations of cocaine had normal motility, morphology, and viability in vitro

(Yazigi et al, 1991). These findings suggest a novel mechanism for the transport of cocaine into the ovum and the potential for production of developmental abnormalities.

Environmental Causes

Concern over the effects of environmental exposure to chemical pollutants has grown over the last 30 years, but research elucidating the effects of such pollutants has lagged far behind. Air pollutants, food additives, ingested contaminants, herbicides and pesticides, and metal compounds all may have the potential to cause implantation failure, but in most cases the possible hazards of such agents are unknown.

Air pollutants commonly include compounds such as oxides of nitrogen and sulfur and incompletely oxidized hydrocarbons such as carbon monoxide that are produced by the combustion of fossil fuels. The most serious source of air pollution for pregnant women may be cigarette smoke because it exposes the embryo to high levels of carbon monoxide, hydrogen cyanide, cadmium, nicotine, and polycyclic aromatic compounds.

Carbon monoxide, which originates from automobile exhaust and industrial combustion as well as cigarettes, is the most studied air pollutant. In animals, exposure to carbon monoxide has been shown to cause pregnancy loss and birth defects. Studies in rats show that reproductive failure via resorption of the embryo and early spontaneous abortion is concentration-dependent (Longo, 1977).

Among food additives two of the better studied are Red Dye No. 2 and cyclamate. Use of these compounds as food additives has been banned or restricted in the United States because studies linked them to cancer. However, their effects on pregnancy also have been studied. Red Dye No. 2 has been linked with a tendency for preimplantation loss and increased embryo resorption, though the variability of results between studies raises question about the significance of these findings (Collins et al, 1976). Studies on cyclamate generally have failed to find embryotoxic or teratogenic effects, but research has shown that cyclohexylamine, a metabolite of 35S-cyclamate, can cause preimplantation loss and early fetal death in Swiss mice (Kroes et al, 1977).

The extensive use of herbicides and pesticides has resulted in various levels of human exposure to these agents. The embryotoxic and teratogenic effects of many of these compounds have been conclusively demonstrated in lower species but not (with the exception of the organomercurials) in humans (Kurzel and Cetrulo, 1985). Though the effects of such agents on implantation and early pregnancy are largely unknown, some studies have suggested detrimental consequences. A study by the Dow Chemical Company on the reproductive performance of female employees exposed to 2,4,5-T (one of the components of Agent Orange) and dioxin showed no significant differences in pregnancy outcomes or birth defects between exposed and unexposed workers. However, exposed workers had almost three times fewer pregnancies, raising the possibility that exposure was causing an increase in preimplantation and early embryo losses.

Immunologic Causes

One of the most important mechanisms in successful implantation is the process by which the blastocyst signals the maternal system, and in particular the surface cells of the uterus, to prevent immunologic rejection of the blastocyst. How the blastocyst does this in humans is unknown, but in sheep and cows it has been shown that the primary embryonic signal to elicit maternal recognition of the blastocyst is an interferon-like protein known as trophoblast protein-1. Known as oTP-1 in sheep and bTP-1 in cows, this protein is produced by the blastocyst's trophoblast cells. In sheep, oTP-1 appears to act locally as a paracrine hormone directed to the uterine lining, where it appears to influence both protein synthesis and production of prostaglandins (Imakawa and Roberts, 1989). Some researchers have speculated that oTP-1 may act on the histocompatibility complex of the uterine cells in a mechanism similar to that of metastatic cancer cells, tricking the complex into recognizing the blastocyst cells as self rather than foreign (Yoshinaga, 1989a).

Embryo Development

The earliest stages of preimplantation embryo growth and development are under the influence of proteins and enzymes derived from maternal mRNA. Consequently, the transfer to embryonic control represents the first significant event that must occur to ensure continued successful development (First and Barnes, 1989). This "maternal to zygotic transition" appears to occur at about the four- to eight-cell stage in humans and is associated with an increased nuclear cytoplasmic ratio and increased cell length (Braude et al, 1979). Although zygotic control of gene expression has been shown in humans and a variety of animal models to require organization of functional nucleoli, the molecular signals required to initiate such organization remain elusive (Tesarik et al, 1987). Indeed, it is unclear whether these signals are external in origin or whether this fundamental developmental milestone is a "preprogrammed" event. The latter is supported by the ability to culture embryos in vitro in relatively simple medium free of the growth factors normally associated with external control of cell processes.

The first "developmental objective" of preimplantation embryos is the establishment of a blastocyst comprising two irreversible cell lines: embryoblast (inner cell mass from which embryo development will continue) and trophectoderm (which will interact with the endometrial lining during placentation). Failure to establish this initial differentiation step, which is normally not completed until the 128-cell stage (Winkel and Pedersen, 1988), is believed to be a common source of implantation loss in humans (Johnson, 1989). In mouse embryos, the model in which early development is best described, blastocyst formation follows "compaction" of the 8-cell embryo. This event, characterized by formation of tight junctions between the outermost-situated cells, establishes cellular polarity and creates "inside and outside" microenvironments that may herald subsequent differentiation (Ducibella et al, 1975; Fleming and Johnson, 1988). The functional counterpart of these events in humans has not been characterized for practical and ethical reasons.

Subsequent development of the preimplantation embryo is oriented toward the goal of implantation and placentation, discussed below.

Endometrial Receptivity

A highly invasive trophoblast allows certain mammalian embryos to implant almost without discretion in a variety of ectopic sites, regardless of host tissue, hormonal milieu, or timing of such an interaction (Kirby, 1965). Conversely, it is overtly conspicuous that the only tissue in which trophoblast invasion does not occur indiscriminately is the natural implantation site, the uterine lining. The endometrium is permissive to blastocyst implantation only during a brief receptive interval, and following appropriate hormonal stimulation, as demonstrated by the failure of transferred blastocysts to result in endometrial attachment except at the time of normal uterine receptivity for that species (Finn, 1977). Any perturbation of the normal progression of hormonal sequelae, or cellular responses to that stimulation (i.e., both excessive and different), results in diminution of reproductive potential from implantation failure. In human IVF programs in which ovarian hyperstimulation is used to obtain multiple ova for fertilization, excessively high serum levels of estradiol and progesterone often result in advanced endometrial histologic maturation or stromal–epithelial asynchrony (Jones, 1985). Such events could contribute to embryo–uterine asynchrony and appear to be partially responsible for the relatively modest success rates of these programs, typically less than 20 percent.

The nature of the embryonic developmental window has been examined by several investigators. Alterations in endometrial surface membrane components during the preimplantation differentiation period (surface negativity, lectin binding, and glycoprotein expression) have been described, with particular reference to the development of adhesive properties (Nilsson et al, 1973; Johnson and Calarco, 1980; Magnuson and Epstein, 1981; Chavez and Enders, 1982).

The regulatory role of the uterine luminal epithelium in embryo–uterine interactions is provided by a growing number of observations (in a wide variety of species). Alterations of surface charge and glycocalyx composition and the temporal and spatial characteristics of stage-specific protein and glycoprotein expression provide potential marker parameters of uterine receptivity in molecular terms. Progesterone-stimulated endometrial effects are discussed elsewhere; here our focus is on stage-specific alterations of the luminal surface and membrane components, cell surface (i.e., morphology and electric charge), lectin binding, and membrane protein expression (see Anderson, 1989, for review).

The Cell Surface

Histologic features of luminal and glandular epithelial cells related to physiologic states of the endometrium (including uterine receptivity) have been well described in rodents, rabbits, nonhuman primates, and humans (Noyes and Dickmann, 1960; Parr, 1982; Williams and Rogers, 1980; Hoffman and Davies, 1973). Surface modifications of luminal epithelial cells have also been reported (Nilsson, 1974; Martel et al, 1987), including the stage-specific appearance of "pinopod" protrusions. Accompanying these morphologic changes, discrete alterations have been identified in the composition and "morphology" of luminal glycocalyx material in several species (Chavez and Anderson, 1985; Anderson et al, 1986; Lee and Damjanov, 1985; Jansen et al, 1985). Common features include decreased surface negativity, increased thickness of periodic acid-Schiff-positive staining material, and a more fibrillar glycocalyx structure by electron microscopy. The reduction of luminal surface negativity prior to the time of implantation is consistent with proposed mechanisms of cellular adhesion involving changes in membrane glycoproteins, and with the general observation that events such as adhesion and fusion are enhanced when repulsive charges on the involved cell surfaces are diminished (Vicker and Edwards, 1972; Weiss et al, 1975).

The significance of increased glycocalyx thickness on the uterine luminal surface at the time of implantation is not clear. Skutelsky and coworkers (1974) proposed that sialic acid plays a "protective role" during differentiation by masking antigenic or recognition sites. Surface negativity of the nonreceptive rabbit uterus appears related to the presence of sialic acid (Anderson and Hoffman, 1984); this is likely true in the primate as well. Anderson and coworkers (1986) found that enzymatic removal of sialic acid from the glycocalyx of nonreceptive rabbit uteri revealed *N*-acetylglucosamine and galactose residues. Similar removal of sialic acid to reveal D-galactose residues on glycoproteins appears to be a recognition mechanism common to a variety of biological systems, including implantation.

Recognition of galactose-containing glycoproteins by apical membrane endogenous lectins appears to mediate adhesion of chick blastoderm cells (Milos and Zalik, 1982) as well as cell–cell or cell–substratum adhesion of some tumor cells in relation to metastatic potential.

Lectin Binding and Membrane Protein Expression

Investigation of apical membrane protein expression by endometrial epithelial cells in relation to the window of uterine receptivity has been limited. Stage-specific proteins in plasma membrane fractions from rabbit endometrial scrapings, and in extracts of rabbit epithelial membranes isolated by intraluminal incubation of detergent solution, have been described (Anderson et al, 1986; Lampelo et al, 1985). Lectin binding to proteins isolated by the latter method and transferred to nitrocellulose has revealed some to be glycoproteins. Of particular interest is a 42-kDa glycoprotein recognized by both WGA and RCA-I lectins. Other proteins, seemingly present in extracts from nonreceptive and receptive uteri, appeared to be glycosylated differentially depending on the hormonal state, exhibiting lectin binding only in the receptive uterus. Steroid hormones appear to influence epithelial apical membrane protein composition both at the transcription level and by posttranslational modification.

Summary

Alterations in glycocalyx thickness and morphology toward the time of implantation appear to reflect a "decondensation" due to reduced negativity and increased saccharide complexity of glycoproteins synthesized under specific hormonal stimulation. Increased availability of D-galactose residues in the glycocalyx at this time may be indicative of a carbohydrate-based cell recognition system during the initial phase of embryo–uterine interaction. The role of stage-specific apical membrane proteins of uterine luminal epithelial cells remains unclear. However, one might predict that they mediate, among other events, signal transduction between the implanting blastocyst and the endometrium, and that any derangement in this elaborate time and hormone-dependent system could result in implantation failure.

Embryo–Uterine Synchrony

The sequence of events culminating in successful implantation must be coordinated such that oviductal transport delivers an appropriately developed blastocyst to a uterus of maximal endometrial receptivity. Evidence from a variety of investigations implicates progesterone excess or deficit as a key factor involved in all these processes.

Progesterone administration during the preimplantation interval has been shown to lead to implantation failure in rabbits, rats, and (albeit debated) humans (Psychoyos, 1976; Schacht and Foote, 1978; Trounson et al, 1986; Jones, 1985). Although the exact mechanism by which implantation failure is effected by progesterone excess has not been elucidated, it is believed to involve advanced endometrial maturity and possibly delayed tubal transport of the preimplantation embryo.

The advent of progesterone antagonists such as RU 486 and ZK 98734 has facilitated our examination of the mechanism of implantation failure due to inadequate progesterone stimulation. In animal models, these compounds have been demonstrated to retard embryo development, delay endometrial maturation, and accelerate tubal transport (Psychoyos, 1989; Psychoyos and Prapas, 1987; Sarantis et al, 1988). Similar perturbation of embryo development, endometrial maturation, and tubal physiology has been described following passive immunization of experimental animals with an antiprogesterone monoclonal antibody (Wang et al, 1984; Rider and Heap, 1986; Rider et al, 1987). It has been suggested that the molecular mechanism may involve altered protein glycosylation by mucosal epithelial cells (Whyte et al, 1987).

It is easy to extrapolate these laboratory findings to the clinical scenario of luteal phase defects in which progesterone stimulation is inadequate. However, out-of-phase endometrial maturation despite adequate progesterone levels is less easily explained. Recent research has provided some insight into the molecular mechanism by which this may occur. Expression of the glycoprotein TAG-72 by endometrial epithelial cells has been examined in vivo throughout the menstrual cycle and in a well-defined endometrial cell culture system under a variety of hormonal stimulations (Osteen et al, 1990; Osteen and Anderson, 1991). Results of these investigations suggest that the action of progesterone on maturation of endometrial epithelial cells may be mediated indirectly through stromal cells via paracrine interactions involving growth factors. Thus, aberrant endometrial stromal–epithelial interactions may be an underlying cause of implantation failure in the face of appropriate serum progesterone levels. Such abnormalities may be manifested by stromal–glandular asynchrony of endometrial biopsies.

Attachment and Invasion

The search for embryo-produced molecules that mediate embryo–maternal interactions has been long-standing, but interest has recently increased dramatically with the desire to predict successful implantation in programs of human IVF. A thorough discussion of all molecular alterations of trophoblastic cells and products of the blastocyst that have been implicated in embryo–uterine interactions is far beyond the scope of this chapter; rather, an outline of different categories of bioactive molecules is provided with notable examples. Such molecules can be classified broadly as those facilitating recognition–attachment, invasion–placentation, or maternal recognition of pregnancy. Note that many of the molecular features described here in relation to embryo implantation parallel what we believe to occur in tumor invasion and metastasis. These similarities cannot be addressed adequately here but are of considerable interest to a growing number of investigators.

Decreased trophoblast cell surface charge and changes in glycocalyx composition have been described in relation to the acquisition of adhesive properties

(Johnson and Calarco, 1980). It is believed that a variety of surface glycoconjugates may be involved in embryo–uterine recognition, and there is evidence to support the involvement of a cell surface galactosyltransferase in initial adhesion of the embryo to the endometrium (Chavez, 1990).

A variety of cell adhesion molecules, including receptors for the extracellular matrix proteins fibronectin, laminin, type IV collagen, and heparin–heparin sulfate have been described (Fisher et al, 1990; Carson et al, 1990; Farach et al, 1987). In addition to the ability of trophoblast cells to adhere to extracellular matrix proteins, it has been demonstrated that connective tissue degrading enzymes (which appear to facilitate stromal invasion by implanting embryos) are elaborated by trophoblast cells (Fisher et al, 1990).

Synthesis and release of lipid-based molecules, such as prostaglandins and PAF have been described in human and experimental animal embryos; a role for inducing vascular permeability has been described (Harper et al, 1989; O'Neill, 1985; Collier et al, 1988). Interestingly, prostaglandins have been shown to stimulate matrix-degrading enzyme production by tumor cells and uterine cervical fibroblasts (Yoshinaga, 1989b).

Numerous lipid, carbohydrate, and protein molecular species elaborated by trophoblastic cells have been postulated as maternal recognition signals. The identification of PAF-binding sites in uterine membranes has been offered as evidence for such a role in the rabbit (Harper et al, 1989). Rizzino (1985) has identified a transforming growth factor–like substance elaborated by mouse embryos, which may have an autocrine role in embryonic growth as well as maternal signaling. Perhaps the most well characterized maternal recognition signals are oTP-1 and bTP-1. These proteins share almost 70 percent sequence homology with interferon-α and have been demonstrated to have an antiluteolytic activity (Roberts et al, 1989). This similarity also raises the possibility of an immunoprotective role for the implanting embryo, as discussed earlier in this chapter.

Angiogenesis Failure

Angiogenesis, or new blood vessel growth, is a key development in the implantation of the embryo. Inadequate angiogenesis resulting in vascular failure, or excessive angiogenesis, such as that seen in the penetration of trophoblastic neoplasia and the normal placenta beyond the endometrium, may be significant sources of pregnancy loss.

At present, little is definitively known about the process of angiogenesis in implantation, but it is believed to be analogous to angiogenesis that occurs in malignant tumors, wound healing, inflammation, and the graft-versus-host reaction. In recent years, research has begun to elucidate some possible factors involved in implantation angiogenesis, raising the possibility that embryonic factors may play an important and perhaps even crucial role in this process.

During the process of implantation, the endometrial surface undergoes several changes, possibly in response to stimulation from the perinidation embryo. Regression of the local epithelium occurs along with proliferation of the surrounding maternal tissue and stimulation of a capillary network underlying the surface epithelium (Barthelmaz, 1951).

During implantation, embryonic factors may stimulate angiogenesis in several ways. Proteolytic enzymes produced by the blastocyst, necessary for the destruction of the endometrial surface epithelium, have been implicated as important mediators of angiogenesis (Moscatelli et al, 1986). These enzymes may participate in the degradation of the capillary basement membrane and/or the extracellular matrix. Such actions could liberate endothelial cell mitogens such as basic fibroblast growth factor, fibronectin, fibrin, or fibrin degradation products. All have been shown to participate in or initiate angiogenesis.

The preimplantation endometrium is rich in prostaglandins, which have been shown to stimulate angiogenesis. Local endometrial regression at implantation may liberate these prostaglandins. Inadequate prostaglandin synthesis is associated with poor vascular permeability and implantation failure.

Although early work in the mechanisms of angiogenesis was largely limited to issues of carcinogenesis, such work has since been extended to other physiologic and pathophysiologic circumstances including embryonic implantation (Hertz, 1978). One of the most interesting findings in implantation angiogenesis has been the documentation of angiogenic growth factors within the ovaries and other areas of the female reproductive tract (for review, see Koos, 1989). The angiogenic growth factors found in the ovary are thought to originate from the granulosa or theca cells directly, but the possibility that these factors are from the oocytes themselves cannot be ruled out.

Recent reports have documented the presence of several growth factors and their RNA transcripts in preimplantation murine embryos (Rappolee et al, 1988; Akhurst et al, 1990). One of these growth factors transforming growth factor beta (TGF-β), is the modulator for fibroblast growth factor (FGF), the known angiogenic growth factor.

In addition to these growth factors, another embryonic factor that appears to play a principal role in angiogenesis is PAF. PAF has been shown to be a potent inducer of vascular permeability (Angle et al, 1986), and it has been hypothesized that PAF may be the initial signal produced by the embryo for maternal recognition of pregnancy. A recent review of research on PAF (Harper, 1989) found several studies showing that in mice and rats PAF appears to be essential for implantation. For human zygotes, PAF production has been used as a measure of zygote viability. Zygotes that resulted in pregnancy produced significantly higher levels of PAF in culture prior to transfer than zygotes that were not viable (O'Neill et al, 1987).

The direct question of whether the early conceptus is responsible for initiation of its own blood supply through the secretion of such factors is still largely uninvestigated. However, the findings of at least one preliminary study suggest that after the eight-cell stage of development, preimplantation monkey blastocysts secrete an angiogenic growth factor crucial to a successful pregnancy (Simon and Hodgen, 1991, unpublished data). Further research will be needed to confirm these observations. However, the hypothesis that the developing embryo acquires the capacity for angiogenic induction and successful implantation at or after the morula stage is supported by the high rates of pregnancy wastage (approximately 30 percent) in human IVF programs (Jones et al, 1983), where embryo transfers usually take place at or before the eight-cell stage.

The embryo's ability to implant outside the endometrium, including some sites distant from hormonally sensitive reproductive tract tissues such as the fallopian tube and ovary, indicates the primacy of embryonic factors over endometrial factors in implantation. This is evident in the large numbers of ectopic pregnancies seen in women (Iffy, 1981) and laboratory animals (Kreitmann-Gimbal et al, 1981).

In the future, a delineation of the growth factors involved in angiogenesis may assist in the understanding and treatment of some types of infertility associated with occult pregnancy and recurrent spontaneous abortion, as well as with gestational trophoblastic neoplasia and early placentation.

REFERENCES

Abel EL, Moore C, Waselewsky D, Zajac C, Russell LD: Effects of cocaine hydrochloride on reproductive function and sexual behavior of male rats and on the behavior of their offspring. *J Androl* 10:17–27, 1989.

Akhurst RJ, Lehnert SA, Faissner et al: TGF beta in murine morphogenetic processes: The early embryo and cardiogenesis. *Development* 108:645–656, 1990.

Anderson TL: Biomolecular markers for the window of uterine receptivity, in Koji Yoshinaga (ed): *Blastocyst Implantation.* Serono Symposia, USA. Boston, Adams Publishing, 1989, pp 219–224.

Anderson TL, Hoffman LH: Alterations in the epithelial glycocalyx of rabbit uteri during early pseudopregnancy and pregnancy, and following ovariectomy. *Am J Anat* 171:321–334, 1984.

Anderson TL, Olson GE, Hoffman LH: Stage-specific alterations in the apical membrane glycoproteins of endometrial epithelial cells related to implantation in rabbits. *Biol Reprod* 36:599–618, 1986.

Angle MJ, McManus LM, Pinckard RN: Age-dependent differential development of leukotactic and vasoactive responsiveness to acute inflammatory mediators. *Lab Invest* 55:616–621, 1986.

Auletta FJ, Flint APF: Mechanisms controlling corpus luteum function in sheep, cow, nonhuman primates, and women especially in relation to the time of luteolysis. *Endocr Rev* 9:88–105, 1988.

Auletta FJ, Paridis DK, Gibson M: Corpus luteum LH receptors during spontaneous and prostaglandin F2alpha-provoked luteolysis in the rhesus monkey (abstract). *Proceedings of the Society for Gynecological Investigation*. Toronto, Canada, 1986.

Baird DD, Weinberg CR, Wilcox AJ, McConnaughey DR, Musey PI, Collins DC: Hormonal profiles of natural conception cycles ending in early, unrecognized pregnancy loss. *J Clin Endocrinol Metab* 72:793–800, 1991.

Balasch J, Vanrell JA: Corpus luteum insufficiency and fertility: A matter of controversy. *Hum Reprod* 2:557–567, 1987.

Beckman G, Beckman L, Nordenson I: Chromosome aberrations in workers exposed to arsenic. *Environ Health Perspect* 19:145–146, 1977.

Ben-David M, Schenker JG: Transient hyperprolactinemia: A correctable cause of idiopathic female infertility. *J Clin Endocrinol Metab* 57:442–444, 1983.

Braude P, Pelham H, Flach G, Lobatto R: Post-transcriptional control in the early mouse embryo. *Nature* 282:102–105, 1979.

Breskin MW, Worthington-Roberts BS, Knopp RH, Brown Z, Plovic B, Mottet NK, Mills JL: First trimester serum zinc concentrations in human pregnancy. *Am J Clin Nutr* 38:943–953, 1983.

Brody BL, Wentz CW: An update on the clinical relevance of luteal phase inadequacy. *Semin Reprod Endocrinol* 7:138–154, 1989.

Buster JE, Simon JA: Placental hormones, hormonal preparation for and control of parturition, and hormonal diagnosis of pregnancy, in DeGroot LJ (ed): *Endocrinology* (vol 3). Philadelphia, WB Saunders, 1989, pp 2043–2073.

Carson DD, Wilson OW, Dutt A: Glycoconjugate expression and interactions at the cell surface of mouse uterine epithelial cells and peri-implantation embryos. *Troph Res* 4:211–241, 1990.

Chasnoff IJ, Burns WJ, Schnoll SH, Burns KA: Cocaine use in pregnancy. *N Engl J Med* 313:666, 1985.

Chavez DJ: Possible involvement of D-galactose in the implantation process. *Troph Res* 4:259–272, 1990.

Chavez DJ, Anderson TL: The glycocalyx of the mouse luminal epithelium during estrus, early pregnancy, the peri-implantation period, and delayed implantation: I. Acquisition of Ricinus communis-I binding sites during pregnancy. *Biol Reprod* 32:1135–1142, 1985.

Chavez DJ, Enders AC: Lectin binding of mouse blastocysts: Appearance of *Dolichos bifloris* binding sites on the trophoblast during delayed implantation and their subsequent disappearance during implantation. *Biol Reprod* 21:545–552, 1982.

Check JH, Vaze MM, Epstein R, Wu CH, Quattrocchi J, Vetter B: 17-Hydroxyprogesterone levels as a marker for corpus luteum function in aborters versus nonaborters. *Int J Fertil* 35(2): 112–115, 1990.

Collier M, O'Neill C, Ammit AJ, Saunders DM: Biochemical and pharmacological characterization of human embryo-derived platelet activating factor. *Hum Reprod* 3:993–998, 1988.

Collins TXF, Ruffles DI, Hosons JF et al: Teratological evaluation of FD&C Red #2—A collaborative government–industry study. *J Toxicol Environ Health* 1:851, 1976.

Cook CL, Rao CV, Yussman MA: Plasma gonadotropin and sex steroid hormone levels during early, midfollicular, and midluteal phases of women with luteal phase defects. *Fertil Steril* 40:45–48, 1983.

Csapo AI, Pulkkinen MO, Wiest WG: Effects of luteectomy and progesterone replacement therapy in early pregnant patients. *Am J Obstet Gynecol* 115:759–765, 1973.

Dierschke DJ, Hutz RJ, Wolf RC: Induced follicular atresia in rhesus monkeys: Strength–duration relationships of the estrogen stimulus. *Endocrinology* 17:1397–1403, 1985.

Ducibella T, Albertini DF, Anderson E, Biggers JD: The preimplantation embryo: Characterization of intercellular junctions and their appearance during development. *Dev Biol* 45:231–250, 1975.

Farach MC, Tang JP, Decker GL, Carson DD: Heparin/heparin sulfate is involved in attachment and spreading of mouse embryos in vitro. *Dev Biol* 123:401–410, 1987.

Finn CA: The implantation reaction, in Wynn RM (ed): *Biology of the Uterus.* New York, Plenum Press, 1977, pp 245–303.

First NL, Barnes FL: Development of preimplantation mammalian embryos, in Yoshinaga K, Mori T (eds): *Development of Preimplantation Embryos and Their Environment.* New York, Alan R Liss, 1989, pp 151–170.

Fisher SJ, Sutherland A, Moss L, Hartmann L, Crowley E, Bernfield M, Calarco P, Damsky C: Adhesive interactions of murine and human trophoblast cells. *Troph Res* 4:115–138, 1990.

Fleming TP, Johnson MH: From egg to epithelium. *Annu Rev Cell Biol* 4:459–485, 1988.

Ghosh D, Sengupta J: Patterns of estrogen and progesterone receptors in rhesus monkey endometrium during secretory phase of normal menstrual cycle and preimplantation stages of gestation. *J Steroid Biochem* 31(2):223–229, 1988.

Golub MS, Gershwin ME, Hurley LS, Baly DL, Hendrickx AG: Studies of marginal zinc deprivation in rhesus monkeys: II. Pregnancy outcome. *Am J Clin Nutr* 39:879–887, 1984.

Harper MJK: Minireview. Platelet-activating factor: A paracrine factor in preimplantation stages of reproduction? *Biol Reprod* 40:907–913, 1989.

Harper MJK, Kudolo GB, Alecozay AA, Jones MA: Platelet-activating factor (PAF) and blastocyst-endometrial interactions, in Yoshinaga K, Mori T (eds): *Development of Preimplantation Embryos and Their Environment.* New York, Alan R Liss, 1989, pp 305–315.

Herbst AL, Senekjian EK, Frey KW: Abortion and pregnancy loss among diethylstilbestrol-exposed women. *Semin Reprod Endocrinol* 7:124–129, 1989.

Hertz R: *Choriocarcinoma and Related Gestational Trophoblastic Tumors in Women*. New York, Raven Press, 1978.

Hoffman LH, Davies J: Studies on the progestational endometrium of the rabbit: I. Light microscopy, day 0 to day 13 of gonadotrophin-induced pseudopregnancy. *Am J Anat* 137:423–446, 1973.

Hurley LS: Teratogenic aspects of manganese, zinc, and copper nutrition. *Physiol Rev* 61:249–295, 1981.

Iffy L: Ectopic pregnancy, in Iffy L, Kaminetzky HA (eds): *Principles and Practices of Obstetrics and Perinatology*. New York, Wiley, 1981.

Imakawa K, Roberts M: Interferons and maternal recognition of pregnancy, in Yoshinaga K, Mori T (eds): *Development of Preimplantation Embryos and Their Environment*. New York, Alan R Liss, 1989.

Jacquet P, Leonard A, Gerber GB: Embryonic death in mouse due to lead exposure. *Experientia* 31:1312–1313, 1975.

Jansen RPS, Turner M, Johannisson E, Landgren B-M, Diczfalusy E: Cyclic changes in human endometrial surface glycoproteins: A quantitative histochemical study. *Fertil Steril* 44:85–91, 1985.

Johnson LV, Calarco PG: Mammalian preimplantation development: The cell surface. *Anat Rec* 196:201–219, 1980.

Johnson MH: How are two lineages established in early mouse development? in Yoshinaga K, Mori T (eds): *Development of Preimplantation Embryos and Their Environment*. New York, Alan R Liss, 1989, pp 189–198.

Jones GES: Some newer aspects of management of infertility. *JAMA* 141:1123–1129, 1949.

Jones HW: Factors affecting the luteal phase in IVF and ET programs, in Testart J, Frydman R (eds): *Human in Vitro Fertilization*. INSERM Symposium No. 24. Amsterdam, Elsevier, 1985, pp 219–235.

Jones HW Jr, Acosta AA, Garcia JE, Sandow BA, Veeck L: On the transfer of conceptuses from oocytes fertilized in vitro. *Fertil Steril* 39:241–243, 1983.

Karande VC, Lester RG, Muasher SJ et al: Are implantation and pregnancy outcome impaired in diethylstilbestrol-exposed women after in vitro fertilization and embryo transfer? *Fertil Steril* 54:287–291, 1990.

Kauppila A, Leinonen P, Vihko R, YI'Ostalo P: Metoclopramid-induced hyperprolactinemia impairs ovarian follicle maturation and corpus luteum function in women. *J Clin Endocrinol Metab* 54:955, 1982.

Keller DW, Wiest WG, Askin FB et al: Pseudo corpus luteum insufficiency: A local defect of progesterone action on endometrial stroma. *J Clin Endocrinol Metab* 48:127, 1979.

Kennedy TG: Evidence for a role for prostaglandins in the initiation of blastocyst implantation in the rat. *Biol Reprod* 16:286–291, 1977.

Kirby DRS: The transplantation of mouse eggs and trophoblast to extrauterine sites, in Daniel JC (ed): *Methods in Mammalian Embryology*. San Francisco, WH Freeman, 1965, pp 146–156.

Koos RD: Potential relevance of angiogenic factors to ovarian physiology, in Speroff L (ed): *Seminars in Reproductive Endocrinology* (vol 7). New York, Thieme–Stratton, 1989, pp 29–40.

Kratzer PG, Taylor RN: Corpus luteum function in early pregnancies is primarily determined by the rate of change of human chorionic gonadotropin levels. *Am J Obstet Gynecol* 163:1497–1502, 1990.

Kreitmann O, Hodgen GD: Retarded cleavage rates of preimplantation monkey embryos in vitro. *JAMA* 246:627–629, 1981.

Kreitmann-Gimbal B, Bayard F, Hodgen GD: Changing ratios of nuclear estrone to estradiol binding in endometrium at implantation: Regulation by chorionic gonadotropin and progesterone during rescue of primate corpus luteum. *J Clin Endocrinol Metab* 52:133–137, 1981.

Kroes R, Peters PW, Berkvens J et al: Long term toxicity and reproduction study with cyclamate, saccharin, and cyclohexylamines. *Toxicology* 8:285, 1977.

Kuenssberg EV, Knox JD: Imipramine in pregnancy. *Br Med J* 2:292, 1972.

Kurzel RB, Cetrulo CL: Chemical teratogenesis and reproductive failure. *Obstet Gynecol Surv* 40: 397–424, 1985.

Lampelo SA, Ricketts AP, Bullock DW: Purification of rabbit endometrial plasma membranes from receptive and non-receptive uteri. *J Reprod Fertil* 75:475–484, 1985.

Lee M-C, Damjanov I: Pregnancy-related changes in the human endometrium revealed by lectin histochemistry. *Histochemistry* 82:275–280, 1985.

Longo LD: The biological effects of carbon monoxide on the pregnant woman, fetus, and newborn infant. *Am J Obstet Gynecol* 129:69, 1977.

McNeilly AS, Kerin J, Swanston IA, Bramley TA, Baird DT: Changes in the binding of human chorionic gonadotropin/luteinizing hormone and prolactin to human corpora lutea during the menstrual cycle and pregnancy. *J Endocrinol* 87:315, 1987.

Magnuson T, Epstein CJ: Characterization of concanavalin A precipitated proteins from early mouse embryos: A two-dimensional gel electrophoresis study. *Dev Biol* 81:193–199, 1981.

Martel D, Frydman R, Glissant M, Maggioni C, Roche D, Psychoyos A: Scanning electron microscopy of postovulatory human endometrium in spontaneous cycles and cycles stimulated by hormone treatment. *J Endocrinol* 114:319–324, 1987.

Miklovich L, van den Berg BJ: An evaluation of the teratogenicity of certain antinauseant drugs. *Am J Obstet Gynecol* 125:244–248, 1976.

Milos N, Zalik SE: Mechanisms of adhesion among the cells of the early chick blastoderm. Role of the beta-D-galactosamine-binding lectin in the adhesion of extra-embryonic endoderm cells. *Differentiation* 21:175–182, 1982.

Moore KL: *The Developing Human: Clinically Oriented Embryology* (4th ed). Philadelphia, WB Saunders, 1988.

Moscatelli D, Presta M, Rifkin DB: Purification of a factor from human placenta that stimulates capillary endothelial cell protease production, DNA synthesis, and migration. *Proc Natl Acad Sci USA* 83:2091–2095, 1986.

Nilsson BO: Changes of the luminal surface of the rat uterus at blastocyst implantation. *Z Anat Entwickl* 144:337–342, 1974.

Nilsson O, Lindkvist I, Ronqvist G: Decreased surface charge of mouse blastocysts at implantation. *Exp Cell Res* 83:421–423, 1973.

Noyes RW, Dickmann Z: Relation of ovular age to endometrial development. *J Reprod Fertil* 1:186–196, 1960.

Odenbro A, Kihlstrom JE: Frequency of pregnancy and ova implantation in triethyl lead-treated mice. *Toxicol Appl Pharmacol* 39:359–363, 1977.

O'Neill C: Partial characterization of embryo-derived platelet-activating factor in mice. *J Reprod Fertil* 75:375–380, 1985.

O'Neill C, Ferrier AJ, Vaughan J, Sinosich MJ, Saunders DM: Causes of implantation failure after in-vitro fertilization and embryo transfer. *Lancet* 2:615, 1985.

O'Neill C, Gidler-Baird AA, Pike IL, Saunders DM: Use of a bioassay for embryo-derived platelet-activating factor as a means of assessing quality and pregnancy potential of human embryos. *Fertil Steril* 47:969–975, 1987.

Osteen KG, Anderson TL: Effect of estrogen on human endometrial epithelial cell growth and differentiation in vitro. *Steroids* 56:279–283, 1991.

Osteen KG, Anderson TL, Hargrove JT, Hill GA, Gorstein F: Epithelial cells from normal human endometrium express a tumor-associated glycoprotein (TAG-72) epitope in vitro. *Am J Obstet Gynecol* 163:479–484, 1990.

Parr MB: Effects of ovarian hormones on endocytosis at the basal membranes of rat uterine epithelial cells. *Biol Reprod* 26:909–913, 1982.

Psychoyos A: Hormonal control of uterine receptivity for nidation. *J Reprod Fertil* (suppl) 25:17–28, 1976.

Psychoyos A: Anti-progestins and egg-implantation, in Yoshinaga K, Mori T (eds): *Development of Preimplantation Embryos and Their Environment*. New York, Alan R Liss, 1989, pp 289–294.

Psychoyos A, Prapas I: Inhibition of egg development and implantation in rats after post-coital administration of the progesterone antagonist RU 486. *J Reprod Fertil* 80:487–491, 1987.

lar press. The heightened awareness of the American population to reproductive concerns, together with a general paucity of knowledge regarding the etiology of specific reproductive disorders, fuels a perception that environmental agents or drugs are responsible for most of the health issues. However, at least some of the increase in reproductive problems in this country can be attributed to the changing demographics of the American population. The median age of the population of this country is increasing. This alone must explain much of the prevalence in age-related reproductive disorders in men and women. For example, one would expect the incidence of prostatic and mammary tumors to increase as a population ages. Moreover, the fact that many women are now delaying childbearing until they are in the mid to late 30s, a period of life when fertility is declining, is also likely to contribute to infertility and heritable defects. However, in addition to these natural biological reasons for reported increases in reproductive problems, technologic advances have enhanced our ability to detect, and fortunately to treat, reproductive disorders. This in effect makes it appear as if reproductive problems are more common now than in the past, when such technologies were unavailable.

On the other hand, concerns over toxicant-induced reproductive health problems are not without justification. Exposure to certain drugs and environmental agents has undoubtedly increased the incidence of infertility, impotence, pregnancy loss, and malformations at birth. For example, certain cancer chemotherapies lead to oocyte loss, premature menopause, and infertility in women (Lyons, 1981), as well as inducing testicular atrophy and sterility in men (Steinberger and Lloyd, 1985). Drugs that target the central nervous system, and many of those that lower blood pressure, can lead to impotence in men (Benson, 1988). Pseudoprecocious puberty in girls can result when they are accidentally exposed to estrogen-containing creams (Hopwood, 1985). Other examples include exposure to lead, which has been known to be an abortifacient in women for centuries (Schardein, 1985; Fahim et al, 1976); occupational exposure to manganese, which can cause impotence and infertility in men (Lauwerys et al, 1985); and exposure to the fungicide dibromochloropropane (DBCP), which led to infertility due to reduced sperm production in factory workers (Whorton et al, 1977).

In addition to effects on the reproductive function of adult men and women, drugs and xenobiotics are known to adversely affect fetal development, with effects ranging from lethal malformations and cancers to subtle disorders of the central nervous system later in life (Schardein, 1985). For example, children of women who used the drug diethylstilbestrol (DES) during pregnancy developed unusual reproductive tract tumors and displayed a number of malformations of the reproductive tract (Steinberger and Lloyd, 1985). Finally, environmental exposure to methyl mercury during gestation caused CNS malformations in the children (Murakami, 1972; Amin-Zaki et al, 1974).

The multiplicity of xenobiotic exposures, lesion sites, and potential reproductive disorders precludes a simple screening of suspected reproductive toxicants in humans. Moreover, the chemicals that cause reproductive dysfunctions or developmental alterations, and the effects themselves, are often impossible to predict because of the complexity of the interactions between toxicants and the reproductive and developmental processes. Thus, it is particularly noteworthy that the chemicals mentioned here as examples of chemicals that cause adverse reproductive outcomes in humans are also reproductive or developmental toxicants in rodents.

Mammalian reproductive toxicology, which will be discussed in this chapter, is so complex that a comprehensive research strategy is necessary to determine if a drug or environmental agent poses a reproductive risk to humans (National Research Council, 1989). Today, major research efforts are underway in many laboratories across the country to develop animal models that might identify the various manifestations of reproductive toxicity in both male and female animals. Animal models that have been used to identify reproductive toxicants will be discussed in this chapter, and an example of a testing strategy for the utilization of these methods is presented in Fig. 8-1. Briefly, reproductive toxicology testing usually begins with long-term, subchronic studies in which both male and female rodents are exposed to the test chemical for at least two generations. Different regulatory groups have slightly different test protocols for these multigenerational studies, but basically, they rely on fertility and histologic measures for the detection of reproductive alterations. Two additional reproductive test protocols under development at the present time will also be discussed.

After identification of a reproductive effect and the affected sex, more-specialized testing might continue on a chemical if additional information is desired. Such testing would focus on the male or female as appropriate. At this level of testing, organs and tissues would first be examined after in vivo administration of the test chemical. This approach allows confirmation of the adverse effect in vivo and leads to identification of the affected target organ(s) or tissue(s). Some information about the subcellular target(s) and mechanism(s) of action of the toxicant may also be forthcoming from such investigations.

Given that sufficient information exists about the metabolism of the test chemical, and if relevant tissue levels are known, then the chemical could be directly administered to the appropriate target tissue or cell in vitro. This would aid in the identification of subcellular targets and the mechanism of action of the chemical. Furthermore, it may then be possible to repeat these in vitro exposures in a similar human tissue or cell culture system, if an appropriate one exists. If a xenobiotic agent perturbs both rodent and human cell systems via homologous mechanisms, then we gain confidence to extrapolate the effects of the compound from the animal model to humans. Some of the in vitro reproductive systems that would be useful in this level of testing are discussed in the chapter. Although we have prioritized in vivo and in vitro testing in Fig. 8-1, we will discuss individual animal models on a system–organ–tissue basis throughout the remaining text. It is beyond the scope of this chapter to discuss, in detail, the nature of all the animal models currently

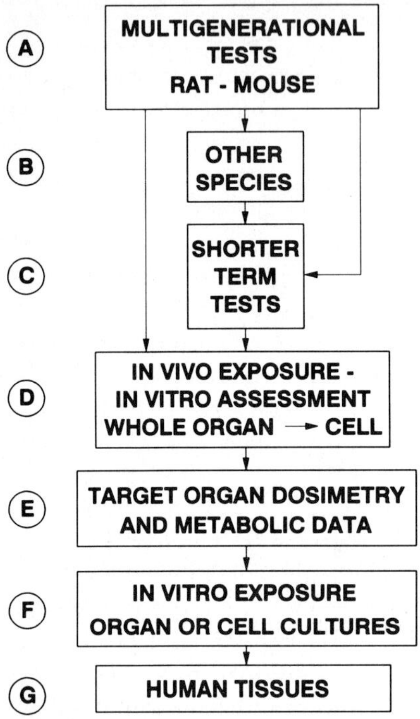

Figure 8-1. A flow diagram of a tier-testing strategy proposed to test a suspected reproductive toxicant. Tier A: A multigenerational study is conducted in the mouse or rat. A subsequent cross-over mating is then done to establish the affected sex. This identifies reproductive dysfunction before, during, or after reproductive development in the male or the female and determines a NOAEL (no-observed-adverse-effect level). Tier B: Other species may be tested for species sensitivity or relevancy information depending on the nature of the reproductive dysfunction indicated in the rat or mouse. For example, the quinea pig may be chosen as an alternate model when a mid–late gestational disorder is indicated. This determines whether the NOAEL is based on the most appropriate species. Tier C: With sex and affected age established for the most appropriate species, a short-term (4 to 90 days) test is done to state or confirm the reproductive target organ and possibly establish a lower NOAEL. This study provides more-sophisticated measures (seminiferous tubule fluid, epididymal sperm protein, etc.) and high-quality histopathologic studies. Tier D: To verify a target tissue or cell, more definitive in vitro measures are obtained following xenobiotic exposure in vivo. For example, measuring hCG-stimulated testosterone production by decapsulated whole testes and purified Leydig cells would most likely be more informative than measuring serum testosterone levels. This again may establish a lower NOAEL. Tier E: The time course of distribution and metabolism of a reproductive toxicant within a target organ provides essential information for subsequent in vitro studies designed to identify a mechanism of action. Tier F: The cellular site and mechanism of action may be identified by a validated in vitro organ and cell culture system. For example, to identify a lesion in the steroidogenic pathway, purified Leydig cells can be cultured with various steroidogenic substrates in the presence of a xenobiotic or its active metabolite. Tier G: Once a cellular site and mechanism of action are established for a reproductive toxicant, an appropriate in vitro system or test may be applied to human reproductive tissue obtained following biopsy or orchidectomy. This may establish whether a homologous mechanism exists for the reproductive toxicant. While tier A is proposed as the starting point for a chemical with unknown toxicity, other points of entry should be considered if sufficient information is already available. For example, one could start at tier C if the affected sex or target organ is known, or at tier B if a species other than the rat is the most appropriate animal model.

in use. We will, however, describe selected examples of adult animal models, both in vivo and in vitro, that are either currently in use or that might be adapted for use in assessing reproductive risk in the human population.

MATING PAIR TESTING

Standard Multigeneration Protocols

In both the male and female, complex interactions between cellular and hormonal events play a role in the reproductive competence of the mating pair. Naturally, the clinician must be concerned with both individuals, the couple, when fertility is discussed. To provide a comprehensive data base for male and female rats, most regulatory groups [Environmental Protection Agency (EPA), Food and Drug Administration (FDA), etc.] require data from a multigenerational reproduction study for risk assessment. Extensive reviews of these protocols have been published (Palmer, 1981; Francis and Kimmel, 1988). Typically, in the first generation (called the P_0 or F_0), male and female rats are exposed to the test chemical beginning shortly after puberty and continuing through adulthood, mating, gestation, and ultimately until necropsy. The chemical is usually administered in the diet at four different dosage levels. After production of one or two litters (called the F_{1a} and F_{1b}), the P_0 are necropsied and reproductive organs are saved for histopathologic evaluations.

In these multigenerational studies, the F_1 and F_2 animals are continuously exposed to the test chemical throughout their life cycle. After puberty, the F_1 animals are mated to produce the F_2 generation. A two-generation study is terminated when the F_2 animals are weaned. Frequently, some of the F_1 and F_2 animals are necropsied at weaning. Unlike their parents, these animals are exposed during their fetal and neonatal life as well as during adulthood. For this reason, developmental reproductive toxicants can be detected by mating to produce the F_2 generation. In addition, the F_1 animals are exposed for a longer period than the P_0, so that chemicals that bioaccumulate will also have a greater effect in the F_1. It is for these reasons that regulatory agencies generally require at least a two-generation reproduction test (Francis and Kimmel, 1988). If the test chemical results in a reproductive disorder and a no-observed-adverse-effect level (NOAEL) was detected, the data can be used for "quantitative" risk assessment (Barnes and Dourson, 1988). In this process, a *reference dose* (RfD) is calculated and human exposure to the chemical at or below the RfD is considered to be safe. The RfD is determined by adjusting the NOAEL by a number of *uncertainty factors* (UFs). The less known about a chemical, the greater the UFs. The UFs are used to account for a number of factors, including the fact that humans may be more sensitive than the test species used in the toxicology study (interspecies UF), and some segments of the population may be more sensitive (i.e., infants, the aged, anemic, asthmatic, etc.) than others (intraspecies UF). Although this method of risk assessment has been criticized on a number of grounds, and other methods of using data for risk assessment are being considered, like the use of the *benchmark* dose (Crump, 1984) in place of the NOAEL, it is not clear at this time if the newer methods actually improve the process.

Other Multigeneration Protocols

As indicated above, standard multigenerational tests rely primarily upon fertility, litter size, and viability, and routine histopathologic examination to detect reproductive disorders. Over the last decade it has become increasingly apparent that fertility *is often* (but not always) an insensitive measure of reproductive toxicity. For example, studies using male rats or mice have found that fertility is normal even when the sperm count or testis weight is reduced by 90 percent (Meistrich and Brown, 1983; Gray et al, 1988b, 1989a). In the female, alterations of puberty and estrous cyclicity may occur at doses well below those that alter fertility (Gray et al, 1988a). Some consideration has been given to using more-sophisticated assessments of testicular histology (Lamb and Chapin, 1985; Russell et al, 1991) and sperm function (Linder et al, 1986a; Overstreet, 1984; Klinefelter et al, 1991b), but effects on other reproductive organs in the male (e.g., pituitary and endocrine function) and additional reproductive measures in females (such as estrous cyclicity) are rarely reported.

The need for alternative reproductive tests was recently recognized by the EPA, and guidelines for their development were presented (Francis and Kimmel, 1988). To detect male and female reproductive and developmental toxicity, these protocols need to span at least two generations, include histologic study of the reproductive organs, and provide information on the other important endpoints just mentioned. These protocols are used to identify reproductive toxicants and prioritize them for additional testing, if warranted.

Currently, there are two independent efforts involved in the development and validation of additional reproductive testing protocols. One of these protocols originated in our reproductive toxicology laboratory (Gray et al, 1988b, 1989b, 1990) at the EPA and is referred to as *alternative reproductive test* (ART). The other protocol (Morrissey et al, 1989) is under development at the National Institute of Environmental Health Sciences and is referred to as *reproductive assessment by continuous breeding* (RACB).

In the ART protocol, exposure is initiated at weaning when male and female rats are 21 days old. The xenobiotic is administered at five doses, and treatment is continued through puberty, young adulthood, mating, gestation, and lactation. This protocol incorporates indices of puberty in the male and female rat along with other morphologic and physiologic indices of reproductive health. In females, the ages of vaginal opening and first estrus, estrous cyclicity, litter size, the number of implants, organ weights, reproductive behavior, hormones, and histology are assessed. The assessment of the male rat includes the age at preputial separation, organ weights, testicular and epididymal sperm counts, a quantitative measure of sperm motility, sperm morphology, testicular and epididymal histology, testicular seminiferous and interstitial fluid evaluation, behavioral evaluation, and endocrine measures including in vitro human chorionic gonadotropin (hCG)-stimulated testosterone production. After about 80 days of treatment, the P_0 are mated for a 16-h period during estrus. At this time the reproductive behavior of the male and female can be observed. If reproductive function is altered during the first mating of the P_0 generation, treated rats are paired with untreated rats and mated again (crossover mating). This allows one to determine the affected sex, and often the effects become more obvious than in the first mating. The growth, viability, and reproductive function of the offspring (F_{1a}) are observed during a 4-month period of continuous breeding. Some of the F_1 are necropsied at weaning, while the breeders are necropsied at the end of the study.

This protocol clearly exceeds the requirements proposed for additional reproductive testing protocols. In order to refine this protocol, a data base is being developed on known reproductive toxicants to determine which endpoints should be utilized for every evaluation and which should be reserved for more specialized evaluations. To date, data have been collected on several chemicals (Gray et al, 1988b) including the estrogenic pesticide methoxychlor; the fungicide carbendazim, which inhibits microtubule formation; a plasticizer, dibutyl phthalate; the miticide lindane; another pesticide, linuron; and the fungicide fenarimol, which inhibits fungal sterol synthesis. From these investigations conducted to date we have found that reproductive organ weights, pubertal landmarks, reproductive behavior, estrous cyclicity, cauda epididymal and testicular sperm counts, testicular histology, and sperm morphology are useful endpoints. A testing protocol that includes these additional endpoints seems to have several advantages over standard multigenerational reproduction tests. The inclusion of these more sensitive endpoints would enable the risk assessor to establish more accurate NOAELs. Moreover, a protocol such as this also requires fewer animals because the statistical precision of these quantitative measures is often greater than that of fertility, an assessment that requires additional animals.

The RACB protocol has been used on a larger number of chemicals than the previously described ART protocol, but dose–response data on measures other than fertility are often lacking. In the RACB (Morrissey et al, 1989), adult male and female mice are exposed to the test chemical for 1 week prior to mating and then during a 14-week period of continuous breeding. This protocol uses larger numbers of animals than is required for other multigenerational studies, with 40 animals per sex in the control group and 20 per sex in each of three dosage groups. If an adverse effect of treatment on fertility is detected during the continuous breeding phase, additional evaluations are considered, including crossover mating and examination of the F_1 animals. In the RACB protocol, only the control and the high-dosage group pups have been examined in the F_1 generation. In RACB, 4 to 5 litters are produced per breeding pair, thereby allowing for the temporal progression of reproductive effects to be noticed and potentially increasing the sensitivity of fertility data. The 14-week exposure in the RACB protocol is sufficient to ensure that all spermatogenic cell types including spermatogonia are exposed, but a shorter duration of treatment prior to the first mating, 7 days in the RACB versus 70 to 80 days in ART, may delay the onset of observed infertility. The fundamental differences between these two protocols include the following: the P_0 generation is used for continuous breeding in RACB, and F_{1a} is used in ART; estrous cycle data are collected for 1–2 weeks in RACB as opposed to at least 40 days in ART; the measurement of organ weights, histologic, and sperm measures is not routinely done at all dosages in RACB, while it is routine in ART; and finally, endocrine, pubertal, and behavioral measures are an integral part of ART, but not the RACB.

MODELS TO ASSESS FEMALE REPRODUCTIVE DYSFUNCTION

General Considerations

Research to date in reproductive toxicology of sexually mature females is much less extensive than comparable efforts in the male animal. This is indeed unfortunate, since it has been estimated that ovulatory problems alone cause the infertility of 40 percent of all infertile couples (Speroff et al, 1983). Exposure to xenobiotics, like lead, in the environment can play a role in certain cases of female-mediated infertility. In addition, there are numerous phytoestrogens and estrogenic fungal mycotoxins present in the environment that may have an additive effect leading to an "estrogen body burden" and affect fertility. Although some of these compounds are only weakly estrogenic, sufficient exposure can render an animal infertile (Gray et al, 1989b). For example, halogenated hydrocarbons commonly used in agriculture are well known for their estrogenic effects in animal studies (Haney, 1985). A classic example, chlordecone, acts in a receptor-mediated manner, as does a metabolite of the pesticide methoxychlor and another pesticide, *o,p′*-DDT. These compounds induce vaginal cornification in female rodents and are uterotropic, increasing uterine weight and protein synthesis. Estrogens also affect behavior, pituitary luteinizing hormone (LH), ovarian steroid production, embryo transport in the oviduct, and uterine decidualization and implantation. For this reason multiple dysfunctions may result. These possible manifestations will be discussed individually later.

Overview of Female Reproductive Physiology

In mammalian species, including rats, mice, hamsters, primates, and humans, in which the adult female is a spontaneous ovulator, the ovarian estrous cycle in rodents or the menstrual cycle in humans and nonhuman primates is controlled, in part, by cyclic fluctuations of the pituitary hormones LH and follicle-stimulating hormone (FSH). The cyclic pulse of LH that induces ovulation is controlled by an LH-releasing hormone (GnRH) signal of hypothalamic origin. The cyclic pacemaker(s) for the GnRH hormone resides within the CNS. The cyclic nature of these hormones in the adult female distinguishes them from males, in which the hormonal secretions are "tonic."

In rodents, developmental exposure to steroid hormones like testosterone or estradiol defeminizes the CNS of the female, such that as adults they are infertile because their hypothalamic hormonal secretions are tonic rather than cyclic. Similar effects have not been obtained in monkeys exposed in utero to androgens (Goy, 1978), or in women exposed prenatally to the estrogen DES. On the other hand, menstrual disturbances are a common complaint from women in the workplace (Mattison, 1985) and of the chemicals that have been tested, many produce alterations of rodent estrous cycles. For ethical and economic reasons, very few chemicals have been tested for their ability to alter menstrual cycles in nonhuman primates. For these reasons and because there is considerable homology between rodent estrous and human menstrual cycle, rodents are the species of choice for the identification of chemicals that may disrupt adult cyclicity.

During a menstrual cycle, oocyte maturation, folliculogenesis, ovulation, and corpus luteum (CL) formation are under the influence of gonadotropins that act on the ovarian cells to initiate morphologic changes, steroidogenesis, and the induction of various receptors. Many of these events are orchestrated similarly in rodents. Factors other than gonadotropins appear to be involved in the initial stimulation of growth of the resting primordial follicle, but once the follicle is committed to growth, the selection of antral, preovulatory follicles is dependent on the LH surge of the preceding cycle (Manson and Kang, 1989). Once follicles are recruited for growth, their maturation is dependent on both FSH and LH stimulation. Androgens, primarily androstenedione, are produced from progesterone by the thecal cells of the ovary in response to LH stimulation. Aromatization of these androgens to estrogen occurs in the granulosa cells under the influence of FSH. Estrogen acts, together with FSH, to differentiate the follicle by increasing gonadotropin receptors and granulosa cell proliferation, as well as the activities of a number of enzymes including aromatase, cholesterol side chain cleavage, and prostaglandin synthetase in the granulosa cell (Richards and Bogvich, 1980). Follicle maturation beyond the preantral stage is dependent on FSH-stimulated formation of an antrum, which consists of mucopolysaccharides, low-molecular-weight plasma proteins, and peptide and steroid hormones

(Mattison and Ross, 1983). The relative amounts and types of constituents can vary depending on ovulatory status.

In response to the LH surge, the primary oocyte in the preovulatory follicle completes meiosis 1 and forms a secondary oocyte blocked at metaphase of meiosis 2. In both humans and rodents, estrogen production increases with growth of the dominant follicle(s) and ultimately stimulates the LH surge responsible for ovulation. The oocyte is ovulated, still at metaphase 2, until fertilization, when the second meiotic division is completed with the extrusion of the second polar body and the formation of both a male and a female pronucleus.

During the onset of luteinization, estrogen levels begin to decline, the estrogen-induced inhibition of the conversion of pregnenolone to progesterone decreases, and continued LH stimulation initiates a rise in serum progesterone levels. These events are followed by vascularization of the granulosa cell layer and granulosa cell transformation into luteal cells, which continue to produce the progesterone needed to prepare the endometrial lining of the uterus for implantation of the conceptus (Haney, 1985). If implantation fails to occur, luteolysis occurs.

An analysis of follicular alterations is difficult because at any one time the sexually mature ovary contains a diverse population of resting, maturing, and mature follicles. This appears especially true if the resting follicles contain damaged primary oocytes, because these may easily go undetected. Unlike the male, in which several cohorts of germ cells form mature sperm, the female has only a single cluster of follicles that is selected for maturity with each cycle. Moreover, after a specified number of mitotic divisions during fetal life, the oogonia enter meiosis, and there is no mechanism to replace oocytes destroyed by toxicants. For this reason, oocyte destruction can shorten the fertile life span of the female. Similarly, if irreparable DNA damage to the oocytes occurs, mutagenic manifestations could occur if fertilization does eventually result. By contrast, if preovulatory or ovulatory follicles are deleteriously affected by xenobiotic exposure during adulthood, the ultimate reproductive effects may be noticed relatively quickly. Moreover, each component of the follicle—the granulosa cell, thecal cell, and oocyte—may be uniquely vulnerable to particular toxicants (Mattison and Ross, 1983; Mattison et al, 1990b). A xenobiotic might increase the progesterone/estradiol or testosterone/estradiol ratios in the granulosa cell stimulated by FSH, or alter steroidogenesis and androgen production by the thecal cells, thereby delaying follicle maturation and ovulation. Finally, the oocyte DNA is susceptible to xenobiotic damage at various times during oocyte maturation and fertilization depending on the agent used (Generoso, 1980). This presumably explains the fact that cancer drugs deplete the pool of oocytes and cause premature menarche in women (Mattison and Ross, 1983).

Although many features of ovarian cyclicity are common to both humans and rodents, certain distinct differences exist that must be considered in toxicologic study (Cooper et al, 1989). For instance, in most mammals, but not in women, follicular growth continues to occur in the luteal phase. Furthermore, luteolysis is thought to be mediated by uterine prostaglandins in rats and nonhuman primates. However, estrogen rather than prostaglandins appears to be the effective luteolytic agent in humans. Other important species distinctions arise during pregnancy (Elger et al, 1990). In rats, progesterone production by the CL maintains pregnancy throughout its duration, while in humans and guinea pigs, placental progesterone plays a significant role in mid and late pregnancy maintenance. Furthermore, progesterone levels must be significantly reduced for parturition to occur in rats; but in humans and guinea pigs, high progesterone levels persist until parturition. Finally, in rats deciduoma formation is highly dependent on the maintenance of a correct progesterone/estrogen ratio, with abortion occurring when the serum estrogen level is increased. Thus, the rat pregnancy is highly vulnerable to estrogen excess or withdrawal, while the guinea pig and human are estrogen-resistant (Elger et al, 1990). By contrast, prostaglandins have abortifacient activity in humans and guinea pigs, but not in rats. Taken together, these findings suggest that the guinea pig may better serve to address certain issues related to toxicant-induced disturbances of mid to late pregnancy.

When the most appropriate species is used, several compartments likely to become altered following xenobiotic exposure can be studied. For instance, impaired follicular growth may be detected by differential morphometry of the ovary (Pederson and Peters, 1968; Mattison et al, 1990a). Follicular fluid may serve as a pool for the bioaccumulation of xenobiotics that may render the oocyte more susceptible to damage (Mattison and Ross, 1983). Steroids and proteins could be profiled following successful aspiration of follicular fluid in rodents. Steroid levels in follicular fluid are currently being profiled for patients undergoing in vitro fertilization (Vanluchene et al, 1991). Following follicular rupture, simultaneous evaluations of oocyte integrity, steroidogenesis, and follicular fluid could be performed on a pool of several adult rat ovaries. CL function is routinely assessed indirectly by the duration of progesterone production in pseudopregnant rats or by the decidual reaction of the uterine endometrium.

In vitro evaluation of ovarian function is complicated by the requisite interaction of the various compartments of the ovary and the change that occurs within the individual compartments during the ovarian cycle. For these reasons, evidence for an effect in vitro must be confirmed using in vivo methodology. Moreover, the absence of an effect in vitro does not preclude the possibility of in vivo toxicity through formation of an active intermediate or metabolite, or through indirect effects mediated via the hypothalamic–pituitary axis.

The hormonal, neural, and molecular mechanisms for female reproductive behaviors in the rat have been the subject of many elaborate investigations and reviews. In fact, this research provides a model system for the study of CNS function and behavioral development. However, our ability to relate this research to human female behavior is limited because of lack of information on the role of these processes in human and nonhuman primate reproductive behaviors.

Female vertebrate reproductive behavior is dependent on the ovarian hormones estrogen and progestins (Beach, 1948; Young, 1961). While most of the basic research has been conducted on rodents, reproductive behavior in female monkeys clearly changes with the menstrual cycle or the steroid hormone injected. While copulation in nonhuman primates is restricted to the period around ovulation, the role of the steroid hormones in women has not been clearly demonstrated, such studies often being fraught with methodologic problems (Pfaff and Schwartz-Giblin, 1988). It can be assumed that in humans steroid hormone behavioral mechanisms are in place, but other social and environmental factors are more important. It is obvious, however, that in human females the onset of sexual behaviors is dependent upon the secretion of steroid hormones during puberty. In women a number of perceptual and sensory behaviors vary with the menstrual cycle (Gandelman, 1983), and some women experience changes in susceptibility to depression, anxiety, and irritability during the premenstrual period (Sanders and Reinisch, 1985). Interestingly, it has been reported that women exposed to DES during prenatal development display reduced feminine sexual behavior (Meyer-Bahlburg et al, 1985). Some other sexually dimorphic behaviors also seem to be masculinized in girls by exposure to androgens during the prenatal stage of life (Ehrhardt and Meyer–Bahlburg, 1981).

In addition to copulatory behaviors, maternal behaviors are also hormonally dependent, in part, in the rat (Numan, 1988), but there have been few such investigations in nonhuman primates. The available evidence suggests that steroid hormones may facilitate maternal responsiveness to alien newborns in monkeys. In humans, the early postpartum period is a time of increased emotional alterations, with mild changes occurring in many women, while a few experience severe disturbances (Sandler, 1978). At present, the human and rodent studies suggest a possible role for serum prolactin on postpartum hostility (Numan, 1988), but this association is far from certain.

We will now present our perspective on the current methodology in the study of toxicology in the mature female, taking this opportunity to propose new experimental approaches.

The Hypothalamic–Pituitary Axis

Toxicants that adversely affect fertility and estrous cyclicity in female rats in vivo have also been shown to alter hypothalamic and pituitary secretions in an in vitro perifusion system. Such systems enable the investigator to examine the specific functions of each organ system independently of the other endocrine organs. They often provide less variable data than those collected in vivo, and they allow one to challenge the system with a variety of stimulants. For example, the triazole fungicide, R151885 delayed the LH surge on proestrus and ovulation by inhibiting ovarian estradiol synthesis in vivo (Middleton et al, 1986). In a follow-up study, Milne and coworkers (1987) studied pituitary function in an in vitro perifusion system after in vivo administration of R151885. They found that pituitaries from rats treated at midday on diestrus 2 produced significantly less LH following GnRH stimulation. In fact, the total LH secretion in response to the GnRH challenges was reduced. This chemical was later found to be an inhibitor of aromatase, preventing the synthesis of estradiol by the ovary.

The Ovary

Rodent ovotoxicity models have been used to study the dose- and stage-dependent and temporal responses associated with oocyte and follicle damage following various xenobiotic exposures. These models utilize morphometry to determine the number and size of oocytes or follicles within ovaries over time following exposure. In studies such as these, serial sections are generated from the ovaries, and sized oocytes (Smith et al, 1990) or follicles (Plowchalk and Mattison, 1991) are enumerated in every 10th or 20th section. The former study identified a species-dependent mechanism of small-oocyte destruction following 4-vinylcyclohexene administration, while the latter study elucidated the active metabolite in cyclophosphamide-induced ovarian toxicity.

Since follicular maturation and pregnancy maintenance in the rat depend on the sequential interaction of ovarian steriod hormones, the profile of steroid hormones produced by ovarian cells in response to LH stimulation might serve to predict lesions in steroidogenesis following in vivo xenobiotic exposure. Recently an effort has been directed toward elucidating these in vivo–induced lesions in cycling rats in vitro by stimulating small pieces of ovarian tissue with hCG over time and quantitating the testosterone, estrogen, and progesterone production (Laskey et al, 1993). This work indicated that the hormone pattern obtained in vitro can be used to characterize in vivo ovarian steroidogenic dysfunction. In addition, these methods provide useful information regarding the potential site of toxicant action. This type of methodology is easily executed, permits the evaluation of a large number of animals rapidly, and appears to be sensitive. Similar technology is now being applied to study the effects of xenobiotics on human ovarian tissue (Weber et al, 1991).

Other examples of experimental models for ovarian study are the in vitro perifused ovary (Peluso, 1990) and the in vitro perfused ovary (Brannstrom and Janson, 1989; Schmidt et al, 1990). Unfortunately these methods have not yet been applied to the study of a reproductive toxicant. Moreover, although elegant and useful in developing our understanding of ovulatory events, these systems have been developed for study of immature, PMSG-primed ovaries, which do not support the full steroidogenic capacity of the mature ovary, that is, estrogen production by aromatization. Thus, further work awaits the applicability of these methods for the study of the mature ovary following toxicant exposure.

Finally, some of the work discussed above was done in the mouse; but it might be advantageous to attempt to develop similar methodology in rats. Rats provide an increased yield of tissue, and this might facilitate research on isolated oocyte, granulosa cell, and luteal cell compartments. If the work of Liu and Hsueh (1986) were extended to the proestrous ovary in adult rats, thecal cells could be isolated by enzymatic disruption of the oocyte- and granulosa cell-free ovary. These thecal cells could then serve as a third cell compartment to include for study in granulosa cell–oocyte cocultures. In addition, thecal cells could be cocultured with granulosa cells to study the periovulatory luteinization of the granulosa cell following toxicant exposure.

Ovulation

Recently, in vivo animal models have been applied to study, both directly and indirectly, ovulatory function. The indirect approach attempts to assess a toxicant-induced delay of ovulation by monitoring the delay in the LH surge in adult female rats (Goldman et al, 1991). In this model female rats are synchronized by ovariectomy followed by estradiol implantation, a procedure known to induce daily afternoon LH surges (Legan and Karsch, 1975). This offers the advantage of selecting for experiment a large cohort of females having similar endocrine status on a given day. Several hours before an anticipated LH surge, the animals were exposed to the pesticides chlordimeform and amitraz, and blood LH levels were assayed over time. Both these compounds produced a pronounced delay in the LH surge, which might be expected to delay ovulation and result in oocyte aging. Both aged (Butcher and Fugo, 1967) and immature (Dodds et al, 1989; Punnonen et al, 1988) oocytes have been associated with infertility and congenital malformations. Thus, any toxicant-induced alteration in the amplitude of the LH surge that hastens or delays ovulation could have significant reproductive consequences. Moreover, exposure of intact females may result in an increased number of resorption sites and/or corpora lutea, suggesting a functional defect from ovulatory delay, and in a sense, making the LH surge a biomarker of toxicant effect in the female.

An adult animal model has been devised in which a population of female rats is first synchronized with an LH-releasing hormone (LHRH) agonist and then implanted with infusion pumps delivering FSH to cause them to superovulate (Hamilton and Armstrong, 1991). The high ovulation rate and oocyte viability achieved with this model make it particularly attractive and provide a direct means of assessing toxicant-induced alterations in ovulatory function in adult rats. Moreover, ovulation in adult rabbits is now being studied using an in vitro-perfused ovary system (Miyazaki et al, 1991). This model seems particularly amenable to study mechanisms of toxicant-perturbed ovulation.

The Oocyte and Fertilization

Oocyte maturation is tightly regulated by an interaction between the oocyte support cells in the follicle, the granulosa and thecal cells, and the oocyte (Buccione et al, 1990). In vivo, oocytes are in meiotic arrest until the onset of the LH surge, when the morphologic change signaling oocyte maturation occurs; that is, germinal vesical breakdown. This is immediately followed by resumption of meiosis 1 and extrusion of the first polar body. Once removed from the follicle, however, immature oocytes frequently progress spontaneously through meiosis 1 to metaphase of meiosis 2. Recently, mature bovine granulosa cells isolated and cocultured with germinal vesicle oocytes successfully inhibited premature germinal vesicle breakdown within the oocyte (Sirard and Bilodeau, 1990). These results exemplify the importance of intimate contact between the granulosa cell layer and the oocyte in normal development of meiotic competence. To date no such model has been applied in rats, but elegant work has evolved to study in vitro-matured mouse oocytes (Eppig and Schroeder, 1989). For this, preantral follicles consisting of oocytes surrounded by two granulosa cell layers are cultured in a bicameral culture system on an extracellular matrix produced from bovine placentas. This system maintains the attachment and integrity of the granulosa cell–oocyte complex and therefore provides the cell–cell interaction required for oocyte growth and meiotic competence. After 10 days of culture in a 5% O_2–5% CO_2 environment, FSH is added to induce germinal vesicle breakdown and meiosis 1 division. The oocytes, surrounded by cumulus cells, are then inseminated with sperm in vitro and either allowed to develop in vitro, or transferred at the two-cell stage to pseudopregnant females. Alternatively, oocytes are allowed to grow in vivo to various sizes and then matured in vitro and inseminated. These approaches lend themselves to either in vitro or in vivo toxicant exposures and offer a means by which the toxicologist could study oocyte maturation, oocyte fertilization ability, and pre-implantation development of the embryo in vitro using relatively few animals. Moreover, xenobiotics shown to be effective in this in vitro model should be predictive of possible problems in the human female since the physiology of these events is similar between the species. This kind of methodology could address more mechanistic concerns such as changes in the distribution of cortical granules in the oocyte cytoplasm (Ducibella et al, 1988) and changes in oocyte chromatin and microtubule assembly (Mattson and Albertini, 1990) during meiotic maturation using fluorescent probes following in vitro xenobiotic perturbation.

Mutagen-induced DNA repair capability in oocytes could be more easily assessed during various stages of oocyte maturation in vitro than in vivo. In female rats, the last scheduled DNA synthesis occurs prior to birth during the premeiotic S phase. Certain mutagens and ultraviolet irradiation can result in an unscheduled DNA synthesis, which can be detected in cells by autoradiography (Pedersen and Brandriff, 1980). This unscheduled synthesis represents DNA excision repair that involves the removal and replacement of damaged bases. Oocytes from the resting or primordial dictyate stage through germinal vesicle breakdown are particularly capable of DNA repair. This is not the case for oocytes involved in meiotic maturation, but the mature oocyte and zygote retain some DNA repair capacity. Thus, the above in vitro model could provide cells from maturational stages with different repair capabilities.

Corpus Luteum

An in vitro ovarian perfusion system has been developed to study the function of the rabbit CL during pseudopregnancy (Dharmarajan et al, 1988). This model is well suited to monitor toxicant-induced alterations in progesterone production by corpora lutea. These ovaries could be exposed to a toxicant in vitro via the perfusing medium or could be perfused following an in vivo toxicant exposure. In addition, after perfuse-fixation, morphologic evaluations can be performed on the toxicant-exposed tissue after perfusion to determine the extent of cytotoxicity, if any. Bovine CL cells are being successfully harvested and cultured (Stirling et al, 1991) but proven rodent models are not yet available.

The Uterus

The blastocyst and the uterus must both be ready for successful implantation to occur. Thus, the synchronous differentiation of the blastocyst and uterus is critical. Transfer of zygotes that are more than a few days out of phase with the uterus can result in a high rate of failure of implantation. The environment in which the preimplantation embryo develops is controlled by hormonal mechanisms. For example, in the absence of gonadal steroids, the embryo of the rat can develop from the morula to the blastocyst after ovariectomy, but progesterone increases survival of the blastocyst prior to implantation and stimulates mitotic activity in the embryo.

Implantation involves the establishment of contact between the trophoblast of the blastocyst and the uterine wall, erosion of the uterine epithelium, and invasion and decidualization of the stroma by the blastocyst, so that it is embedded in the uterine wall. The preparation of the endometrium for implantation is under hormonal control and is dependent on a complicated interaction between progesterone and estrogen. In rodents, mating causes maintenance of the CL and progesterone secretion. On day 2 following fertilization, estradiol concentrations increase and remain elevated until day 8. Progesterone is required for preparation of the uterus for implantation and in rats, the presence of estrogen, subsequent to the progesterone, causes a short period of receptivity for the blastocyst. Treatments that alter the internal hormonal environment, inhibit protein synthesis, or inhibit mitosis or cell differentiation can block implantation. Implantation on day 4 induces stromal differentiation, the so-called decidual cell reaction, or decidualization.

Experimental models to assess toxicant-induced alterations in implantation have been the focus of much effort in the last few years. As indicated above, successful implantation depends on correct orchestration of placental hormone production, synchronous embryo arrival, and the decidual reaction. Model compounds have been used to demonstrate the role toxicants may play in each of these processes (Cummings and Gray, 1987, 1989; Cummings, 1990). First an early pregnancy protocol was used to establish that dose-dependent effects in pregnancy maintenance in the rat ensued from a short exposure regimen from days 1 through 8 of pregnancy. Thus, this dosing period includes implantation and development of the decidua, the maternal component of the placenta. On day 9 the animals are necropsied and multiple endpoints are evaluated, including the ovarian and uterine weights, as well as the number of corpora lutea and implantation sites, and serum hormone levels. Later, this protocol was modified to distinguish between preimplantation and postimplantation pregnancy losses. This simply involves forming two groups of animals following mating. One group is dosed during days 1 to 3 of pregnancy to evaluate preimplantation losses. Another group is dosed during days 4 to 8 of pregnancy to assess postimplantation pregnancy losses. Again, on day 9 all relevant endpoints are measured. An animal model was also developed to study the effect of xenobiotics on the decidual response. For this, the decidual reaction was induced by uterine trauma on day 4 of pseudopregnancy in rats stimulated to become pseudopregnant on day 0, estrus. Finally, alterations in the synchronicity of embryo arrival and uterine receptivity can also be assessed by dosing females during the first 3 days of pregnancy and performing necropsy at regular intervals throughout these first 3 days. The number of embryos recovered from the oviduct and uterus over time are enumerated.

Recently, implantation delay and antideciduogenic activities were studied following exposure to the antiandrogen, hydroxyflutamide (Chandrasekhar et al, 1990). This model, however, used immature rats that were induced with PMSG to ovulate synchronously. It is well known that this procedure supports normal embryonic implantation and development (Nuti et al, 1975). It offers the obvious advantage of facilitating synchrony among a large group of animals. Again, supplemental hormones were given during the course of the study, and uterine trauma was used. Implantation sites were visualized by Evans blue dye localization. These areas of increased vascularity were counted, excised, and weighed. The results indicated that hydroxyflutamide delayed implantation and fetal development in pregnant rats and inhibited the formation of the decidua in pseudopregnant rats.

The Oviduct

Descriptions of the pathologic processes of the oviducts of animals other than humans are not common. Developmental abnormalities include agenesis, segmental aplasia, and hypoplasia. Oviductal agenesis is often associated with agenesis or segmental aplasia of the ipsilateral uterine horn. Hypoplasia of otherwise well formed oviducts results most commonly from a lack of estrogen stimulation, and for this reason, this condition may not be recognized until after puberty. Cysts occasionally occur in the oviducts. Paraovarian cysts occur with some frequency in laboratory species and arise in the vestigial remnants of the mesonephric tubules. Functional disorders include atrophy due to the lack of estrogens and hyperplasia of the oviductal epithelium from prolonged exogenous estrogenic stimulation. Hyperplastic oviducts are enlarged and secrete increased quantities of fluid. Ectopic pregnancy, in the oviduct, is a rare event in all mammals except humans. It has been estimated that one ectopic pregnancy occurs in every 64 live human births and that 96 percent of these occur in the oviduct.

After mating, vaginal and uterine contractions are critical in the transport of spermatozoa from the vagina. Vaginal contractions and, in rodents, the copulatory plug facilitate the movement of sperm through the cervix. It seems that prostaglandins, possibly from the seminal fluid, stimulate uterine activity. It is also known that in the rat, if the female does not receive sufficient cervical stimulation from mating, then sperm transport is also inhibited. During transport sperm undergo critical maturational changes such as capacitation and the acrosome reaction, requisite processes of fertilization. In rodents only motile sperm can ascend the oviduct. In the oviduct, sperm are distributed in two ways. A few sperm are rapidly transported to the ampullary region, while a second group, which actually fertilize the eggs, remain in the proximal isthmus until the time of ovulation.

The transport of the preimplantation embryo in most mammalian species is fairly constant, taking 3 to 4 days. Transport of the embryo through the oviduct is controlled by the contractility of the smooth musculature, the activity of the cilia, and the flow of oviductal secretion. A great many agents that affect motility of the oviduct and sperm transport have been described. These include steroids, weakly estrogenic pesticides, cyclic nucleotides, prostaglandins, catecholamines, and assorted peptides. In women, low doses of administered estrogens accelerate ovum transport, while high doses of estrogens or progestins delay transport and increase the incidence of ectopic pregnancies (Harper, 1988).

Fertilization and early embryonic development begin in the oviduct, and a reliable means of confirming ovulation is the retrograde flushing of the oviducts and collection of the eggs. The eggs can then be evaluated for viability to determine if fertilization took place, and the embryos can be staged. Recently an in vitro system was developed to coculture bovine oviductal epithelial cells with sperm (Ellington et al, 1991). This system was capable of producing capacitated, acrosome-reacted sperm capable of fertilizing bovine oocytes in vitro. A model such as this should facilitate a better understanding of the intimate interaction between oviductal epithelial cells and sperm requisite for normal fertilization and might be used in a toxicologic study to better identify and characterize preimplantation losses.

The Embryo

Work that questions the DNA repair capacity of the zygote is provided by Generoso and coworkers (1988). In this study, one group of female mice was treated with a mutagen before mating to assess effects on preovulatory oocytes. These females then were mated within 4 days of treatment. Another group of females was treated 1, 6, 9, and 25 h following a restricted, 30-min, mating period, coinciding with the time of sperm entry, early pronuclear stage, pronuclear DNA synthesis, and two-cell zygotes, respectively. Implants were categorized according to the time of death: implantation, midgestation, and late gestation. The results indicated that all stages of exposure, including zygotic, are susceptible to embryopathy, and that both the affected stage and the time of death vary among mutagens. Ethyl nitrosourea and triethylene melamine, for example, caused preimplantation death regardless of the stage at which females were exposed. Ethylene oxide and ethyl methanesulphonate, on the other hand, caused midgestational and late fetal death when female mice were exposed at the time of fertilization or during the early pronuclear stage. The site of action of this second group of mutagens was later investigated using the techniques of reciprocal zygote transfer and cytogenetic analysis (Katoh et al, 1989). The former technique is used to determine whether there is a direct effect on the zygote or whether the effect is mediated through the pseudopregnant foster mother. For this, separate groups of zygotes are transferred into untreated and treated foster mothers. The latter technique is performed on pronuclear metaphase chromosomes, and metaphase chromosomes of early cleavage embryos, to identify any structural or numerical aberrations. Their results suggested a direct zygotic effect and a novel, nonmutational alteration in gene expression as an underlying embryotoxic mechanism for these two compounds. Recently, doxorubicin (Adriamycin) and platinol, compounds that induce dominant lethality (genetic postimplantation loss) when females are treated but not when males are treated, were similarly studied (Katoh et al, 1990). These compounds cause increased embryonic mortality when females are exposed during meiosis 1 prior to mating. Reciprocal transfer eliminated maternal toxicity as a factor in the embryotoxicity, and cytogenetic analysis revealed distinct metaphase chromosome aberrations. Thus, this methodology provided confirmation of the suspected dominant lethal oocyte lesion.

Typically, embryos from the rat exhibit a block at the two-cell stage in culture. This obviously precludes in vitro study of preimplantation development. Recently, rat embryos have been maintained without degeneration through the eight-cell stage by coculturing two-cell embryos with uterine epithelial and stromal cells (Pampfer et al, 1990). Numerous laboratories have been working to develop a model that will confer a high incidence of blastocyst formation and posthatch development. The published methodology has been both complicated and diverse. In addition, two-cell embryos have been cocultured with oviductal cells in mice (Lavranos and Seamark, 1989), sows (White et al, 1989), cows (Ellington et al, 1990), and rabbits (Carney et al, 1990). One study has examined the effects of both oviductal and uterine cells on cocultured embryos (Sakkas and Trounson, 1990). From the results of these studies, the ultimate model might involve coculture of two-cell embryos, first with plaques of oviductal cells and then with plaques of uterine cells, each on an extracellular matrix in a bicameral environment, in a reduced-oxygen atmosphere, while providing adequate hormonal stimulation.

Again, these more elaborate experimental approaches are recommended only after appropriate in vivo testing has indicated that such measures are feasible to further elucidate the active metabolite or mechanism and site of action of a xenobiotic. Once fully developed for use, this model could have great utility in the in vitro identification of preimplantation defects. Using various fluorescent probes such as rhodamine 123, antitubulin antibodies, and rhodamine–phalloidine, the embryonic distribution of mitochondria, microtubules, and microfilaments can be ascertained (Batten et al, 1987; Albertini, 1987). Thus, subtle toxicant-induced embryonic defects during the preimplantation period might be better identified. If these techniques were used with highly sophisticated computer-based image analysis technology incorporating confocal microscopy and laser-generated fluorescence, the sensitivity and resolution of these measurements would be increased greatly. Although an extremely costly resource, it might prove to be an invaluable means of studying the course of events from the immature oocyte through the expanded blastocyst from either in vivo–derived or in vitro–generated tissue.

MODELS TO ASSESS REPRODUCTIVE DYSFUNCTION IN MALES

General Considerations

Like the female, the male requires highly coordinated events of the central and peripheral nervous systems, the hypothalamic–pituitary axis, and both gonadal and extragonadal systems to reproduce successfully. Sperm production in a healthy human is only 20 percent as efficient, on a per-gram-testis basis, as sperm production in rodents, rabbits, dogs, and monkeys (Amann, 1986). However, the total number of sperm in the ejaculate is similar in the rat and the human. In men with low sperm counts, a slight reduction in the ejaculated sperm number may result in infertility. It is well known that the number of sperm ejaculated by men is influenced by drug treatment or pesticide exposure, and many men treated with chemotherapeutic agents are azoospermic (Schilsky et al, 1980). Other xenobiotics, such as the environmental pesticide DBCP, can render industrial workers infertile (Levine et al, 1981). Taken together this information suggests that a considerable potential exists for toxicant-induced infertility in men.

In addition to direct gonadal effects, many commonly used drugs cause erectile dysfunction and alter libido in men. Workers exposed to manganese oxides in the air suffer from impotence and, consequently, have greater difficulty having children (Lauwerys et al, 1985). In fact, Benson (1988) proposed that "medication may be the single most common cause of impotence in the United States." Virtually every class of drugs has been reported to cause impotence through effects on the CNS, elevated serum prolactin levels, antiandrogenic effects, and anticholinergic or adrenergic effect. Other drugs, like yohimbine or testosterone, may be useful in the clinical treatment of impotence under certain conditions. Thus, a male reproductive toxicant can exert its effect either by perturbing the CNS, the peripheral nervous system, or the hormonal balance of the reproductive system, or by acting directly on germ cells or sperm to effect a change in spermatogenesis or sperm maturation.

Overview of Male Reproductive Physiology

The process of spermatogenesis involves coordination of endocrine, autocrine, and paracrine influences in both the interstitium and the seminiferous epithelium of the testis. This in turn mediates the temporal and spatial proliferation and differentiation of spermatogonia into spermatids and their release into the lumen of the seminiferous tubule every 60 days in the rat and every 74 days in humans. For recent discussions of this complex process see Amann (1989b), Zenick and Clegg (1989), and Johnson (1991). Briefly, the interstitium between the seminiferous tubules contains the Leydig cells, which produce the testosterone needed to maintain spermatogenesis and sperm maturation. The seminiferous epithelium is divided into basal and adluminal compartments by tight junctional complexes formed between adjacent Sertoli cells. In the basal compartment, spermatogonia and preleptotene (DNA-synthesizing) spermatocytes are found. In the adluminal compartment, the remaining meiotic spermatocytes and spermatids are found. Sertoli cells promote the differentiation of germ cells in this latter compartment, and tight junctions between Sertoli cells form a blood–testis barrier that denies access of many compounds to the adluminal compartment. Thus, effects of xenobiotics that affect spermatids and spermatocytes might be mediated through the Sertoli cell.

Spermatogenesis begins with the mitotic differentiation from A_1 (stem) spermatogonia through B spermatogonia. The series of mitotic divisions resulting from a single stem spermatogonium in a particular region of the epithelium remains in synchrony through intercellular connecting bridges. Thus, a cohort of identical cells is formed with the mitotic division of each generation, and each generation appears in succession in the adluminal direction over time. Furthermore, as some stem spermatogonia commit to differentiation and begin mitotic division, others divide to repopulate the stem cell population. In addition, in rodents and perhaps humans, there exists a subpopulation of A_0 (reserve) nondividing spermatogonia that appear to be better protected from xenobiotic damage and are capable of dividing to replenish the stem cell population upon toxic insult. Thus, unlike the situation in the female ovary, where toxic insult at an early age may completely destroy the oocyte population, spermatogenesis is rarely irreversibly destroyed.

The time it takes for A_1 spermatogonia in the most basal layer of the epithelium to complete the mitotic divisions and form B spermatogonia, which appear in the next adluminal layer, marks the duration of a spermatogenic cycle, 12.9 and 16 days in rat and human, respectively. At this interval, within a specific region of the seminiferous epithelium, the successive cohorts of more mature generations are established from the basement membrane to the lumen of the seminiferous epithelium. Thus, aside from this regional separation in a cross section of the epithelium, these cohorts are temporally separated from one another by a constant 16 days in humans. The profile of generations seen in a given cross section of the seminiferous tubule represents a particular cellular association or stage. Just downstream from this region of the seminiferous tubule, one finds a different profile in cross section representing a different stage. This difference in spatial relationship reflects the fact that A_1 to B spermatogonial differentiation is not occurring simultaneously along the length of the seminiferous tubule. Think of it as the staggering of the start and subsequent positioning in a relay race. The difference here is that the elapsed time is identical for every cell that starts the race regardless of positioning, 74 days in humans. This represents 4½ 16-day generations (legs of the race). In the rodent, the entire circumference of a particular cross section depicts a single stage. In humans, however, several stages are found along the circumference of a tubule cross section, much like slices in a pie. Regardless of their spatial arrangement, the existence of these spatially different stages at a given point in time has toxicologic implications. Although an entire cohort of cells, pachytene spermatocytes, for example, may be altered by a particular toxicant, just downstream from a stage containing pachytene spermatocytes, a different stage can be found that does not contain these susceptible cells. Thus, if the exposure to the toxicant is acute, one may find only a slight, transient decline in spermatogenesis. With increasing duration of exposure, the effect, although still transient, will increase in magnitude.

The last spermatogonial division results in the formation of preleptotene spermatocytes, which undergo the DNA synthesis to initiate prophase of meiosis 1. This cell type is followed by other primary spermatocytes in various phases of prophase. These are the leptotene, zygotene, pachytene, and diplotene spermatocytes, which exemplify chromatin condensation, pairing, recombination, and chiasmata, respectively. The pachytene spermatocyte has the longest life of these primary spermatocytes. This cell type appears to have a unique reliance on lactate for energy and is responsible for much of the RNA and structural protein synthesis that occurs during spermatogenesis. These features appear to render the pachytene spermatocyte uniquely susceptible to many toxicants. Following diakinesis (synaptonemal dissolution and chromosome recondensation) there is cell division to separate homologous chromosomes to form the secondary spermatocyte, which contains a haploid number of duplicated chromosomes. With the second meiotic division, the sister chromatids are separated and spermatids containing a haploid number of single chromosomes are formed.

The morphologic differentiation of spermatids, spermiogenesis, then begins. Morphogenesis includes nuclear condensation and elongation, acrosome development, and flagellum formation. During nuclear condensation and elongation, histone proteins are replaced with protamines. Together, these events act to compact the genetic material and cause transcription to cease. The Golgi apparatus of the spermatid differentiates into the acrosome of the spermatozoan released into the lumen of the seminiferous epithelium upon spermiation. This transformation has been categorized into four morphologically distinct phases: Golgi, cap, acrosomal, and maturational. These phases of spermatid differentiation can be found in distinct stages of the spermatogenic cell cycle and are routinely used to stage cross sections of seminiferous tubules. In the rat and human, the number of stages that can be identified based on the morphologic differentiation of the acrosome is 14 and 6, respectively. As discussed later, this is an invaluable means of assessing spermatogenic function following a toxic insult to the testis. As mentioned above, spermatogenesis in the human male is inherently less efficient than spermatogenesis in the rat. This may be due in part to the fact that the germ cell astresia that occurs during spermatogenesis in humans is much greater than that which occurs in the rat (Johnson, 1986). Thus, it would seem that any xenobiotic that decreases spermatogenesis in the rat would almost certainly impair spermatogenesis in humans.

When newly formed spermatozoa leave the testis and enter the epididymis, they are incapable of progressive motility, fertility, and the ability to induce normal embryonic development. Sperm acquire these functional attributes during epididymal transit through a variety of biochemical changes that are collectively referred to as the process of sperm maturation. To achieve full maturity during its epididymal sojourn, the sperm must modify intracellular constituents involved in energy transduction during progressive motility. In addition, the plasma membrane must be modified for motility, zona recognition, and acrosomal stabilization and destabilization. Finally, nuclear chromatin is further stabilized by additional disulfide bonding. The reader is encouraged to refer to recent discussions of epididymal sperm maturation (Robaire and Hermo, 1988; Eddy, 1988).

Sperm maturation occurs in the proximal regions of the epididymis in all mammals studied. This process appears to be mediated to a large extent by androgen-dependent proteins that are secreted by the epididymal epithelium in the proximal epididymis (Orgebin-Crist and Jahad, 1978) and the subsequent association of these secretions with the luminal sperm (Klinefelter and Hamilton, 1985). This could certainly account for the sperm membrane changes mentioned above. Aside from these biochemical alterations, the translocation of the cytoplasmic droplet from the neck of immotile sperm to the distal midpiece of progressively motile sperm is a morphologic indicator of normal sperm maturation. Sperm transit through the proximal (caput and corpus) epididymis in 2 and 3 days in humans and rats, respectively. In both species the sperm that reach the distal region of the epididymal duct, the cauda, have the capacity for progressive motility and fertility. Sperm are stored and maintained in the cauda until they are removed by copulation or voided in the urine (Amann, 1970). Therefore, sperm recovered from the cauda epididymis following xenobiotic exposure, that were immotile and possessed a cytoplasmic droplet in the neck would be considered infertile.

Accessory sex gland function is also androgen-dependent, and a xenobiotic that interferes with this process would most likely affect seminal characteristics. Unfortunately, ejaculated sperm and seminal fluid are difficult to obtain from rodents used in most routine toxicology tests. For this reason, the rabbit has been proposed as the animal model of choice to evaluate seminal characteristics (Williams et al, 1990) since rabbit ejaculates are readily obtained. Typically, however, rodent epididymal or uterine sperm are used for morphology and motility assessments. We will discuss the utility of these sperm endpoints in detail later.

As indicated earlier, fertility in rodents is often an insensitive endpoint in conventional animal models, and more sensitive approaches are not routinely practiced. Indirect methods of assessing fertilized capacity of sperm are difficult to establish due to the continuous process of modification sperm undergo in both the male and the female reproductive tract to fertilize an egg (Amann, 1988; Eddy and O'Brien, 1989). Although in vitro methods of assessing the fertilizability of sperm bypass many of the environments sperm normally encounter, they provide at least relative indications of the capacity of untreated and xenobiotic-exposed sperm to undergo the acrosome reaction, binding to the vitelline membrane of the egg, and the formation of male and female pronuclei.

Research on mammalian reproductive behavior, including the roles of the central and peripheral nervous systems, is based primarily on studies in rodents. Current ethics and economics make research on nonhuman primates and other larger animals more difficult to conduct, and for this reason, rodents will continue to provide much of the data in this field in the future. There are, however, sufficient data on reproductive behavior in men and nonhuman primates to demonstrate that even though the patterns of copulation differ among mammalian species, many of the underlying mechanisms are similar. For example, bilateral lesions of the medial preoptic–anterior hypothalamic area of the brain eliminate mating behavior in sexually experienced males in every mammalian species that has been studied, including rodents and rhesus monkeys (Sachs and Meisel, 1988). In fact, such lesions also block mating behavior in lizards and fish. Manipulation of the monoamine systems with lisuride overcomes the effects of these lesions, while administration of LHRH or naloxone are without effect (Hansen and Gummesson, 1982). Electric stimulation of the medial preoptic area in monkeys elicits penile erection and occasionally ejaculation (MacLean et al, 1963). In addition, castration inhibits mating behavior in male rodents and nonrodents, while androgen replacement in hypogonadal men increases penile erectability and stimulates sexual activity (Davidson et al, 1979).

In other cases, effects seen in rodents may not be as important in nonrodent species. Olfactory bulbectomy inhibits mating in rats but has no effect on copulation in male cats, dogs, or rhesus monkeys. Furthermore, male rats treated with an inhibitor of aromatization fail to mate (Gray et al, 1991), while men exposed to aromatase inhibitors are unaffected (Gooren, 1985).

In addition to the effects of damage to the brain on male mating behavior, damage to the spinal cord interferes with the afferent and efferent paths between the cord and peripheral structures, eliminates components of mating behavior and prevents penile erection. The general plan of the penis is similar in mammalian species, and the mechanical basis for erection is a combination of vascular and muscle actions dependent upon innervation from the pudendal, pelvic, and hypogastric nerves. However, unlike rodents, humans do not possess an os penis or a retractor penis muscle. The complexity of the interactions between the reproductive structures and the central and peripheral nervous systems, required for successful copulation, renders it highly susceptible to chemical disruption. As indicated earlier, Benson (1988) reported that "medication may be the single most common cause of impotence in the United States," and almost every class of drugs has been reported to cause impotence. However, the lack of a universally accepted animal model for penile erection and a lack of information about the basic physiology of this process in humans hinders research in this area. In general, adrenergic agonists can suppress or eliminate copulation, while antagonists can activate mating. Increased dopaminergic activity accelerates copulation, while, in contrast, increased serotonergic activity inhibits it (Sachs and Meisel, 1988).

We will now begin to discuss current animal models in use in male reproductive toxicology, as well as some we will propose for use, beginning with the endocrine system, which drives, as discussed above, the processes of spermatogenesis, sperm maturation, and sexual behavior.

The Hypothalamic–Pituitary Axis

The endocrine regulation of the male reproductive system has been the subject of recent review (Everett, 1988). The paracrine and autocrine influences believed to play various roles in the normal function of the testis were also discussed (Heindel and Treinen, 1989). We also refer the reader to a review by Chapin and Williams (1989), for current theoretic details of endocrine regulation in the testis. For some time now we have known that following adrenergic stimulation, hypothalamic pulses of GnRH are released into the hypothalamic–pituitary portal blood, and these pulses in turn stimulate the pituitary to release FSH and LH (Barraclough and Wise, 1982). Recently chlordimeform, an insecticide formerly used in the United States, was shown to decrease serum gonadotropin and testosterone levels by perturbing the hypothalamic release of GnRH (Goldman et al, 1990). In addition to measuring circulating blood hormone levels, researchers in this study cultured hypothalamic and pituitary fragments obtained from treated rats, with maximally stimulating concentrations of catecholamine and GnRH, respectively, to assess function. Since hypothalamic but not pituitary hormone production was compromised, their results indicate that this xenobiotic interferes with the adrenergic stimulation of GnRH release. Similarly, perifusion systems have been utilized to study the release of inhibin and FSH from pituitary cells incubated with GnRH and Sertoli cell-conditioned medium, respectively (Jakubowiak et al, 1989, 1990; Janecki et al, 1984). These perifusion studies were carried out on immature pituitary cells, but this technology could also be applied to mature cells; and conceptually the latter study lends itself to a coculture system that could provide answers to questions regarding cell–cell interactions. Initially this notion may seem abstract, but surely the key to understanding the mechanism of action of many toxicants lies in improving our knowledge of cell–cell interactions, and we will continue to allude to examples of this concept throughout our discussion.

The Leydig Cell

Although the etiology of idiopathic oligospermia in men varies (Nankin et al, 1977), an endocrine imbalance is usually associated with the observed decrease in spermatogenesis. The endocrine requirements for the quantitative maintenance of spermatogenesis may be different in rats and men. Testosterone alone maintains spermatogenesis in rats (Awoniyi et al, 1989), whereas both testosterone and FSH appear to be required for quantitative maintenance of spermatogenesis in men (Matsumoto and Bremner, 1989). Regardless of the exact differences that may exist, the need for testosterone is unequivocal. Thus, the measurement of serum testosterone levels and assessment of Leydig cell function may be pivotal to the diagnosis of toxicant-induced infertility. Indeed, a correlation between Leydig cell number and steroidogenic capacity was shown to exist in oligospermic men (Oshima et al, 1977).

Subsequent to this, Ewing et al (1981) advanced our capability to detect xenobiotic effects on Leydig cell structure and function in rats and rabbits. An in vitro perfused testis model was adopted that maintains the normal cytoarchitecture of the testis during vascular perfusion while maximally stimulating the Leydig cell with LH. During a 3-h perfusion, effluent medium is collected and assayed by high-performance liquid chromatography (HPLC) for testosterone and precursor substrates produced. Known inhibitors of specific steroidogenic reactions were used to validate the method. Immediately following the perfusion, the testis can be perfuse-fixed for subsequent morphologic examination. This methodology was recently used in a study to determine the mechanism by which 2,3,7,8-tetrachlorodibenzo-*p*-dioxin (TCDD) decreases testosterone production (Kleeman et al, 1990). This study suggested that the decreased serum testosterone levels in response to TCDD treatment in vivo result from an inability of the hCG-stimulated testis to form pregnenolone.

Although the in vitro perfused testis model offers several distinct advantages, it is a more labor-intensive procedure than many investigators care to adopt. Frequently, radiolabeling studies done on testicular homogenates from the rat (Kelce et al, 1990) or microsomal preparations from humans (Albertson et al, 1988; Higashi et al, 1990) can provide desired mechanistic data since homologous steroidogenic enzymes exist in the Leydig cells of both rodents and animals. The loss of Leydig cell integrity inherent in these procedures, however, raises natural concerns.

Today, the more common methods to assess Leydig cell viability and steroidogenic capacity following xenobiotic exposure involve in vitro preparations of interstitial (Phelps and Laskey, 1989; Laskey and Phelps, 1991) or highly purified (Klinefelter et al, 1991a) Leydig cells. In the former study, cadmium produced age-related damage to Leydig cells. An in vitro and in vivo correlation was made measuring hCG-stimulated interstitial cell testosterone production and hCG-stimulated serum testosterone, respectively. The in vitro assessment of Leydig cell function has often proved to be more sensitive than the serum testosterone measurement. Leydig cells purified to 98 percent homogeneity by a multistep procedure (Klinefelter et al, 1987) were used to determine the dose–response, site-of-action, and morphologic integrity following in vitro exposure to the Leydig cell toxicant ethane dimethanesulphonate (EDS). After determining the dose–response of Leydig cells in an interstitial cell preparation following in vivo exposure, purified Leydig cells were obtained from animals given the estimated ED_{50} dose and then were similarly evaluated for the site-of-action and morphologic integrity. The results clearly demonstrated that for both in vitro and in vivo exposures, the decrease in hCG-stimulated testosterone production results from a lesion between the second messenger cyclic AMP and the enzyme that converts cholesterol to pregnenolone. Thus, these results are similar to those obtained in the TCDD study above. Furthermore, at the EC_{50} and ED_{50} doses, for in vitro and in vivo exposures, respectively, there appears to be no apparent morphologic change in Leydig cell integrity.

Leydig cells purified as above were also used recently to study the interaction of Leydig cells and Sertoli cells during in vitro exposure to the testicular toxicant tri-*o*-cresyl phosphate (Chapin et al, 1990). This study demonstrated that metabolism of the xenobiotic by the Leydig cell was requisite for the subsequent Sertoli cell toxicity. These results are illustrative of the complex interaction that exists between Leydig cells and Sertoli cells in a normal adult testis (Sharpe et al, 1990).

The Seminiferous Tubule

A xenobiotic that perturbs any component of the seminiferous epithelium, Sertoli cell, or germ cell may produce infertility at some point in time following exposure. There seems to be no standard animal model to determine testicular effects. The existing diversity of experimental approaches makes it impossible for us to present a comprehensive, clear picture. Therefore, we will present the approach that was used for 1,3-dinitrobenzene (DNB), a chemical often encountered in the manufacturing of plastics, dyes, and explosives. This compound is highly toxic and readily absorbed through human skin, and in rats it produced testicular atrophy upon exposure in drinking water. An initial study to determine whether testicular toxicity was associated with infertility (Linder et al, 1986b) examined the effects at four gavage doses, including vehicle, over a subchronic 12-week exposure. Infertility was found in the high-dose group, but a significant decrease in homogenization-resistant testicular spermatids was observed in the intermediate-dose group. This was the only endpoint affected beneath the high dose and serves to highlight the utility of this long-standing, simple measure (Amann and Lambiase, 1969; Robb et al, 1978; Meistrich, 1989). In an acute study in which a single dose of DNB was used, quantitative histologic changes (Hess et al, 1988) were compared with changes in sperm quality and fertilizing ability (Linder et al, 1988). The sequelae of effects that follow a single exposure allow the pathogenesis of the lesion to be characterized (Perreault et al, 1989).

Histologic criteria used to assess spermatogenesis have been reviewed (Berndtson, 1977; Amann and Berndtson, 1986; Hess et al, 1990). In the acute DNB study, seminiferous tubule diameter was measured for only stage VII tubules, since this is the most prevalent stage. In this study, 10 tubules were counted per male. More recently, 15 tubular cross sections have been recommended (Berndtson et al, 1989). Two hundred tubules from each animal were qualitatively classified as normal or abnormal (regressive, atrophied, etc.). In addition, the number of step 19 spermatids in stages later than VIII, an indication of delayed spermiation, were enumerated. One hundred tubules per testis were evaluated for stage frequency, and stages were grouped (I–IV, V–VIII, IX–XIII, XIV) for more accurate identification of the toxicant-treated testes. However, since then, Hess and coworkers (1990) have recommended evaluating 200 tubule cross sections per testis if the frequency of all stages of spermatogenesis is to be determined. Finally, cell-specific alterations were determined by counting the number of cells of each type in 3 tubules from each stage of spermatogenesis in three rats. Counts are corrected for section thickness and expressed per Sertoli cell nucleolus within a tubule cross section.

The major morphologic effects observed within the first 2 days after dosing were pachytene spermatocyte degeneration, exfoliation of round spermatids, and retention of step 19 spermatids. Furthermore, the frequency of spermatogenic stages was altered and, if normalcy was not established in 3 weeks, tubule atrophy occurred. After 5 weeks, when tubular atrophy was evident, the fertilizing ability of available sperm diminished as evidenced by a decreased number of fertilized ova that could be flushed from the oviduct of mated females. Coincident with this was the recovery of morphologically abnormal sperm from the cauda epididymis. Fertilizing ability eventually returned to a control level in many, but not all, of the animals. Presumably these were animals that regained the normal kinetics of spermatogenesis.

More recently, the effects of DNB on pachytene spermatocytes and round and elongating spermatids have also been established by flow cytometry in both rats and mice (Evenson et al, 1989a,b). In these studies, testicular cell preparations are stained with the metachromatic fluorochrome acridine orange, which binds to nucleic acid. The intensity of the red and green fluorescence emitted following blue laser light excitation is a reflection of the amounts of single- and double-stranded nucleic acid, respectively. The relative amount of fluorescence is a measure of the RNA, single-stranded DNA, or double-stranded DNA content in a cell, which serves to distinguish specific testicular cell types (for a thorough review refer to Evenson, 1989). Similar technology has also been applied recently to the fungicide methylbenzimidazole-2-yl carbamate (Evenson et al, 1987), the chemotherapeutic agent thiotepa (Evenson et al, 1986a), and the immunosuppressive agent cyclophosphamide (Toppari et al, 1990). This last study compared the results obtained with flow cytometry with those obtained using morphometry and found that while morphometry proved to be a more sensitive method of evaluation, flow cytometry gave similar results and was much less labor-intensive.

Clearly, a definitive, accurate determination of the affected cell type(s) in the seminiferous epithelium is possible only with careful morphologic evaluation of the seminiferous epithelium over time. Although time-consuming and tedious, this type of work is feasible in an animal such as the rat. If a given toxicant alters the kinetics of either spermatogenesis and/or epididymal transit, the affected cell type might be quite different than one would predict based on published cell cycle and transit times. Furthermore, an accurate prediction of the affected cell is useful for pursuing mechanistic studies on the toxicant.

The effect of DNB on pachytene spermatocytes and spermatids suggested that certain aspects of Sertoli cell nurturing might be compromised. Initial in vitro work with DNB identified Sertoli cell vacuolization and germ cell exfoliation in coculture (Foster et al, 1986, 1987). Subsequent work using

Sertoli cell cultures showed that the production of lactate and pyruvate, substrates required for germ cell survival, was altered in a dose-dependent manner following DNB exposure (Williams and Foster, 1988). Recently, data suggesting the pathway of metabolic activation of DNB have been provided by the use of these Sertoli cell–germ cell cocultures (Cave and Foster, 1990).

The interaction between Sertoli cells and germ cells has been intensively studied in vitro (Onoda and Djakiew, 1990; Janecki et al, 1988; O'Brien et al, 1986; Hadley et al, 1985). Virtually all the in vitro models in use today, including those discussed above in reference to DNB, utilize Sertoli cells from immature animals. Castellon et al (1989) demonstrated age-dependent differences in the secretion of both transferrin and androgen-binding protein by Sertoli cells from 10-, 18-, and 26-day-old rats. More recently, it has been shown that lactate production in cultures of human Sertoli cells is a function of the age of the men from whom the Sertoli cells were derived (Drosdowsky et al, 1991). If such age-related differences in the in vitro response of Sertoli cells exist, one might expect that with the acquisition of fully differentiated cell function upon sexual maturation, Sertoli cell responsivity would undergo further change. A recent toxicologic study demonstrated support for this notion when immunoactive inhibin production from cultures of seminiferous tubules from adult rats was greater and less variable in response to DNB exposure than cultures of Sertoli cells from immature rats (Allenby et al, 1991). This is in line with in vivo data demonstrating that DNB has no testicular effect when given to prepubertal mice (Evenson et al, 1989a) and has less effect on young adult rats than on old adult rats (Linder et al, 1990). Finally, in an article describing the only protocol used to date to successfully culture Sertoli cells from sexually mature rats (Wright et al, 1989), data are presented that demonstrate a dramatic increase in transferrin secretion following sexual maturity. Since clinical testing suggests that seminal transferrin levels may predict seminiferous tubule dysfunction (Mallea et al, 1988), transferrin production should be optimized in Sertoli cell cultures designed to monitor toxicity.

Although there is considerable variability in the specific nature of the in vitro system one might use, the culture of Sertoli cells on a semipermeable filter (Janecki and Steinberger, 1986) coated with an extracellular matrix (Hadley et al, 1988) appears to offer distinct advantages. Unfortunately, this model has been applied to only a single toxicology study (Chapin et al, 1990). In addition, multiple biochemical endpoints are used to evaluate the functional status of the Sertoli cell in vitro. Recently, Allenby and coworkers (1990) evaluated multiple endpoints including morphology, pyruvate and lactate production, and inhibin secretion to test the effects of DNB on Sertoli cells in vitro. With the identification of cyclic protein 2 (Wright, 1988; Wright et al, 1989), another biochemical marker for Sertoli cell function exists that awaits application to in vitro toxicology.

Likewise, since this protein is secreted primarily by Sertoli cells associated with stage VI and VII seminiferous tubules, and since many toxicants affect stage VII seminiferous tubules, antiserum to this protein may be used to detect toxicant-induced decreases in cyclic protein 2 in seminiferous fluid following in vivo exposures. This analysis could be performed in conjunction with androgen-binding protein assays currently used to assess alterations in seminiferous tubule fluid (Rehnberg et al, 1988) and assay of immunoactive inhibin in the interstitial fluid compartment of the adult rat testis (Allenby et al, 1991).

The early identification of the type of germ cell affected in testicular toxicity is fundamental in determining the subsequent duration of subfertility or infertility. Naturally, not all xenobiotics perturb meiotic and postmeiotic germ cells, as was the case for DNB. For example, radiation and some cancer chemotherapeutic agents destroy the rapidly dividing, differentiating spermatogonia (Meistrich, 1986). The alkylating agent busulfan affects spermatogenesis by killing both the reserve and the differentiating spermatogonia but exerts a greater effect on the reserve stem cell population (Bucci and Meistrich, 1987). Typically, the reserve stem cell is more toxicant-resistant than the differentiating spermatogonium; but once this cell type is affected by a toxicant, the period of infertility is related to the degree of stem cell killing and the time it will take for repopulation of the seminiferous tubules.

As discussed earlier, a dominant lethal test is used to determine a toxicant's ability to cause germ cell mutagenesis. Aneuploidy or chromosomal deletion typically results in early embryonic death in rodents. This model detects significant chromosome breakage, not point mutations unless they are lethal, in both male and female germ cells. To identify an affect in the male, for instance, the male is treated acutely with a putative mutagen and bred to untreated females for 8 to 10 weeks, which encompasses the duration of spermatogenesis plus epididymal transit in the mouse and rat, respectively. This period of time is more than sufficient since this test detects effects on only those germ cells capable of surviving the mutagenic insult, the non-dividing meiotic and postmeiotic germ cells. A couple of weeks later, the females are killed and a dominant lethal effect is defined as a decrease in the number of implants and/or an increase in fetal death, which represents pre- and postimplantation loss, respectively. Preimplantation loss can be a reflection of fertilization failure attributed to toxicant-induced sperm dysfunction. This accounted for most of the dominant lethality associated with methyl chloride exposure (Working, 1989).

Unscheduled DNA synthesis has been observed in both meiotic and postmeiotic male germ cells using known mutagens such as ethyl methanesulphonate, methyl methanesulphonate, cyclophosphamide, and x-rays in mice (Sega, 1980). If a mutagen causes damage to deoxynucleotides and/or single-strand breaks, unscheduled DNA synthesis constituting base repair may occur in primary spermatocytes beyond the preleptotene stage (the last scheduled DNA synthesis), including round and elongating spermatids according to the work of Bentley and Working (1988). It is believed that the

DNA synthesis in this case might constitute repair of the damaged genetic material. This type of repair would not be expected to correct specific locus mutations or double-strand breaks, and therefore despite repair efforts, dominant lethal tests may prove positive. Regardless, unscheduled DNA synthesis is a good indicator of a testicular insult and mutagenic potential (Working, 1989). To test for unscheduled DNA synthesis, tritiated thymidine is injected into the testis immediately following toxicant administration. The DNA synthesis can then be identified in the affected cells directly by autoradiography, or indirectly by liquid scintillation counting of sperm taken from the epididymis over time, estimating the affected cell type from the kinetics of spermatogenesis. With the latter method one must assume that the toxicant does not alter the kinetics of spermatogenesis or epididymal transit in addition to causing sperm DNA damage. Autoradiography, on the other hand, is costly, labor-intensive, and time-consuming and therefore does not seem practical until the site of toxicant action becomes important. Therefore this test may best serve as an indicator of an effect. Elongating spermatids and epididymal spermatozoa are much less sensitive to DNA damage as protamines replace sperm histones and nuclear compaction proceeds. Likewise, spermatogonia, which undergo scheduled DNA synthesis, are not candidates for study in this animal model. Despite these limitations, this method has been extremely predicative of mutagenic effects. Recently, Bentley and Working (1988) have increased the sensitivity of the method by isolating exposed rat seminiferous tubules in specific stages of spermatogenesis by trans-illumination (Parvinen and Ruokonen, 1982) and incubating these tubules in vitro with labeled thymidine. This method then provides increased accuracy in assigning both stage-specific and cellular effects.

In recent years there has been considerable interest in the development of molecular probes to assess testicular dysfunction. The utility of various DNA probes to detect toxicant-induced events in the testis has been reviewed (Hecht, 1987), and the evidence of their utility is beginning to emerge. The inherent advantage of molecular probing is the increase in sensitivity to observe an effect as minor as a small-base-pair substitution. On the other hand, the inherent disadvantage is that any given probe usually represents a very small percentage of the total genomic DNA and therefore limits the amount of DNA that can be assessed in a single hybridization assay. Recently, a cDNA probe was used to describe stage- and cell-specific changes in the mRNAs for ornithine decarboxylase by Northern blot and in situ hybridization of RNA from stage-isolated seminiferous tubules and testis sections, respectively, in the rat (Kaipia et al, 1990). Since the levels of mRNA for this enzyme were found to accumulate in late pachytene spermatocytes and early spermatids, this model might have provided simple, early detection of the effects of DNB discussed above. By utilizing molecular probes and current testicular cell isolation and culture systems, the new computerized fluorescent imaging systems might provide a means of quicker, more sensitive detection on a per cell basis. Multiple probe might be used to screen more of the genome in an individual isolation or culture. Molecular probes have been used to detect alterations in the relative amounts of mRNA for mitochondrial NADH dehydrogenase and human protamine 1 in human testis poly(A+) RNA from individuals in which testicular dysfunction ranged from benign cysts to Leydig cell hyperplasia (Krawetz et al, 1989). As expected, the relative amount of protamine transcript decreased with increasing spermatogenic dysfunction. Similar strategy might be applied clinically if toxicant-induced infertility is suspected.

The Epididymis

It is now an accepted fact that immature sperm acquire factors, some of which are proteins, during epididymal transit and that this in turn confers functional attributes necessary for normal fertilizability such as progressive motility, binding to the zona pellucida, ability to undergo the acrosome reaction, and binding to the vitelline membrane of the egg. Despite the unique importance of this organ, few compounds have been associated with epididymal dysfunction to date. However, a number of estrogenic chemicals, including the weakly estrogenic pesticide methoxychlor, reduce epididymal sperm numbers dramatically at dosage levels that have no effect on testicular sperm production (Gray et al, 1989a). Epididymal sperm are routinely evaluated in toxicologic study for alterations in number, motility, and morphology (Blazak, 1985); but when effects are observed, it is often impossible to distinguish whether the effect is the result of an epididymal lesion, a testicular lesion, or a lesion in both organs. Methods for the assessment of epididymal function include histologic evaluation, efferent duct ligation, semen analysis for epididymal secretions, and culture of epididymal cells.

A dominant lethal study testing the effects of inhaled methyl chloride, both a testicular and an epididymal toxicant, was the first of a series of studies that elucidated an epididymal role in the postimplantation losses observed 1 week following exposure (Working et al, 1985). The authors hypothesized that the DNA damage that occurred to the sperm was secondary to an inflammatory neutrophil invasion of the cauda epididymides, associated with the formation of sperm granulomas shortly after exposure (Chapin et al, 1984). Indeed, treatment with an anti-inflammatory agent prevented both methyl chloride-induced sperm granuloma formation in the epididymis and postimplantation loss (Chellman et al, 1986).

Correlation between reduced fertility and impaired sperm motion parameters and/or morphology has also been described for sulphonamides and related drugs (Wong et al, 1987), epichlorhydrin (Toth et al, 1989a), and ornidazole (McClain and Downing, 1988). Unfortunately, the length of dosing or time until postexposure evaluation in these studies precludes any determination of an effect that might be mediated through the epididymis. However, a single inhalation exposure of rats to epichlorhydrin resulted in significant decrease in sperm velocity 1 day later, suggestive of a direct epididymal effect (Slott et al, 1990).

To better characterize toxicant-induced epididymal effects, Klinefelter et al (1990) used a 4-day postexposure evaluation to identify effects of the Leydig cell toxicant EDS on the epididymis of adult rats. Four days approximates the time required for sperm to transit from the androgen-dependent caput epididymis to the cauda epididymis; and since EDS depletes serum testosterone in 24 h, effects on the androgen-sensitive epididymis seemed likely. Techniques of testosterone implantation and ligation of the efferent ducts coursing from the testis to the epididymis were used to control for effects of serum testosterone and testicular fluid, respectively. Conclusions were based upon the evaluation of multiple endpoints such as testicular spermatid counts, organ weights, serum testosterone levels, light and electron microscopy of the testis and epididymis, computer-assisted sperm motion analysis of proximal cauda epididymal sperm, and sperm membrane protein characterization of both caput and proximal cauda sperm. Sperm from the proximal cauda epididymis represent a more homogeneous age group compared with distal cauda sperm and were therefore deemed more sensitive. To a certain extent, the observed decreases in progressive motility and sperm proteins were attributed to the cytotoxic effect of the compound on the Leydig cell and the resulting decrease in circulating testosterone levels. A proportion of the treatment effects, however, could not be prevented by replacement of serum testosterone levels. Likewise, efferent duct ligation failed to ameliorate or prevent any of the epididymal effects. Thus, certain effects appeared to be mediated directly through the epididymal epithelium.

The anticancer and immunosuppressive drug cyclophosphamide was shown to affect the ultrastructural integrity of caput and cauda sperm midpieces (Trasler et al, 1988). Ultrastructural changes have been seen in sperm obtained by testicular biopsy from infertile men (Baccetti et al, 1988; Siew et al, 1977); but in the Trasler study, morphologic changes were seen in both the sperm and epithelial compartments of the rat epididymis, which prompted further investigation. In a manner reminiscent of the EDS study, a short exposure protocol and efferent duct ligation were used to suggest a direct epididymal component to the postimplantation losses that result from cyclophosphamide treatment (Qiu et al, 1992).

Significant decrease in epididymal-derived constituents of seminal plasma such as glycerolphosphocholine, α-glucosidase, and L-carnitine have been found in oligospermic patients (Cooper et al, 1988). One of these seminal markers, α-glucosidase, has recently been localized in the epididymal epithelium of both humans and rats (Yeung et al, 1990) and might serve to detect clinically relevant toxicant-induced alterations. A system of culturing both human (Cooper et al, 1990) and rat (Klinefelter, 1992a) epididymal epithelial cells has been described, but α-glucosidase secretion in response to toxicant perturbation awaits testing.

At least a portion of our current understanding of epididymal sperm maturation has come from organ cultures of rat (Klinefelter and Hamilton, 1985) and human (Tezon and Blaquier, 1981) epididymal tubules. Some of the proteins secreted in the latter system were immunoprecipitated using an antiserum raised against ejaculated sperm proteins. This antiserum, which recognizes the acrosomal cap of the sperm head, has been used recently to detect changes in the pattern of immunolocalization among sperm from fertile and infertile patients (Blaquier et al, 1987). This antiserum may also detect alterations on sperm from patients exposed to xenobiotics.

Recently, a system for the coculture of rat epididymal epithelial cells and sperm has been developed that enables identification of direct-acting epididymal toxicants. Epididymal epithelial cell plaques are cultured on an extracellular matrix on semipermeable supports beneath another filter insert of greater porosity that contains the sperm. This system was used to determine that ethane dimethanesulphonate decreases protein secretion by the epididymal epithelium, protein acquisition by cocultured sperm, and the progressive motility of cocultured sperm (Klinefelter et al, 1992a). These results confirm the above supposition that many of the effects this compound has on the epididymis, which are unrelated to serum testosterone levels, are mediated through the epididymal epithelium by a direct action.

Finally, flow cytometry has been used to relate changes in epididymal epithelial cell characteristics with seasonal alterations in androgen status in the ram (Kunze and Hammerstedt, 1991). This offers another exciting possibility for the detection of toxicant-induced change in the epididymis.

Sex Accessory Glands

Sex accessory tissues such as the prostate and seminal vesicles vary considerably between species, and for this reason it is difficult to generalize about structure and function (Coffey, 1988). It can be stated that these glands secrete fluids and chemicals that constitute the major volume of the ejaculate, and some chemicals (prostaglandins, fructose, polyamines, citric acid, and Zn, for example) are found in very high concentrations in the seminal plasma. It is known that these secretions facilitate male fertility, although they may not be essential for fertility in all species. It is also clear that mammalian sex accessory tissues are dependent, in part, upon androgens to sustain their function. Other important growth factors include GnRH, LH, and FSH, while prolactin, corticotropin, and estrogens also affect accessory gland structure and function. In these tissues, testosterone is metabolized to dihydrotestosterone (DHT) as in the epididymis. DHT binds to the androgen receptor, which is in turn activated in the nucleus and binds to the DNA, a process that initiates growth and development in these tissues. Chemicals that prevent the synthesis of androgens, or antagonize their action, inhibit accessory gland growth and function. For example, subchronic administration of the weakly estrogenic pesticide methoxychlor dramatically reduces the size of the seminal vesicles in the male rat (Gray et al, 1988b). Administration of antiandrogens, like cyproterone acetate, also inhibits sex accessory gland growth.

As indicated above, there are significant differences in sex accessory gland anatomy and physiology in different mammalian species. Rats and men have large seminal vesicles, but they are absent in cats and dogs. Even though the prostate is present in all mammals, its structure, function, and, importantly, its pathologic susceptibilities vary considerably among species. The rat prostate is divided into several distinct lobes, each with separate functions, while humans and dogs have a single prostate. Detailed descriptions of the anatomy, physiology, and biochemistry of the sex accessory tissues are provided in reviews by Coffey (1988) and Setchell and Brooks (1988). Pathologic conditions of the prostate are among the most common problems in men. Hyperplasia of the prostate occurs in over 80 percent of aged men, and about 25 percent of these will require surgery to correct this overgrowth (Berry et al, 1984). This hyperplasia is referred to as benign prostatic hyperplasia (BPH). Prostatic cancer is rarer but is the second leading cause of cancer deaths in males in the United States. Due to similarities between the dog and human in prostatic anatomy and function, dogs have been the animal model of choice to study the underlying mechanisms of BPH. However, recently a number of laboratories are attempting to develop rodent models of BPH, which would facilitate the progress of toxicologic research in this area.

Sperm

The analysis of sperm measures from ejaculated samples has been routinely conducted in infertility clinics for decades. Routine sperm analysis typically includes a determination of sperm concentration, the percentages of both motile and progressively motile sperm, and sperm morphology. While it is the intent to relate these human data to fertility, such a relationship is often difficult to demonstrate. In 1983 Meistrich and Brown proposed a mathematical model to estimate the incidence of human infertility based on alterations in sperm counts, using three published data sets. Meistrich (1989b) reported that this model could also estimate the incidence of human infertility from alterations in other sperm measures, like motility, but sufficient data on these sperm measures from humans were lacking. Such information would allow one to predict the increase in infertility in the human population that would result from an environmentally induced alteration in a sperm measure like sperm count.

Rabbits have been used as an animal model to assess the effects of a number of toxicants on ejaculated sperm measures, including sperm counts, sperm morphology, sperm motility, and semen characteristics (Desjardins et al, 1968; Williams et al, 1990). For example, Rao et al (1982) found that exposure via inhalation to DBCP, a fungicide that produced hypospermatogenesis and infertility in men, reduced ejaculated sperm counts at 1 ppm and produced infertility at 10 ppm. By contrast, epididymal sperm numbers can be reduced 90 percent in a rat without observing a significant decrease in fertility (Aafjes et al, 1980). While it is generally accepted that ejaculated sperm counts cannot be collected in rodents, recent advances in the field have overcome this barrier by collecting ejaculated sperm from the uterus shortly after mating (Zenick et al, 1984; Oudiz et al, 1984; Toner and Adler, 1985; Gray et al, 1991). Subchronic oral exposure to ethoxyethanol at 1872 mg/kg for 5 days reduced ejaculated sperm numbers in treated male rats. Studies by Gray et al (1991) indicate that a single injection of the testicular toxicant cadmium chloride during pubertal development reduces ejaculated sperm counts and fertility without affecting mating behavior. Doses of 2.0 mg/kg were effective in reducing sperm number and fertility, while the 0.5-, 1.0-, and 1.5-mg/kg doses were without effect.

Over the years human sperm morphology, typically head shape, has been a less variable endpoint than sperm count in the prediction of fertility (Sherins et al, 1977). Since the variability is less, a smaller sample size is needed to detect a change. Furthermore, this measure appears to be less sensitive to abstinence. The fact that approximately 50 percent of the sperm in the ejaculate of a healthy, fertile man can be classified as abnormal (Wyrobek et al, 1982) suggests that as with sperm counts, a relatively small change in the percentage of sperm with abnormal morphology may result in decreased fertility. However, epididymal sperm from the mouse are sensitive enough to generate dose–response data. Moreover, data from the mouse suggest that alterations in sperm head morphology are related to the ability of a toxicant to act as a germ cell mutagen (Wyrobek et al, 1982). The determination of sperm morphology has been routinely performed from air-dried smears that are subsequently fixed in ethanol and stained for DNA by the Feulgen or Papanicolaou method. This method is semiobjective in the sense that the individual performing the evaluation periodically refers to a set of reference slides that are coded for sperm head shape. Katz and coworkers (1986) used a semiautomated system to measure the morphometric parameters of human sperm and found that although the absolute dimensions of fixed and air-dried sperm were different than those in a wet preparation, the sperm head length/width ratio, an indicator of the shape of the sperm head, was constant. More recently, automated image analysis has been used successfully to classify air-dried sperm from healthy men into clinically familiar shape categories (Moruzzi et al, 1988). This technology will most certainly soon be applied to mouse sperm since mouse and human sperm have tested similarly for exposure to several chemicals (Wyrobek et al, 1983). The successful adaptation of this method to rat sperm may not be possible since it is difficult to establish parameters for the sickle-shaped head of the rat sperm, but nevertheless the task is being undertaken.

A thorough semen evaluation, which includes the assessment of motility, morphology, structural stability, membrane status, and metabolic activity, was originally proposed by Eliasson (1978). It is now commonly agreed that several sperm endpoints need to be measured for an accurate assessment of sperm quality. One of today's more popular endpoints is sperm motility. A study comparing sperm from fertile and infertile men suggests that objective motility assessments are more sensitive than morphologic evaluations (Morales et al, 1988). Quite recently, rabbit sperm motility has been shown to be perturbed by ethylene dibromide (Williams et al, 1991) in the same fashion described previously for human sperm (Schrader et al, 1988). Sperm morphology, on the other hand, was affected in humans but not in rabbits. Routine, objective clinical evaluations of human sperm motility has been made possible by the advent of computer-automated sperm analysis, but standardized methods of sample preparation and analysis have not yet been agreed upon or adopted (Boyers et al, 1989; Davis and Katz, 1989). The importance of carefully optimizing individual components of sperm motion analysis was illustrated recently in a study investigating the effects of ethane dimethanesulphonate on rat sperm motion parameters (Klinefelter et al, 1991b). This study provided evidence that the method used to collect sperm from the epididymis has a significant influence on the outcome of motion analysis. A method that provides accurate, precise, and sensitive data is described. Similarly, Toth and coworkers (1989b) carefully evaluated automated analysis of rat sperm following epichlorhydrin administration. Their research detailed important statistical considerations and the derivation of optimal instrument settings. Although recently the limitation of computerized sperm motility in fertility assessment has been questioned (Grunert et al, 1989), recent studies suggest that such motion analysis can be predictive of success in in vitro fertilization in humans (Check et al, 1990; Liu et al, 1991).

Other sperm endpoints may be more uniquely capable of predicting fertilizing capacity. For example, sperm from infertile patients that tested normal by routine classic clinical criteria were found to have increased susceptibility to DNA denaturation (Evenson, 1986). The technique, now referred to as the sperm chromatin structure assay (Evenson, 1989), is similar to that which we described for testicular cells earlier. Briefly, the acridine orange stain will intercalate into native DNA and fluoresce green, while binding to the single strands in denatured DNA results in red fluorescence. An index of DNA denaturation is provided on a per-cell basis by the fluorescence ratio of red to red plus green. The initial treatment of the sperm with detergent at low pH permeabilizes the cell membrane, and this allows the acridine orange stain to gain access to structurally altered DNA that was partially denatured by the acid. Most recently, this assay has been used to demonstrate that significant differences in denaturable sperm DNA exist between normal individuals, but that there is little within-individual variation (Evenson et al, 1991). When applied to cattle, this assay also proved to be highly correlated with the heterospermic performance test of fertility but only weakly correlated with the other classic sperm measures (Ballachey et al, 1988). Taken together, these results suggest that this assay offers a unique, objective endpoint that should be used in conjunction with other endpoints in the detection of toxicant-altered sperm. The sperm chromatin structure assay has been used to detect an effect of zeranol on the DNA of sperm from the vas deferens in the bull (Ballachey et al, 1986), as well as on DNA of epididymal sperm following ethylnitrosourea (Evenson et al, 1985) and triethylenemelamine (Evenson et al, 1989c) exposure. In studies with mice, the chromatin structure assay results appear to be highly correlated with the effects of these two mutagens on sperm head morphology.

Fluorescent probes have been used to assess the integrity of other constituents of the sperm during the last several years. Toxicant-induced alteration in the normal oxidation of free sulfhydryl groups that occurs during epididymal transit may be studied in rats and other laboratory animals with 7-diethylamino-3-(4′-maleimidylphenyl)-4-methylcoumarin (Evenson et al, 1989d). Flow cytometric analysis of mitochondrial function of sperm in human semen has been studied using the green fluorescence of rhodamine 123 in conjunction with the red fluorescence of impermeant, viability dyes such as ethidium bromide (Evenson, 1982) and propidium iodide (Auger et al, 1989). The latter study also correlated sperm fluorescence patterns with computerized sperm motility data and found good correlation between mitochondrial activity and sperm motility. Inhibitors of mitochondrial activity were used in a recent flow cytometric study of bull sperm to illustrate the utility of rhodamine 123 staining (Graham et al, 1990). In theory, a reproductive toxicant could be equally effective in reducing mitochondrial function, and therefore rhodamine 123 would be sensitive to a detectable change.

Most recently, a technique that utilizes routine fluorescent microscopy was applied to ram and boar semen (Harrison and Vickers, 1990). Briefly, live sperm are immobilized with a low concentration of formaldehyde for viewing. Carboxyfluorescein diacetate passes through intact membranes and is deesterified to carboxyfluorescein, an impermeable dye, by the action of nonspecific esterases. Intact sperm will fluoresce green throughout their length, sperm with damaged plasma membranes will fluoresce only in the acrosomal and/or mitochondrial compartments, and completely damaged sperm will not fluoresce green at all. Simultaneous staining with propidium iodide, which binds DNA, verifies the specificity of compartmental damage. This procedure did not appear to compromise the viability of the sperm, since sperm motility returned to normal after washing, and the staining over time of sperm allowed to sit in staining solution mimicked the staining over time of unstained sperm that were left undisturbed. Finally, the incidence of sperm that stained over time in capacitation buffer corresponded to the observed incidence of acrosome-reacted sperm. A monoclonal antibody directed against the acrosomal ground substance was generated and used in conjunction with indirect immunofluorescence to evaluate the membrane status of sperm in stallion semen (Blach et al, 1988). In this instance, acrosomal staining indicated that damage to the plasma membrane had occurred. A similar probe for rat sperm awaits development.

Flow cytometric probes that might provide a means for the detection of alterations in acrosomal integrity and/or function have recently evolved. Purvis and coworkers (1990) have used a fluorescent conjugate of the plant lectin *Arachis hypogea* agglutinin to study the acrosome labeling in normospermic men. Simultaneous incubation with the calcium ionophore A 23187 suggests that this staining may be used to detect a change in functional status of the acrosome such as the acrosome reaction. Another flow cytometric study used the fluorescently labeled lectin *Pisum sativum* agglutinin, either alone or in conjunction with various phospholipids, to assess acrosome intact and acrosome-reacted bull sperm, respectively (Graham et al, 1990). Modest concentrations of dilauroylphosphatidylcholine liposomes were found to cause acrosomal fluorescence in sperm without loss of rhodamine 123 staining or uptake of propidium iodide. Thus, the functional integrity of the acrosome, the mitochondria, and the plasma membrane can be assessed simultaneously. This type of multicompartment evaluation has tremendous implication in toxicology. Moreover, as we have suggested on numerous occasions, an extension of this work might be the direct application to the new fluorescent imaging systems with confocal microscopy capability. This would provide direct morphologic interpretation of the quantitative fluorescence data.

Finally, quite recently flow cytometry has been used to identify protamine variants in human semen. Briefly, sperm nuclei were decondensed with alkyltrimethylammonium bromide and dithiothreitol and incubated with fluorescein-conjugated monoclonal antibodies specific for the human protamine 1 and 2 proteins (Wyrobek et al, 1990). Relying on the fact that the ratio of the protamine variants is constant within a species, protamines from both fertile individuals and infertile patients were analyzed by gel electrophoresis and microdensitometry (Balhorn et al, 1988). The results indicated that sperm from infertile men have a different distribution of protamines than sperm from fertile men. Using molecular probes for mouse protamine 1 and protamine 2, the expression of these protamines was compared in sperm from the rat and the hamster (Bower et al, 1987), but differential protamine content in sperm from rodents has not yet been measured following toxicant exposure.

Fertilization

We have pointed out that a reduction in sperm number is less frequently associated with a change in fertility in rodents than in humans. The increased percentage of morphologically abnormal sperm in a normal man's ejaculate, together with a relatively poor spermatogenic efficiency in humans, and alterations in other endpoints such as sperm membrane protein composition, sperm motility, and nuclear chromatin integrity, may account for this. Thus, the underlying reason for the apparent insensitivity of rodent fertility is that the rodent offers a greater number of qualitatively normal sperm per ejaculate, and therefore functional alteration, infertility, is less readily detected. With this logic in mind we can now discuss the current trend to assess fertility, following both in vivo and in vitro xenobiotic exposure, in the rodent.

Hurtt and Zenick (1986) demonstrated a dramatic shift in the dose–response curve following ethoxyethanol spermatotoxicity by mating male rats more frequently to reduce the epididymal sperm reserves both prior to and during the dosing period. All the sperm endpoints were affected at lower doses in the mated group of animals. Although fertility was not assessed in this study, it seem likely that any toxicant-induced alteration in fertility would be maximized by this design. To routinely assess fertility following toxicant exposures is costly and labor-intensive. Thus, research efforts are underway to correlate the number of ejaculated sperm in the rat with fertility (Gray et al, 1991). This would greatly increase our capability to predict fertility when a significant decrease in ejaculated sperm number is observed following toxicant exposure. Artificial insemination should further increase the sensitivity of toxicant-induced decreases in sperm quality (Amann, 1986). Recently there has been progress in achieving a high rate of fertilization following in utero insemination of adult, cycling or LHRH-synchronized female rats with cauda epididymal sperm (Klinefelter et al, 1992b). The problems associated with correlating fertility data with in vitro techniques, which are only indicative of fertilizing capacity, have been appropriately highlighted (Amann, 1989a). Nevertheless, if accurate, repeatable in vitro data can be derived to assess fertilizing capacity, the better one can attempt to predict the potential for normal fertilization in humans and possibly gain information regarding the mechanism of toxicant action. We want to emphasize again that in vitro approaches should be applied to the study of reproductive toxicology in laboratory animals after demonstrating an effect on fertilizing capacity in vivo. The in vitro research to evaluate fertilizing capacity has been described in recent reviews (Perreault, 1989; Aitken, 1990).

The hamster egg penetration test popularized by Yanagimachi et al (1976) has been used over the years to screen sperm quality prior to in vitro fertilization and embryo transfer procedures. In this procedure acrosome-reacted sperm bind at their equatorial segment to plasma membrane of the zona-free hamster oocyte. The test has been less predictive in patients with a male factor, and it is suggested that other sperm endpoints be measured in conjunction with hamster egg penetration, including the sperm chromatin structure assay (Ibrahim et al, 1989). The toxicologist is well-advised to implement similar strategies. The assay is attractive since it can be performed with sperm from a variety of laboratory animals and may be very sensitive if the number of acrosome-reacted sperm are maximized using ionophores. The results of this test have correlated positively with both in vitro and in vivo results of fertilization, and the test has been used to demonstrate effects of membrane stabilizing agents such as propranolol on sperm–oocyte fusion (Aitkin et al, 1988). Moreover, since many sperm are capable of binding to zona-free oocytes, functional variants can be monitored simultaneously. For example, Blazak et al (1985) coincubated both control and treated sperm labeled with different fluorescent probes to characterize potential differences in oocyte binding.

The species specificity of sperm–zona binding has limited the study of sperm–zona interaction in toxicology to rodents. Still, the binding of human sperm to the zona pellucida has been tested using nonviable human eggs (Hembree and Overstreet, 1977; Gould et al, 1983). We must take into account that a given toxicant may act at multiple sites during the fertilization process. For example, Perreault et al (1980) first removed the cumulus layer and then the zona to demonstrate the two sites of action of a hyaluronidase inhibitor on the fertilizing capacity of hamster sperm. It occurs to us that the information available from tests such as the hamster egg penetration test and the zona penetration test can be gained during a single in vitro fertilization assay. Unfortunately, only a few laboratories have reported successful in vitro fertilization rates using the rat as an animal model to date (Evans and Armstrong, 1984; Holloway et al, 1990). The latter study confirmed the results of other DNB studies, finding that sperm fertilizing capacity was impaired. An example of the experimental strategy that is currently being tested in our laboratory follows.

As discussed earlier, EDS affects not only the histologic appearance of the epididymis and epididymal sperm motility, but it also causes a diminution in several proteins associated with the sperm membrane. Furthermore, these effects are coincident with the transient infertility observed following exposure to this compound (Jackson, 1964). As pointed out in our discussion of epididymal sperm maturation, certain sperm-associated proteins may play a role in both sperm motility and fertilizing ability. Figure 8-2 is a representation of what might occur as sperm membrane protein composition is altered as a result of toxicant exposure. An increment of change in sperm membrane protein, either qualitative or quantitative, might be correlated with a reduction in fertility of artificially inseminated cauda epididymal sperm (Klinefelter et al, 1992b). The alteration in sperm membrane protein may cause a change in sperm motility that may account for all, or part, of the observed decrease in fertility. If the fertilizing capacity of artificially inseminated epididymal sperm is changed, it would be feasible to test sperm function by in vitro fertilization and again determine to what extent, if any, the changes in sperm membrane protein influence fertilizing capacity during in vitro fertilization. An observed change in the in vitro fertilization rate could be studied further to assess the ability of the sperm to undergo the acrosome reaction, bind to the zona pellucida, and fuse with the oocyte plasma membrane. Finally, the detection of alterations in the membrane protein profile of sperm exposed in vitro during coculture might be tested by artificial insemination and in vitro fertilization as well, using radiolabeling techniques to identify direct, toxicant-induced protein changes.

Successful fusion with the oocyte membrane is followed by the decondensation of the nuclear chromatin of the sperm and the formation of the male pronucleus. If colchicine is added following sperm and hamster egg

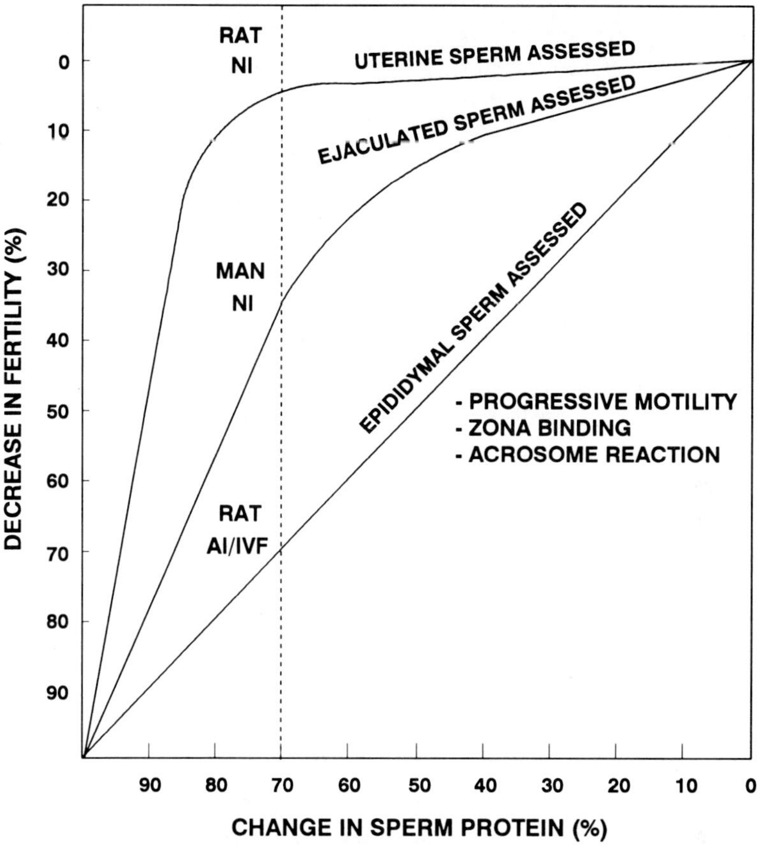

Figure 8-2. A hypothetical relationship between fertility, assessed by both naturial insemination (NI; human and rat) and artificial insemination (AI/IVF; rat) and a sperm endpoint such as sperm membrane protein composition. The broken line represents a 70 percent change in sperm protein composition. Sperm number is constant over the course of each individual line irrespective of the percentage change in protein composition. Since the rodent offers greater numbers of qualitatively normal sperm upon ejaculation, a homologous toxicologic mechanism might best be established by comparing ejaculated human and epididymal rat sperm. Increased sensitivity is offered by artificially inseminating, in utero, the minimum number of rat epididymal sperm that routinely yields 100 percent fertility. Once dose–response data can be generated, in vitro fertilization techniques may be coupled with assessments of sperm motion, zona binding, and the acrosome reaction to address more mechanistic questions.

fusion, it is possible to obtain sperm chromosomes appearing in metaphase for cytogenetic analysis (Rudak, 1978). This has been a valuable technique to assess human chromosome alterations (Brandriff et al, 1985) and might be utilized in an animal model following either an in vivo or in vitro exposure to predict the incidence of fertility or heritable defects. This might be included in the investigation of a toxicant that produced defects in head morphology or tested positive in the sperm chromatin structure assay (Figure 8-3).

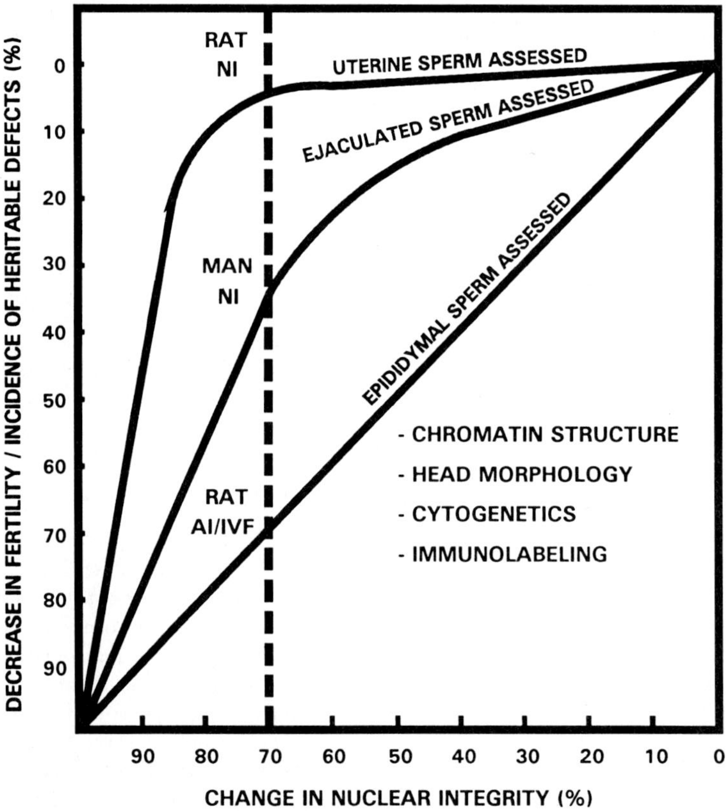

Figure 8-3. A hypothetical relationship between fertility and/or heritable defects and a sperm endpoint such as nuclear integrity. The curves are shaped as in Fig. 8-2 since the same reasoning applies and again the number of sperm will be constant throughout each line. The results of several measures discussed in the text are applied in this figure. For example, the sperm chromatin structure assay and/or morphometric analysis of the sperm nucleus could be correlated with fertility and/or heritable defects. Cytogenetic analysis and the labeling of sperm nuclei with monospecific nuclear protamine antibodies might provide mechanistic data if results correlated well with fertility and/or heritable defects.

FUTURE DIRECTIONS

Throughout this chapter we have attempted to highlight the most recent work as well as offer suggestions for future directives in reproductive toxicology. We would now like to summarize our vision for the future of the field as it relates to human reproductive failure. Clearly, animal models will gain importance as species-homologous biomarkers are developed and tested for the recognition of clinical reproductive disorders following xenobiotic exposure.

The most efficient means to the end, however, will involve a significant level of cooperative spirit among existing and budding academic and regulatory reproductive toxicologists alike. We hope that the future brings a certain standardization and tier testing of animal protocols that are routinely followed during the testing of a suspected reproductive toxicant. Amann (1982) presented physiologic justification for studies that span six cycles of the seminiferous epithelium to identify testicular toxicity; but today many short-term protocols are implemented before the appropriate subchronic data are derived. This means that false negatives that arise from short-term testing are likely to go undetected. Moreover, subtle lesions that may be present during a short-term test, that is, early stages of spermatogonial degeneration, may not be apparent even with sophisticated morphometric analysis or biochemistry. It must also be recognized that these sophisticated procedures require considerable time, money, and expertise, resources that cannot be squandered in science today. Similar logistics need to be applied to studies in the female. In the rat, other than the assessment of vaginal estrous cyclicity, few endpoints have been proposed for use in the nonpregnant female. Additional short-term testing models need to be developed for use. For instance, guinea pigs are currently not used when mid to late gestational problems are indicated.

At some point in a tier-testing strategy, histopathologic evaluation needs to become incorporated into the evaluation process. Although time-consuming and tedious, high quality corroborative morphologic data are reliable means of accurately identifying the cellular site of action of a compound at a given point in time. This information is critical prior to the development of biochemical, immunologic, or molecular probes for the dysfunction of a particular cell. Although probes never before tested following toxicant exposure, such as cyclic protein 2, should be utilized in appropriate testing strategies provided a viable hypothesis exists, it seems clear that a probe should be developed against a molecule for which data are available for a number of xenobiotics, many of which are shown to affect that molecule. Unfortunately, relatively few proven probes exist at the present time.

The current in vitro systems highlighted in the text provide excellent opportunities for identifying cell-specific events, including molecular changes, following toxicant exposure. However, before the utility of these in vitro systems can be realized, tier testing must first establish that the xenobiotic exerts an in vivo effect on a specific tissue or cell either by itself or via a commercially available active metabolite. In many cases insufficient data are available to determine if the proposed animal model is appropriate. An excellent example of proper implementation of the final tiers of our strategy has been provided recently in the study of 4-vinylcyclohexene (VCH) by Smith and coworkers (1991). After species differences in ovotoxicity were identified by morphometry, the active blood metabolite was identified, and finally species differences were evaluated in vitro by comparing mouse, rat, and human hepatic microsome metabolism. The results clearly demonstrated that if the rate of VCH epoxidation determines the degree of ovotoxicity, the rat, but not the mouse, is appropriate for human extrapolation. Another example is ethane dimethanesulphonate, first shown to exert effects on the epididymis in vivo, and subsequently in vitro. Research is now underway to identify at least one common sperm membrane protein that changes following both routes of exposure. If successful identification is achieved, this protein will be monitored following exposures to other suspected epididymal toxicants. Attempts will then be made as described above to correlate the membrane alteration to an alteration in sperm function, motility, and/or fertility. Provided this, too, is successful, an antibody to the sperm membrane protein will serve to probe rat epididymal sperm following exposure to a suspected toxicant: Animals with functionally defective sperm will not bind the antibody. It seems feasible that a functional sperm membrane protein in the rat would be derived from a homologous human gene. Therefore, a cDNA probe could be derived, and ultimately an antibody could be generated to probe human sperm.

While the quest for clinical methods to assess toxicant effects in both men and women continues, the variety of animal models available today can provide extensive information regarding the effects of toxicant exposure in animals. Today, whether proof of homology exists or not, a biologically relevant effect in a scientifically proven animal model should be reason for clinical concern. In the future, the effects of an increasing number of xenobiotics in animals will be thoroughly characterized and experimentally extrapolated to humans. Subsequently, structurally homologous compounds will be tested in biologically based dose–response models designed to correlate the target tissue concentration of active xenobiotic with the target tissue response. From this, a structure/activity data base will emerge facilitating the recognition of compounds that might pose clinical problems.

REFERENCES

Aafjes JH, Vels JM, Schenk E: Fertility of rats with artificial oligozoospermia. *J Reprod Fertil* 48:345–351, 1980.

Aitkin RJ: Development of in vitro tests of human sperm function: A diagnostic tool and model system for toxicological analyses. *Toxic in vitro* 4:560–569, 1990.

Aitkin RJ, Clarkson JS, Hulme MJ, Henderson CJ: Analysis of calmodulin acceptor proteins and the influence of calmodulin antagonists on human spermatozoa. *Gamete Res* 21:93–112, 1988.

Albertini DF: Cytoplasmic reorganization during the resumption of meiosis in cultured preovulatory rat oocytes. *Dev Biol* 120:121–131, 1987.

Albertson BD, Frederick KL, Maronian NC, Feuillan P, Schorer S, Dunn JF, Loriaux DL: The effect of ketoconzaole on steroidogenesis. *Res Commun Chem Pathol Pharmacol* 61:17–26, 1988.

Allenby G, Sharpe RM, Foster PMD: Changes in Sertoli cell function in vitro induced by nitrobenzene. *Fundam Appl Toxicol* 14:364–375, 1990.

Allenby G, Foster PMD, Shapre RM: Evaluation of changes in the secretion of immunoactive inhibin by adult rat seminiferous tubules in vitro as an indicator of early toxicant action on spermatogenesis. *Fundam Appl Toxicol* 16:710–724, 1991.

Amann RP: Sperm production rates, in Johnson AD, Gomes WR, Van Demark NL, (eds). *The Testis*. New York, Academic, 1970, pp 433–482.

Amann RP: Use of animal models for detecting specific alterations in reproduction. *Fundam Appl Toxicol* 2:13–26, 1982.

Amann RP: Detection of alterations in testicular and epididymal function in laboratory animals. *Environ Health Perspect* 70:149–158, 1986.

Amann RP: Maturation of spermatozoa, in *Eleventh International Congress of Animal Reproduction* (vol 5). 1988, pp 320–328.

Amann RP: Can the fertility potential of a seminal sample be predicted accurately? *J Androl* 10:89–98, 1989a.

Amann RP: Structure and function of the normal testis and epididymis. *J Am Coll Toxicol* 8:457–471, 1989b.

Amann RP, Berndtson WE: Assessment of procedures for screening agents for effects of male reproduction. Effects of dibromochloropropane (DBCP) on the rat. *Fundam Appl Toxicol* 7:244–255, 1986.

Amann RP, Lambiase JT: The male rabbit: III. Determination of daily sperm production by means of testicular homogenates. *J Anim Sci* 28:369–374, 1969.

Amin-Zaki L, Elhassani S, Majeed MA, Clarkson TW, Doherty RA, Greenwood M: Intrauterine methylmercury poisoning in Iraq. *Pediatrics* 54:587–595, 1974.

Auger J, Ronot X, Dadoune J: Human sperm mitochondrial function related to motility: A flow and image cytometric assessment. *J Androl* 10:439–448, 1989.

Awoniyi CA, Santulli R, Chandrashekar V, Schanbacher BD, Zirkin, BR: Quantitative restoration of advanced spermatogenic cells in adult male rats made azoospermic by active immunization against luteinizing hormone or gonadotropin-releasing hormone. *Endocrinology* 125:1303–1309, 1989.

Baccetti B, Burrini AG, Collodel G, Magnano AR, Piomboni P, Renieri T: Immunocytochemistry and sperm pathology. *J Submicrosc Cytol Pathol* 20:209–224, 1988.

Balhorn R, Reed S, Tanphaichitr N: Aberrant protamine 1/protamine 2 ratios in sperm of infertile human males. *Experientia* 44:52–55, 1988.

Ballachey BE, Miller HL, Jost LK, Evenson DP: Flow cytometry evaluation of testicular and sperm cells obtained from bulls implanted with zeranol. *J Anim Sci* 63:995–1004, 1986.

Ballachey BE, Evenson DP, Saacke RG: The sperm chromatin structure assay. Relationship with alternate tests of semen quality and heterospermic performance of bulls. *J Androl* 9:109–115, 1988.

Barnes DG, Dourson M: Reference dose (RfD): Description and use in health risk assessments. *Regul Toxicol Pharmacol* 8:471–486, 1988.

Barraclough CA, Wise PM: The role of catecholamines in the regulation of pituitary LH and FSH secretion. *Endocr Rev* 3:91–119, 1982.

Batten BE, Albertini DF, Ducibella T: Patterns of organelle distribution in mouse embryos during preimplantation development. *Am J Anat* 178:204–213, 1987.

Beach F: *Hormones and Behavior.* New York, Hoeber, 1948.

Benson GS: Male sexual function: Erection, emission, and ejaculation, in Knobil E, Neill J et al (eds): *The Physiology of Reproduction.* New York, Raven, 1988, pp 1121–1139.

Bentley KS, Working PK: Use of seminiferous tubule segments to study stage specificity of unscheduled DNA synthesis in rat spermatogenic cells. *Environ Mol Mutagen* 12:285–297, 1988.

Berndtson WE: Methods for quantifying mammalian spermatogenesis: A review. *J Anim Sci* 44:818–833, 1977.

Berndtson WE, Neefus C, Foote RH, Amann RP: Optimal replication for histometric analyses of testicular function in rats or rabbits. *Fundam Appl Toxicol* 12:291–302, 1989.

Berry SJ, Coffey DS, Walsh PC, Ewing LL: The development of human benign prostatic hyperplasia with age. *J Urol* 132:474–479, 1984.

Blach EL, Amann RP, Bowen RA, Sawyer HR, Hermenet MJ: Use of a monoclonal antibody to evaluate integrity of the plasma membrane of stallion sperm. *Gamete Res* 21:233–241, 1988.

Blaquier J, Cameo M, Stephany D, Piazza A, Tezon J, Sherins R: Abnormal distribution of epididymal antigens on spermatozoa from infertile men. *Fertil Steril* 47:302–309, 1987.

Blazak WR, Ernst TL, Stewart BE: Potential indicators of reproductive toxicity: Testicular sperm production and epididymal sperm number, transit time, and motility in Fischer 344 rats. *Fundam Appl Toxicol* 5:1097–1103, 1985.

Bloom AD (ed): *Guidelines for Studies of Human Populations Exposed to Mutagenic and Reproductive Hazards*. Proceedings of a conference held Jan. 26–27, 1981, Washington, DC, March of Dimes Birth Defect Foundation. White Plains, NY, March of Dimes, 1981.

Bower PA, Yelick PC, Hecht NB: Both P1 and P2 protamine genes are expressed in mouse, hamster, and rat. *Biol Reprod* 37:479–488, 1987.

Boyers SP, Davis RO, Katz DF: Current problems in obstetrics, gynecology and fertility: Automated semen analysis (vol 12). Chicago, Year Book Medical, 1989.

Brandriff B, Gordon L, Ashworth L, Watchmaker G, Moore D 2d, Wyrobek AJ, Carrano AV: Chromosomes of human sperm: Variability among normal individuals. *Hum Genet* 70:18–24, 1985.

Brannstrom M, Janson PO: Progesterone is a mediator in the ovulatory process of the in vitro–perfused rat ovary. *Biol Reprod* 40:1170–1178, 1989.

Bucci LR, Meistrich ML: Effects of busulfan on murine spermatogenesis: Cytotoxicity, sterility, sperm abnormalities, and dominant lethal mutations. *Mutat Res* 176:259–268, 1987.

Buccione R, Schroeder AC, Eppig JJ: Interactions between somatic cells and germ cells throughout mammalian oogenesis. *Biol Reprod* 43:543–547, 1990.

Butcher RL, Fugo NW: Overripeness and the mammalian ova: II. Delayed ovulation and chromosome anomalies. *Steril Fertil* 18:397–402, 1967.

Carney EW, Tobback C, Ellington JE, Foote RH: Co-culture of rabbit 2-cell embryos with rabbit oviduct epithelial cells and other somatic cells. *Mol Reprod Dev* 27:209–215, 1990.

Castellon E, Janecki A, Steinberger A. Influence of germ cells on Sertoli cell secretory activity in direct and indirect co-culture with Sertoli cells from rats of different ages. *Mol Cell Endocrinol* 64:169–178, 1989.

Cave DA, Foster PMD: Modulation of *m*-dinitrobenzene and *m*-nitrosonitrobenzene toxicity in rat Sertoli–germ cell cocultures. *Fundam Appl Toxicol* 14:199–207, 1990.

Chandrasekhar Y, Armstrong DT, Kennedy TG: Implantation delay and anti-deciduogenic activity in the rat by the anti-androgen, hydroxyflutamide. *Biol Reprod* 42:120–125, 1990.

Chapin RE, Williams J: Mechanistic approaches in the study of testicular toxicity: Toxicants that affect the endocrine regulation of the testis. *Toxicol Pathol* 17:446–451, 1989.

Chapin RE, White RD, Morgan KT, Bus JS: Studies of lesions induced in the testis and epididymis of F-344 rats by inhaled methyl chloride. *Toxicol Appl Pharmacol* 76:328–343, 1984.

Chapin RE, Phelps JL, Somkuti SG, Heindel JJ, Burka LT: The interaction of Sertoli and Leydig cells in the testicular toxicity of tri-*o*-cresyl phosphate. *Toxicol Appl Pharmacol* 104:483–495, 1990.

Check JH, Bollendorf A, Lee MA, Nazari A, Nowroozi K: Correlation of computerized semen analysis with successful fertilization of oocytes in an in vitro fertilization program. *Arch Androl* 24:229–234, 1990.

Chellman GJ, Morgan KT, Bus JS, Working PK: Inhibition of methyl chloride toxicity in male F-344 rats by the anti-inflammatory agent BW755C[1]. *Toxicol Appl Pharmacol* 85:365–379, 1986.

Coffey DS: Androgen action and the sex accessory tissues, in Knobil E, Neill J et al (eds): *The Physiology of Reproduction.* New York, Raven, 1988, pp 1081–1119.

Cooper RL, Goldman JM, Rehnberg GL: Regulation of ovarian function, in Working PK (ed). *Toxicology of the Male and Female Reproductive Systems.* New York, Hemisphere Publishing, 1989, pp 217–229.

Cooper TG, Yeung CH, Nashan D, Nieschlag E: Epididymal markers in human infertility. *J Androl* 9:91–101, 1988.

Cooper TG, Yeung CH, Meyer R, Schulze H: Maintenance of human epididymal epithelial cell function in monolayer culture. *Reprod Fertil* 90:81–91, 1990.

Crump K: A new method for determining allowable daily intakes. *Fundam Appl Toxicol* 4:854–871, 1984.

Cummings AM: Toxicological mechanisms of implantation failure. *Fundam Appl Toxicol* 15:571–579, 1990.

Cummings AM, Gray LE: Methoxychlor affects the decidual cell response of the uterus but not other progestational parameters in female rats. *Toxicol Appl Pharmacol* 90:330–336, 1987.

Cummings AM, Gray LE: Antifertility effect of methoxychlor in female rats: Dose- and time-dependent blockade of pregnancy. *Toxicol Appl Pharmacol* 97:454–462, 1989.

Davidson JM, Camargo CA, Smith ER: Effects of androgen on sexual behavior in hypogonadal men. *J Clin Endocrinol Metab* 8:955–958, 1979.

Davis RO, Katz DF: Computer-aided sperm analysis (CASA): Image digitization and processing. *Biomater Artif Cells Artif Organs* 17:93–116, 1989.

Desjardins C, Kirton KT, Hafs HD: Sperm output of rabbits at various ejaculation frequencies and their use in the design of experiments. *J Reprod Fertil* 15:27–32, 1968.

Dharmarajan AM, Yoshimura Y, Sueoka K, Atlas SJ, Dubin NH, Ewing LL, Zirkin BR, Wallach EE: Progesterone secretion by corpora lutea of the isolated perfused rabbit ovary during pseudopregnancy. *Biol Reprod* 38:1137–1143, 1988.

Dodds WG, Awadalla SG, Hixon C, Roh SI, Friedman CI, Moon KH: Atypical luteinizing hormone rise and associated fertilization failure in non-male factor in vitro fertilization patients. *Obstet Gynecol* 73:191, 1989.

Drosdowsky MA, Foucault P, Curreau S: Some functional features of human Sertoli cells in vitro. *J Androl* (suppl) Jan/Feb 1991.

Ducibella T, Anderson E, Albertini DF, Aalberg J, Rangarajan S: Quantitative studies of changes in cortical granule number and distribution in the mouse oocyte during meiotic maturation. *Dev Biol* 130:184–197, 1988.

Eddy EM: The spermatozoon, in Knobil E, Neill J et al (eds): *The Physiology of Reproduction.* New York, Raven, 1988, pp 27–68.

Eddy EM, O'Brien DA: Biology of the gamete: Maturation, transport, and fertilization, in Working PK, (ed): *Toxicology of the Male and Female Reproductive Systems.* New York, Hemisphere Publishing, 1989, pp 31–71.

Ehrhardt AA, Meyer-Bahlburg HFL: Effects of prenatal sex hormones on gender-related behavior. *Science* 211:1312–1318, 1981.

Elger W, Beier S, Faehnrich M: Interference with hormonal control of rodent reproduction and its implications for human risk assessment, in Valans GN, Sims J, Sullivan FM, Turner P (eds): *Proceedings of the Fifth International Congress of Toxicology.* Philadelphia, Taylor & Francis, 1990, pp 445–456.

Eliasson R: Semen analysis. *Environ Health Perspect* 24:81–85, 1978.

Ellington JE, Carney EW, Farrel PB, Simkin ME, Foote RH: Bovine 1–2-cell embryo development using a simple medium in three oviduct epithelial cell coculture systems. *Biol Reprod* 43:97–104, 1990.

Ellington JE, Schlafer DH, Foote RH: Ultrastructural features of bull sperm cocultures with bovine uterine tube epithelial cells. *J Androl* (suppl) Jan/Feb 1991.

Eppig JJ, Schroeder AC: Capacity of mouse oocytes from preantral follicles to undergo embryogenesis and development to live young after growth, maturation, and fertilization in vitro. *Biol Reprod* 41:268–276, 1989.

Evans G, Armstrong DT: Reduction in fertilization rate in vitro of oocytes from immature rats induced to superovulate. *J Reprod Fertil* 70:131–135, 1984.

Evenson DP: Simultaneous measurement of flow cytometry of sperm cell viability and mitochondrial membrane potential related to cell motility (letter to the editor). *J Histochem Cytochem* 30:279–280, 1982.

Evenson DP: Male germ cell analysis by flow cytometry: Effects of cancer, chemotherapy, and other factors on testicular function and sperm chromatic structure, in Andreeff (ed): *Clinical Cytometry.* New York, New York Academy of Science, 1986, pp 350–367.

Evenson DP: Flow cytometry evaluation of male germ cells, in Yen A (ed): *Flow Cytometry: Advanced Research and Clinical Applications* (vol 1). Boca Raton, FL, CRC Press, 1989, pp 217–246.

Evenson DP, Higgins PJ, Grueneberg D, Ballachey BE: Flow cytometric analysis of mouse spermatogenic function following exposure to ethylnitrosourea. *Cytometry* 6:238–253, 1985.

Evenson DP, Baer RK, Jost LK, Gesch RW: Toxicity of thiotepa on mouse spermatogenesis as determined by dual-parameter flow cytometry. *Toxicol Appl Pharmacol* 82:151–163, 1986.

Evenson DP, Janca FC, Jost LK: Effects of the fungicide methyl-benzimidazol-2-yl carbamate (MBC) on mouse germ cells as determined by flow cytometry. *J Toxicol Environ Health* 20:387–399, 1987.

Evenson DP, Janca FC, Baer RK, Jost LK, Karabinus DS: Effect of 1,3-dinitrobenzene on prepubertal, pubertal, and adult mouse spermatogenesis. *J Toxicol Environ Health* 28:67–80, 1989a.

Evenson DP, Janca FC, Jost LK, Baer RK, Karabinus DS: Flow cytometric analysis of effects of 1,3-dinitrobenzene on rat spermatogenesis. *J Toxicol Environ Health* 28:81–98, 1989b.

Evenson DP, Baer RK, Jost LK: Long-term effects of triethylenemelamine exposure on mouse testis cells and sperm chromatic structure assayed by flow cytometry. *Environ Mol Mutagen* 14:79–89, 1989c.

Evenson DP, Baer RK, Jost LK: Flow cytometric analysis of rodent epididymal spermatozoal chromatin condensation and loss of free sulfhydryl groups. *Mol Reprod Dev* 1:283–288, 1989d.

Evenson DP, Jost LK, Baer RK, Turner TW, Schrader S: Individuality of DNA denaturation patterns in human sperm as measured by the sperm chromatic structure assay. *Reprod Toxicol* 5:115–125, 1991.

Everett JW: Pituitary and hypothalamus: Perspectives and overview, in Knobil E, Neil J et al (eds): *The Physiology of Reproduction*. New York, Raven, 1988, pp 1143–1160.

Ewing, LL, Zirkin BR, Chubb C: Assessment of testicular testosterone production and Leydig cell structure. *Environ Health Perspect* 38:19–27, 1981.

Fahim MS, Fahim Z, Hall DG: Effects of subtoxic lead levels of pregnant women in the state of Missouri. *Res Commun Chem Pathol Pharmacol* 13:309–331, 1976.

Foster PMD, Sheard CM, Lloyd SC: 1,3-Dinitrobenzene: A Sertoli cell toxicant? *Excerpta Med Int Congr Ser* 718:281–288, 1986.

Foster PMD, Lloyd SC, Prout MS: Toxicity and metabolism of 1,3-dinitrobenzene in rat testicular cell cultures. *Toxicol in Vitro* 1:31–37, 1987.

Francis EZ, Kimmel GL: Proceedings of the workshop on one- vs two-generation reproductive effects studies. *J Am Coll Toxicol* 7:911, 1988.

Gandelman R: Gonadal hormones and sensory function. *Neurosci Biobehav Rev* 7:1–17, 1983.

Generoso WM: Repair in fertilized eggs of mice and its role in the production of chromosomal aberrations, in Generoso WM, Shelby MD, De Serres FJ (eds): *DNA Repair and Mutagenesis in Eukaryotes*. New York, Plenum, 1980, pp 389–410.

Generoso WM, Rutledge JC, Cain KT, Hughes LA, Downing DJ: Mutagen-induced fetal anomalies and death following treatment of females within hours after mating. *Mutat Res* 199:175–181, 1988.

Goldman JM, Cooper RL, Laws SC, Rehnberg GL, Edwards TL, McElroy WK, Hein JF: Chlordimeform-induced alterations in endocrine regulation within the male rat reproductive system. *Toxicol Appl Pharmacol* 104:25–35, 1990.

Goldman JM, Cooper RL, Edwards TL, Rehnberg GL, McElroy WK, Hein JF: Suppression of luteinizing hormone surge by chlordimeform in ovariectomized, steroid-primed female rats. *Pharmacol Toxicol* 68:131–136, 1991.

Gooren LJG: Human male sexual functions do not require aromatization of testosterone: A study using tamoxifen, testolactone, and dihydrotestosterone. *Arch Sex Behav* 14:539–548, 1985.

Gould JE, Overstreet JW, Yanagimachi H, Yanagimachi R, Katz D, Hanson FW: What functions of the sperm cell are measured by in vitro fertilization of zona free hamster eggs? *Fertil Steril* 40:344–452, 1983.

Goy RW: Development of play and mounting behavior in female rhesus virilized prenatally with esters of testosterone or dihydrotestosterone, in Chivers DJ, Herbert J (eds): *Recent Advances in Primatology*. New York, Academic, 1978, pp 449–462.

Graham JK, Kunze E, Hammerstedt RH: Analysis of sperm cell viability, acrosomal integrity, and mitochondrial function using flow cytometry. *Biol Reprod* 43:55–64, 1990.

Gray LE Jr, Ostby JS, Ferrell JM, Sigmon ER, Goldman JM: Methoxychlor induces estrogen-like alterations of behavior and the reproductive tract in the female rat and hamster: Effects on sex behavior, running wheel activity, and uterine morphology. *Toxicol Appl Pharmacol* 96:525–540, 1988a.

Gray LE Jr, Ostby J, Sigmon R, Ferrell J, Rehnberg G, Linder R, Cooper R, Goldman J, Laskey J: The development of a protocol to assess reproductive effects of toxicants in the rat. *Reprod Toxicol* 2:281–287, 1988b.

Gray LE Jr, Ostby J, Ferrell J, Sigmon R, Cooper R, Linder R, Rehnberg G, Goldman J, Laskey J: Correlation of sperm and endocrine measures with reproductive success in rodents, in Burger EJ Jr, Tardiff RG, Scialli AR, Zenich H (eds): *Sperm Measures and Reproductive Success: Institute for Health Policy Analysis Forum on Science, Health, and Environmental Risk Assessment*. New York, Alan R Liss, 1989a, pp 193–209.

Gray LE Jr, Ostby J, Ferrell J, Rehnberg G, Linder R, Cooper R, Goldman J, Slott V, Laskey J: A dose–response analysis of methoxychlor-induced alterations of reproductive development and function in the rat. *Fundam Appl Toxicol* 12:92–108, 1989b.

Gray LE Jr, Ostby J, Linder R, Goldman J, Rehnberg G, Cooper R: Carbendazim-induced alterations of reproductive development and function in the rat and hamster. *Fundam Appl Toxicol* 15:281–297, 1990.

Gray LE Jr, Ostby J, Sigmon R, Linder R: A fungicide (fenarimol) that inhibits fungal sterol synthesis also reduces mating behavior and fertility in male rats. *Biol Reprod* 44:132, 1991.

Grunert JH, De Geyter C, Bordt J, Schneider HPG, Nieschlag E: Does computerized image analysis of sperm movement enhance the predictive value of semen analysis for in-vitro fertilization results? *Int J Androl* 12:329–338, 1989.

Hadley MA, Byers SW, Suarez-Quian CA, Kleinman HK, Dym M: Extracellular matrix regulates Sertoli cell differentiation, testicular cord formation, and germ cell development in vitro. *J Cell Biol* 101:1511–1522, 1985.

Hadley MA, Byers SW, Suarez-Quian CS, Djakiew D, Dym M: In vitro models of differentiated Sertoli cell structure and function. *In Vitro Cell Dev Biol* 24:550–557, 1988.

Hamilton GS, Armstrong DT: The superovulation of synchronous adult rats using follicle-stimulating hormone delivered by continuous infusion. *Biol Reprod* 44:851–856, 1991.

Haney AF: Effects of toxic agents on ovarian function, in Thomas JA et al (eds): *Endocrine Toxicology.* New York, Raven, 1985, pp 181–210.

Hansen S, Gummesson BM: Participation of the lateral midbrain tegmentum in the neuroendocrine control of sexual behavior and lactation in the rat. *Brain Res* 251:319–325, 1982.

Harper MJK: Gamete and zygote transport, in Knobil E, Neill J et al (eds): *The Physiology of Reproduction.* New York, Raven, 1988, pp 103–134.

Harrison RAP, Vickers SE: Use of fluorescent probes to assess membrane integrity in mammalian spermatozoa. *J Reprod Fertil* 88:343–352, 1990.

Hatch MC: Assessing female fertility. *Reprod Toxicol* 2:241–245, 1988.

Hecht NB: Detecting the effects of toxic agents on spermatogenesis using DNA probes. *Environ Health Perspect* 74:31–40, 1987.

Heindel JJ, Treinen KA: Physiology of the male reproductive system: Endocrine, paracrine and autocrine regulation. *Toxicol Pathol* 17:411–445, 1989.

Hembree WC, Overstreet JW: Defects in human sperm function in clinical infertility reduced zona pellucida penetration in vitro, in Troen P, Nankin HR (eds): *The Testis in Normal and Infertile Men.* New York, Raven, 1977, pp 513–517.

Hess RA, Linder RE, Strader LF, Perreault SD: Acute effects and long-term sequelae of 1,3-dinitrobenzene on male reproduction in the rat: II. Quantitative and qualitative histopathology of the testis. *J Androl* 9:327–342, 1988.

Hess RA, Schaeffer DJ, Eroschenko VP, Keen JE: Frequency of the stages in the cycle of the seminiferous epithelium in the rat. *Biol Reprod* 43:517–524, 1990.

Higashi Y, Yoshida K, Oshima H: In vitro inhibition by ketoconazole of human testicular steroid oxidoreductases. *J Steroid Biochem* 36:667–671, 1990.

Holloway AJ, Moore HDM, Foster PDM: The use of in vitro fertilization to detect reductions in the fertility of male rats exposed to 1,3-dinitrobenzene. *Fundam Appl Toxicol* 14:113–122, 1990.

Hopwood NJ: Pathogenesis and management of abnormal puberty. *Spec Top Endocrinol Metab* 7:175–236, 1985.

Hurtt ME, Zenick H: Decreasing epididymal sperm reserves enhance the detection of ethoxyethanol-induced spermatotoxicity. *Fundam Appl Toxicol* 7:348–353, 1986.

Ibrahim ME, Moussa MMA, Pederson H: Efficacy of zona-free hamster egg sperm penetration assay as a predictor of in vitro fertilization. *Arch Androl* 23:267–274, 1989.

Jackson H: The effects of alkylating agents on fertility. *Br Med Bull* 20:107–114, 1964.

Jakubowiak A, Janecki A, Steinberger A: Action kinetics of inhibin in superfused pituitary cells depend on gonadotropin-releasing hormone treatment. *Endocrinology* 127:211–217, 1990.

Jakubowiak A, Janecki A, Steinberger A: Similar effects of inhibin and cycloheximide on gonadotropin release in superfused pituitary cell cultures. *Biol Reprod* 41:454–463, 1989.

Janecki A, Steinberger A: Polarized Sertoli cell functions in a new two-compartment culture system. *J Androl* 7:69–71, 1986.

Janecki A, Lukaszyk A, Sobieszczyk S: Stimulatory effect of Sertoli cell culture medium on the FSH release by pituitary halves in vitro. *Horm Metab Res* 16:581–584, 1984.

Janecki A, Jakubowiak A, Steinberger A: Effect of germ cells on vectorial secretion of androgen binding protein and transferrin by immature rat Sertoli cells in vitro. *J Androl* 9:126–132, 1988.

Johnson L: Spermatogenesis and aging in the human. *J Androl* 7:331–354, 1986.

Johnson L: Spermatogenesis, in Cupps PT (ed): *Reproduction in Domestic Animals* (4th ed). New York, Academic, 1991, pp 173–219.

Kaipia A, Toppari J, Mali P, Kangasniemi M, Alcivar AA, Hecht NB, Parvinen M: Stage- and cell-specific expression of the ornithine decarboxylase gene during rat mouse spermatogenesis. *Mol Cell Endocrinol* 73:45–52, 1990.

Katoh M, Cacheiro NLA, Cornett CV, Cain KT, Rutledge JC, Generoso WM: Fetal anomalies produced subsequent to treatment of zygotes with ethylene oxide or ethyl methanesulfonate are not likely due to the usual genetic causes. *Mutat Res* 210:337–344, 1989.

Katoh MA, Cain KT, Hughes LA, Foxworth LB, Bishop JB, Generoso WM: Female-specific dominant lethal effects in mice. *Mutat Res* 23:205–217, 1990.

Katz DF, Overstreet JW, Samuels SJ, Niswander PW, Bloom TD, Lewis EL: Morphometric analysis of spermatozoa in the assessment of human male fertility. *J Androl* 7:203–210, 1986.

Kelce WR, Raisbeck MF, Ganjam VK: Gonadotoxic effects of 2-hexanone and 1,2-dibromo-3-chloropropane on the enzymatic activity of rat testicular 17α-hydroxylase/$C_{17,20}$-lyase. *Toxicol Lett* 42:331–338, 1990.

Kleeman JM, Moore RW, Peterson RE: Inhibition of testicular steroidogenesis in 2,3,7,8-tetrachlorodibenzo-*p*-dioxin-treated rats: Evidence that the key lesion occurs prior to or during pregnenolone formation. *Toxicol Appl Pharmacol* 106:112–126, 1990.

Klinefelter GR, Hamilton DW: Synthesis and secretion of proteins by perifused caput epididymal tubules, and association of secreted proteins with spermatozoa. *Biol Reprod* 33:1017–1027, 1985.

Klinefelter GR, Hall PF, Ewing LL: Effect of luteinizing hormone deprivation in situ on steroidogenesis of rat Leydig cells purified by a multistep procedure. *Biol Reprod* 36:769–783, 1987.

Klinefelter GR, Laskey JW, Roberts NR, Slott V, Suarez JD: Multiple effects of ethane dimethanesulfonate on the epididymis of adult rats. *Toxicol Appl Pharmacol* 105:271–287, 1990.

Klinefelter GR, Laskey JW, Roberts NL: In vitro/in vivo effects of ethane dimethanesulfonate on Leydig cells of adults rats. *Toxicol Appl Pharmacol* 107:460–471, 1991a.

Klinefelter GR, Gray LE Jr, Suarez JD: The method of sperm collection significantly influences sperm motion parameters following ethane dimethanesulphonate administration in the rat. *Reprod Toxicol* 5:39–44, 1991b.

Klinefelter GR, Roberts NL, Suarez JD: Direct effects of ethane dimethanesulphonate in adult rats: An in vitro demonstration *J Androl* (in press) 1992a.

Klinefelter G, Perreault S, Ferrell J, Jeffay S, Suarez J, Roberts N: Altered membrane protein, rather than sperm motility, may be an underlying cause of the infertility of sperm from ethane dimethanesulphonate (EDS)-treated rats. *Biol Reprod* (suppl) 1992b.

Krawetz SA, Cannon PD, Joshi R, Dixon GH: Molecular probes for general testicular and specific spermatogenic function. *Mol Cell Probes* 3:109–115, 1989.

Kunze EL, Hammerstedt RH: Flow cytometric characterization of ram ductal epididymal cells using light scatter and nile red staining. Eleventh North American Testis Workshop: The Male Germ Cell: Spermatogonium to Fertilization. Montreal, April 24–27, 1991.

Lamb JC IV, Chapin RE: Experimental models of male reproductive toxicology, in Thomas JA, Korach KJ, McLachlan JA (eds): *Experimental Models of Male Reproductive Toxicology*. New York, Raven, 1985, pp 85–115.

Laskey JW, Phelps PV: Effect of cadmium and other metal cations in vitro Leydig cell testosterone production. *Toxicol Appl Pharmacol* 108:296–306, 1991.

Laskey JW, Berman E, Gray LE Jr, Ferrell J, Carter H: Steroidogenic assessment using ovary culture in cycling rats: Effects of bis(2-diethylhexyl) phthalate on ovarian steroid production. *Reprod Toxicol* 7:1993.

Lauwerys R, Roels H, Genet P, Toussaint G, Bouchaert A, De Cooman S: Fertility of male workers exposed to mercury vapor or to manganese dust: A questionnaire study. *Am J Ind Med* 7:171–176, 1985.

Lavranos TC, Seamark SF: Addition of steroids to embryo–uterine monolayer coculture enhances embryo survival and implantation in vitro. *Reprod Fertil Develop* 1:41–46, 1989.

Legan SJ, Karsch FJ: A daily signal for the LH surge in the rat. *Endocrinology* 96:57–62, 1975.

Levine RJ, Symons MJ, Balogh SA, Milby TH, Whorton MD: A method for monitoring the fertility of workers: Validation of the method among workers exposed to dibromochloropropane. *J Occup Med* 23:183–188, 1981.

Linder RE, Strader LF, McElroy WK: Measurement of epididymal sperm motility as a test variable in the rat. *Bull Environ Contam Toxicol* 36:316–324, 1986a.

Linder RE, Hess RA, Strader LF: Testicular toxicity and infertility in male rats treated with 1,3-dinitrobenzene. *J Toxicol Environ Health* 19:477–489, 1986b.

Linder RE, Hess RA, Perreault SD, Strader LF, Barbee RR: Acute effects and long-term sequelae of 1,3-dinitrobenzene on male reproduction in the rat: I. Sperm quality, quantity, and fertilizing ability. *J Androl* 9:317–326, 1988.

Linder RE, Strader LF, Barbee RR, Rehnberg GL, Perreault SD: Reproductive toxicity of a single dose of 1,3-dinitrobenzene in two ages of young adult male rats. *Fundam Appl Toxicol* 14:284–298, 1990.

Liu YX, Hsueh AJW: Synergism between granulosa and theca-interstitial cells in estrogen biosynthesis by gonadotropin-treated rat ovaries: Studies on the two-cell, two-gonadotropin hypothesis using steroid antisera. *Biol Reprod* 34:27–36, 1986.

Liu DY, Clarke GN, Baker HWG: Relationship between sperm motility assessed with the Hamilton-Thorn motility analyzer and fertilization rates in vitro. *J Andrology* 12:231–239, 1991.

Lyons MF: Sensitivity of various germ-cell stages to environmental mutagens. *Mut Res* 84:323–345, 1981.

McClain RM, Downing JC: The effect of ornidazole on fertility and epididymal sperm function in rats. *Toxicol Appl Pharmacol* 92:488–496, 1988.

MacLean PD, Dua S, Denniston RH: Cerebral localization for scratching and seminal discharge. *Arch Neurol* 9:485–497, 1963.

Mallea L, Mas J, Padron RS, Diaz JW: Transferrin in seminal plasma of fertile and infertile men. *Andrologia* 20:15–20, 1988.

Manson JM, Kang YJ: Test methods for assessing female reproductive and developmental toxicology, in Hayes AW (ed): *Principles and Methods of Toxicology* (2d ed). New York, Raven, 1989, pp 311–359.

Matsumoto AM, Bremner WJ: Endocrine control of human spermatogenesis. *J Steroid Biochem* 33:789–790, 1989.

Mattison DR: Clinical manifestations of ovarian toxicity, in Dixon RL (ed): *Reproductive Toxicology.* New York, Raven, 1985, pp 109–130.

Mattison DR, Ross GT: Laboratory methods for evaluating and predicting specific reproductive dysfunctions: Oogenesis and ovulation, in Vouk VB, Sheehan PJ (eds): *Methods for Assessing the Effects of Chemicals on Reproductive Functions.* New York, Wiley, 1983, pp 217–246.

Mattison DR, Bogumil RJ, Chapin R, Hatch M, Hendricks A, Jarrell J, LaBarbera AL, Schrader SM, Selevan S: Reproductive effects of pesticides, in Baker SR, Wilkinson CF (eds): *The Effects of Pesticides on Human Health.* Princeton, NJ, Princeton Scientific Publishing, 1990a, pp 297–389.

Mattison DR, Plowchalk DR, Meadows MJ, Al-Juburi AZ, Gandy J, Malek A: Reproductive toxicity: Male and female reproductive systems as targets for chemical injury. *Med Clin North Am* 74:391–411, 1990b.

Mattson BA, Albertini DF: Oogenesis: Chromatin and microtubule dynamics during meiotic prophase. *Molec Reprod Dev* 25:374–383, 1990.

Meistrich ML: Critical components of testicular function and sensitivity to disruption. *Biol Reprod* 34:17–28, 1986.

Meistrich ML: Evaluation of reproductive toxicity by testicular sperm head counts. *J Am Coll Toxicol* 8:551–567, 1989a.

Meistrich ML: Calculation of the incidence of infertility in human populations from sperm measures using the two-distribution model, in Burger EJ Jr, Tardiff RG, Scialli AR, Zenick H (eds): *Sperm Measures and Reproductive Success: Institute for Health Policy Analysis—Forum on Science, Health, and Environmental Risk Assessment,* New York, Alan R Liss, 1989b, pp 275–290.

Meistrich ML, Brown CC: Estimation of the increased risk of human infertility from alterations in semen characteristics. *Fertil Steril* 40:220–230, 1983.

Meyer-Bahlburg HFL, Ehrhardt AA, Feldman JF, Rosen LR, Veridiano NP, Zimmerman I: Sexual activity level and sexual functioning in women prenatally exposed to diethylstilbestrol. *Psychosom Med* 47:496–511, 1985.

Middleton MC, Milne CM, Moreland D, Hasmall RL: Ovulation in rats is delayed by a substituted triazole. *Toxicol Appl Pharmacol* 83:230–239, 1986.

Milne CM, Hasmall RL, Russell A, Watson SC, Vaughan Z, Middleton MC: Reduced estradiol production by a substituted triazole results in delayed ovulation in rats. *Toxicol Appl Pharmacol* 90:426–435, 1987.

Miyazaki T, Sueoka K, Dharmarajan AM, Atlas SJ, Bulkley GB, Wallach EE: Effect of inhibition of oxygen free radical on ovulation and progesterone production by the in vitro perfused rabbit ovary. *J Reprod Fertil* 91:207–212, 1991.

Morales P, Katz DF, Overstreet JW, Samuels SJ, Chang RJ: The relationship between the motility and morphology of spermatozoa in human semen. *J Androl* 9:241–247, 1988.

Morrissey RE, Lamb JC IV, Morris RW, Chapin RE, Dushyant KG, Heindel JJ: Results and evaluations of 48 continuous breeding reproduction studies conducted in mice. *Fundam Appl Toxicol* 13:747–777, 1989.

Moruzzi JF, Wyrobek AJ, Mayall BH, Gledhill BL: Quantification and classification of human sperm morphology by computer-assisted image analysis. *Fertil Steril* 50:142–152, 1988.

Murakami U: The effect of organic mercury on intrauterine life, in Klingberg MA, Abramovici A, Chemke J (eds): *Drugs and Fetal Development*. New York, Plenum, 1972, pp 301–336.

Nankin HR, Castaneda E, Troen P: Endocrine profiles in oligospermic men, in Troen P, Nankin HR (eds): *The Testis in Normal and Infertile Men*. New York, Raven, 1977, pp 529–537.

National Research Council: *Biologic Markers in Reproductive Toxicology*. Washington, DC, National Academy Press, 1989.

Numan M: Maternal behavior, in Knobil E, Neill J et al (eds): *The Physiology of Reproduction*. New York, Raven, 1988, pp 1569–1645.

Nuti KM, Sridharan BN, Meyer RK: Reproductive biology of PMSG primed immature female rats. *Biol Reprod* 13:38–44, 1975.

O'Brien DA, Froysa A, Rockett D: Sertoli cell-conditioned medium increases viability and ATP levels of pachytene spermatocytes and round spermatids in vitro. *J Cell Biol* 103:485a, 1986.

Office of Technology Assessment: *Reproductive Health Hazards in the Workplace*. OTA-BA-266. Government Printing Office, December 1985.

Onoda M, Djakiew D: Modulation of Sertoli cell secretory function by rat round spermatid protein(s). *Mol Cell Endocrinol* 72:35–44, 1990.

Orgebin-Crist MC, Jahad N: The maturation of rabbit epididymal spermatazoa in organ culture: Inhibition by antiandrogens and inhibitors of ribonucleic acid and protein synthesis. *Endocrinology* 103:46–53, 1978.

Oshima H, Nankin HR, Troen P, Yoshida KI, Ochi-ai KI: Leydig cell number and function in infertile men, in Troen P, Nankin HR (eds): *The Testis in Normal and Infertile Men*. New York, Raven, 1977, pp 445–455.

Oudiz DJ, Zenick H, Niewenhuis RJ, McGinnis PM: Male reproductive toxicity and recovery associated with acute ethoxyethanol exposure in rats. *J Toxicol Environ Health* 13:763–775, 1984.

Overstreet JW: Laboratory tests for human male reproductive risk assessment. *Teratogenesis Carcinog Mutagen* 4:67–82, 1984.

Palmer AK: Regulatory requirements for reproductive toxicology: Theory and practice, in Kimmel CA, Buelke-Sam J (eds): *Developmental Toxicology*. New York, Raven, 1981, pp 259–287.

Pampfer S, Vanderheyden I, Miciels B, De Hertogh R: Co-culture of two-cell rat embryos on cell monolayers. *In Vitro Cell Dev Biol* 26:944–948, 1990.

Parvinen M, Ruokonen A: Endogenous steroids in rat seminiferous tubules. Comparison of different spermatogenic stages isolated by transillumination-assisted microdissection. *J Androl* 3:211–220, 1982.

Pederson RA, Brandriff B: Radiation and drug-induced DNA repair in mammalian oocytes and embryos, in Generoso WM, Shelby MD, De Serres FJ (eds): *DNA Repair and Mutagenesis in Eukaryotes*. New York, Plenum, 1980, pp 389–410.

Pederson T, Peters H: Proposal for a classification of oocytes and follicles in the mouse ovary. *J Reprod Fertil* 17:555–557, 1968.

Peluso JJ: Role of the amplitude of the gonadotropin surge in the rat. *Fertil Steril* 53:150–154, 1990.

Perreault SD: Impaired gamete function: Implications for reproductive toxicology, in Working PK (ed): *Toxicology of the Male and Female Reproductive Systems*. New York, Hemisphere Publishing, 1989, pp 217–229.

Perreault SD, Zaneveld LJD, Rogers BJ: Inhibition of fertilization in the hamster by sodium aurothiomalate, a hyaluronidase inhibitor. *J Reprod Fertil* 60:461–467, 1980.

Perreault SD, Linder RE, Strader LF, Slott V: The value of multiple endpoint data in male reproductive toxicology: Revelations in the rat, in Burger EJ Jr, Tardiff RG, Scialli AR, Zenick H (eds): *Sperm Measures and Reproductive Success: Institute for Health Policy Analysis—Forum on Science, Health, and Environmental Risk Assessment*. New York, Alan R Liss, 1989, pp 179–192.

Pfaff DW, Schwartz-Giblin S: Cellular mechanisms of female reproductive behaviors, in Knobil E, Neill J (eds): *The Physiology of Reproduction*. New York, Raven, 1988, pp 1487–1568.

Phelps PV, Laskey JW: Comparison of age-related changes in in vivo and in vitro measures of testicular steroidogenesis after acute cadmium exposure in the Sprague-Sawley rat. *J Toxicol Environ Health* 27:95–105, 1989.

Plowchalk DR, Mattison DR: Phosphoramide mustard is responsible for the ovarian toxicity of cyclophosphamide. *Toxicol Appl Pharmacol* 107:472–481, 1991.

Punnonen R, Ashorn R, Vilja P, Heinonen PK, Kuhansuu E, Tuohimaa P: Spontaneous luteinizing hormone surge and cleavage of in vitro fertilized embryos. *Fertil Steril* 49:479–482, 1988.

Purvis K, Hallgeir R, Scholberg A, Hesla S, Clausen OPF: Application of flow cytometry to studies on the human acrosome. *J Androl* 11:361–366, 1990.

Qiu J, Hales B, Robaire B: Adverse effects of cyclophosphamide (CPA) on progeny outcome can be mediated through the epididymis in rats. *Biol Reprod* 46:926–931, 1992.

Rao KS, Burek JD, Murray FJ, John JA, Schwetz BA, Beyer JE, Parker CM: Toxicologic and reproductive effects of inhaled 1,2-dibromo-3-chloropropane in male rabbits. *Fundam Appl Toxicol* 2:241–251, 1982.

Rehnberg GL, Linder R, Goldman JW, Hein JF, McElroy WK, Cooper RL: Changes in testicular and serum hormone concentrations in the male rat following treatment with *m*-dinotrobenzene. *Toxicol Appl Pharmacol* 95:255–264, 1988.

Richards JS, Bogvich K: Development of gonadotropin receptors during follicular growth, in Maresh VB, Muldoon TG, Saxena BB, Sadler WA (eds): *Functional Correlates of Hormone Receptors in Reproduction.* Amsterdam, Elsevier North Holland, 1980, pp 223–244.

Robaire B, Hermo L: Efferent ducts, epididymis, and vas deferens: Structure, functions, and their regulation, in Knobil E, Neill J et al (eds): *The Physiology of Reproduction.* New York, Raven, 1988, pp 999–1080.

Robb GW, Amann RP, Killian GJ: Daily sperm production and epididymal sperm reserves of pubertal and adult rats. *J Reprod Fertil* 54:103–107, 1978.

Rudak E, Jacobs PA, Yanagimachi R: Direct analysis of the chromosome constitution of human spermatozoa. *Nature* 274:911, 1978.

Russell LD, Ettlin RA, Hikim APS, Clegg ED: *Histological and Histopathological Evaluation of the Testis.* Clearwater, FL, Cache River Press, 1991.

Sachs, BD, Meisel RL: The physiology of male sexual behavior, in Knobil E, Neill J (eds): *The Physiology of Reproduction.* New York, Raven, 1988, pp 1393–1485.

Sakkas D, Trounson AO: Co-culture of mouse embryos with oviduct and uterine cells prepared from mice at different days of pseudo-pregnancy. *J Reprod Fertil* 90:109–118, 1990.

Sanders SA, Reinisch JM: Behavioral effects on humans of progesterone-related compounds during development and in the adult, in Ganten D, Pfaff D (eds): *Current Topics in Neuroendocrinology* (vol. 5). Berlin, Heidelberg, Springer-Verlag, 1985, pp 175–205.

Sandler M (ed): *Mental Illness in Pregnancy and the Puerperium.* New York, Oxford University Press, 1978.

Schardein JL (ed): *Chemically Induced Birth Defects.* New York, Marcel Dekker, 1985.

Schilsky RL, Lewis BJ, Sherins RJ, Young RC: Gonadal dysfunction in patients receiving chemotherapy for cancer. *Ann Intern Med* 93:109–114, 1980.

Schmidt G, Jorgenson J, Kannisto P, Liedberg F, Ottesen B, Owman C: Vasoactive intestinal polypeptide in the PMSG-primed immature rat ovary and its effect on ovulation in the isolated rat ovary perfused in vitro. *J Reprod Fertil* 90:465–472, 1990.

Schrader SM, Turner TW, Ratcliffe JM: The effects of ethylene dibromide on semen quality: A comparison of short-term and chronic exposure. *Reprod Toxicol* 2:191–198, 1988.

Sega GA: Relationship between unscheduled DNA synthesis and mutation induction in male mice, in Generoso WM, Shelby MD, De Serres FJ (eds): *DNA Repair and Mutagenesis in Eukaryotes.* New York, Plenum, 1980, pp 373–387.

Setchell BP, Brooks DE: Anatomy, vasculature, innervation, and fluids of the male reproductive tract, in Knobil E, Neill J (eds): *The Physiology of Reproduction.* New York, Raven, 1988, pp 753–836.

Sharpe RM, Maddocks S, Kerr JB: Cell–cell interactions in the control of spermatogenesis as studied using Leydig cell destruction and testosterone replacement. *Am J Anat* 188:3–20, 1990.

Sherins RJ, Brightwell D, Sternthal PM: Longitudinal analysis of semen of fertile and infertile men, in Troen P, Nankin HR (eds): *The Testis in Normal and Infertile Men.* New York, Raven, 1977, pp 473–488.

Siew S, Troen P, Nankin HR: Ultrastructure of testicular biopsies from infertile men, in Troen P, Nankin HR (eds): *The Testis in Normal and Infertile Men.* New York, Raven, 1977, pp 137–152.

Sirard MA, Bilodeau S: Effects of granulosa cell co-culture on in-vitro meiotic resumption of bovine oocytes. *J Reprod Fertil* 89:459–465, 1990.

Slott VL, Suarez JD, Simmons JE, Perreault SD: Acute inhalation exposure to epichlorhydrin transiently decreases rat sperm velocity. *Toxicol Appl Pharmacol* 15:597–606, 1990.

Smith BJ, Sipes IG: Epoxidation of 4-vinylcyclohexene by human hepatic microsomes. *Toxicol Appl Pharmacol* 109:367–371, 1991.

Smith BJ, Mattison DR, Sipes IG: The role of epoxidation in 4-vinylcyclohexene-induced ovarian toxicity. *Toxicol Appl Pharmacol* 105:372–381, 1990.

Speroff L, Glass RH, Kase NG: Investigation of the infertile couple, in *Clinical Gynecologic Endocrinology and Infertility* (3d ed). Baltimore, Williams & Wilkins, 1983, pp 467–492.

Steinberger E, Lloyd JA: Chemicals affecting the development of reproductive capacity, in Dixon RL (ed): *Reproductive Toxicology.* New York, Raven, 1985, pp 1–20.

Stirling D, Waterman MR, Simpson ER: Expression of mRNA encoding basic fibroblast growth factor (bFGF) in bovine corpora lutea and cultured luteal cells. *J Reprod Fertil* 91:1–8, 1991.

Tezon JG, Blaquier JA: The organ culture of human epididymal tubules and their response to androgens. *Molec Cell Endocrinol* 21:233–242, 1981.

Toner JP, Adler NT: Potency of rat ejaculations varies with their order and with male age. *Physiol Behav* 35:113–115, 1985.

Toppari J, Bishop PC, Parker JW, Ahmad N, Girgis W, DiZerega GS: Cytotoxic effects of cyclophosphamide in the mouse seminiferous epithelium: DNA flow cytometric and morphometric analysis. *Fundam Appl Toxicol* 15:44–52, 1990.

Toth GP, Zenick H, Smith MK: Effects of epichlorhydrin on male and female reproduction in Long–Evans rats. *Fundam Appl Toxicol* 13:16–25, 1989a.

Toth GP, Stober JA, Read EJ, Zenick H, Smith MK: The automated analysis of rat sperm motility following subchronic epichlorohydrin administration: Methodologic and statistical considerations. *J Androl* 10:401–415, 1989b.

Trasler JM, Hermo L, Robaire BB: Morphological changes in the testis and epididymis of rats treated with cyclophosphamide: A quantitative approach. *Biol Reprod* 38:463–479, 1988.

Vanluchene E, Hinting A, Dhont M, DeSutter P, Van Maele G, Vanderckhove D: Follicular fluid steroid levels in relation to oocyte maturity and in vitro fertilization. *J Steroid Biochem Molec Biol* 38:83–87, 1991.

Weber MM, Will A, Adelmann B, Engelhardt D: Effect of ketoconazole on human ovarian $C_{17,20}$-desmolase and aromatase. *J Steroid Biochem Molec Biol* 38:213–218, 1991.

White KL, Hehnke K, Rickords LF, Southern LL, Thompson DL, Wood TC: Early embryonic development in vitro by coculture with oviductal epithelial cells in pigs. *Biol Reprod* 41:425–430, 1989.

Whorton D, Krauss RM, Marshall S, Milby TH: Infertility in male pesticide workers. *Lancet* 2:1259–1261, 1977.

Wilcox AJ: Surveillance of pregnancy loss in human populations. *Am J Indust Med* 4:285–291, 1983.

Williams J, Foster PMD: The production of lactate and pyruvate as sensitive indices of altered rat Sertoli cell function in vitro following the addition of various testicular toxicants. *Toxicol Appl Pharmacol* 94:160–170, 1988.

Williams J, Gladen BC, Schrader SM, Turner TW, Phelps JL, Chapin RE: Semen analysis and fertility assessment in rabbits: Statistical power and design considerations for toxicology studies. *Fundam Appl Toxicol* 15:651–665, 1990.

Williams J, Gladen BC, Turner TW, Schrader SM, Chapin RE: The effects of ethylene dibromide on semen quality and fertility in the rabbit: Evaluation of a model for human seminal characteristics. *Fundam Appl Toxicol* 16:687–700, 1991.

Wong PYD, Lau SKD, Fu WO: Antifertility effects of some sulphonamides and related compounds and their accumulation in the epididymis of male rats. *J Reprod Fertil* 81:259–267, 1987.

Working PK: Mechanistic approaches in the study of testicular toxicity: Agents that directly affect the testis. *Toxicol Pathol* 17:452–456, 1989.

Working PK, Bus JS, Hamm TE Jr: Reproductive effects of inhaled methyl chloride in the male Fischer 344 rat. *Toxicol Appl Pharmacol* 77:133–143, 1985.

Wright WW: Germ cell–Sertoli cell interactions: Analysis of the biosynthesis and secretion of cyclic protein-2. *Dev Biol* 130:45–56, 1988.

Wright WW, Zabludoff SD, Erickson-Lawrence M, Karzai AW: Germ-cell–Sertoli cell interactions. *Ann NY Acad Sci* 564:173–185, 1989.

Wyrobek AJ, Gordon LA, Watchmaker G, Moore DH II: Human sperm morphology testing: Description of a reliable method and its statistical power, in Bridges BA, Butterworth BE, Weinstein IB (eds): *Banbury Report: Indicators of Genotoxic Exposure.* Cold Spring Harbor, Cold Spring Harbor Laboratory, 1982, pp 527–541.

Wyrobek AJ, Gordon LA, Burkhart JG, Francis MW, Kapp RW Jr, Letz G, Malling HV, Topham JC, Whorton MD: *An Evaluation of Human Sperm as Indicators of Chemically Induced Alterations of Spermatogenic Function: A Report of the US Environmental Protection Agency Gene-Tox Program.* New York, Elsevier Science Publishers, 1983.

Wyrobek AJ, Segraves M, Stilwell J, Barton D, Stanker L, Balhorn R: Fluorescence cytometric detection and analysis of rare human sperm nuclei that don't bind monoclonal antibodies for human protamines 1 or 2. *J Cell Biol* 111:110a, 1990.

Yanagimachi R, Yanagimachi H, Roger BJ: The use of zona-free animal ova as a test-system for the assessment of the fertilizing capacity of human spermatozoa. *Biol Reprod* 15:471–476, 1976.

Yeung CH, Cooper TG, Senge T: Histochemical localization and quantification of α-glucosidase in the epididymis of men and laboratory animals. *Biol Reprod* 42:669–676, 1990.

Young WC: The hormones and mating behavior, in Young WC (ed): *Sex and Internal Secretions* (3d ed; vol 2). Baltimore, Williams & Wilkins, 1961, pp 1173–1239.

Zenick H, Clegg ED: Assessment of male reproductive toxicity: A risk assessment approach, in Hayes AW (ed): *Principles and Methods of Toxicology* (2d ed). New York, Raven, 1989, pp 275–309.

Zenick H, Blackburn K, Hope E, Oudiz D, Goeden H: Evaluation of male reproductive toxicity in rodents: A new animal model. *Teratogenesis Carcinog Mutagen* 4:109–128, 1984.

CHAPTER 9

Assessing Reproductive Potential in Men

Steven M. Schrader

Toxicants can attack the male reproductive system at one of several sites or at multiple sites. These sites and assays associated with their respective functions are discussed individually. This does not necessarily indicate, however, that there exists an absolute one-to-one relationship between a particular measurement and the associated site of action. These sites have been described in detail in Chaps. 1, 4, and 5 and include the neuroendocrine system, the testes, the excurrent duct system including the accessory sex glands, and sexual function.

The establishment of a male reproductive profile for assessing reproductive potential for both individual and population investigations is necessary. The same profile can be used for both types of studies, but there are some basic differences in methodology. The assessment profile illustrated here is that being used by the National Institute for Occupational Safety and Health (NIOSH) to assess populations exposed to potential reproductive toxicants. Differences between assessing the individual and assessing the population will be noted. A summary of assessments and specific methodologies follow. If individual data (versus population comparisons) are to be used, care should be taken to compare the results with the normal range of results of the laboratory conducting the analysis and not published values. If a population-based study is being conducted, a concurrent comparison cohort must be used and the analyses should be blind to exposure status.

NEUROENDOCRINE SYSTEM

As discussed in Chap. 4, the endocrine system, in concert with the nervous system, coordinates function of the various components of the reproductive axis, drawing upon external (e.g., sexual cues, temperature) and internal (e.g., checks and balances between endocrine tissue function, metabolic status) inputs. The reproductive endocrine status of the male is best established by measuring the hormones in the circulation, urine, or saliva. The hormones of interest are luteinizing hormone (LH), follicle-stimulating hormone (FSH), testosterone, and prolactin.

Since the circulating profile of LH is pulsatile, the status of this hormone for the individual, if measured in blood, should be estimated in serial samples. The pooled results of three samples collected at 20-min intervals will provide a reasonable estimate of mean concentration (Sokol, 1988). Alternatively, an integral of the pulsatile LH secretion rate may be obtained by measuring this gonadotropin in urine. If a population is being evaluated, a single blood sample per individual may suffice.

Circulating FSH levels are not as variable as those of LH. This is attributable in part to a longer circulating half-life for FSH compared with LH. Thus, analysis of a single blood sample for an individual will provide a more reliable estimate of FSH than of LH. FSH can also be measured in urine for the sake of convenience. Prolactin secretion is variable (Désir et al, 1982) and affected by several factors including a circadian rhythm (Sassin et al, 1972; Quigley et al, 1981), eating (Ishizuka et al, 1983), and physical and psychological stress (Frantz et al, 1972). These variables should be controlled whenever possible. This can be a challenge, however, since the stressful anticipation of providing a blood sample by venipuncture in itself may elevate serum prolactin levels in some individuals. Prolactin has not been demonstrated to be present in saliva, nor are specific immunoassays presently available for measuring intact or fragmented prolactin in urine. Therefore, venipuncture is required for estimating circulating prolactin levels. Because the effects of the stress of venipuncture on prolactin secretion may produce the greatest fluctuations, the most reliable measurement is by calmly but expediently collecting a single blood sample.

Approximately 2 percent of circulating testosterone is free, whereas the remainder is bound to sex hormone–binding globulin (SHBG), albumin, and other serum proteins. The free circulating testosterone is the active component and therefore provides a more accurate marker of physiologically available testosterone than total circulating testosterone does under conditions when SHBG concentration or binding is altered (Sokol, 1988). Circulating testosterone levels, like those of LH, fluctuate considerably over time. Estimates of free and total testosterone levels can be determined in single blood samples but are greatly improved by assaying multiple blood samples and pooling the results. Alternatively, a single measurement of a testosterone metabolite in urine (e.g., androsterone, etiocholanolone, or testosterone glucuronide) provides a convenient index of total testosterone (Bardin, 1986). Quantifying testosterone in saliva affords a convenient alternative to blood sampling while providing a measure of the unbound, biologically active component of circulating testosterone (Raid-Fahmy et al, 1982). Protein hormones such as the gonadotropins are not exuded into the saliva.

If steroid hormone metabolites in urine are measured, consideration must be given to the potential that the exposure being studied may alter the metabolism of excreted metabolites. This is especially pertinent since most metabolites are formed by the liver, a target of many toxicants. Lead, for example, reduces the amount of sulfated steroids that are excreted into the urine (Apostoli et al, 1989).

Blood levels of both gonadotropins become elevated during sleep as the male enters puberty, while testosterone levels maintain this diurnal pattern through adulthood in men (Plant, 1988). Thus, blood, urine, or saliva samples should be collected at approximately the same time of day to avoid variations due to diurnal secretory patterns.

In summary, Table 9-1 provides the hormones to be measured. FSH, LH, testosterone, and prolactin can all be evaluated in a population-based study by assessing the hormone levels in a single blood sample from each man (preferable at about the same time of day). In the study of an individual, three blood samples should be collected 20 min apart, or urinary assessment of testosterone metabolites and gonadotropins should be conducted.

SEMEN ANALYSIS

Semen analysis provides a useful profile of the function of the male reproductive system. The various measurements that are routinely used in the assessment of occupational exposure are presented in Table 9-2. Exact instructions should be provided to each man to ensure that the semen sample is collected by masturbation after a set time of abstinence (usually 2 days) and delivered to the laboratory within 1 h from the time of ejaculation. The men should be instructed to maintain the semen at room temperature avoiding any temperature shock to the sperm cells. NIOSH produced a videotape with these instructions, and the tape is shown to all participants of a semen study (NIOSH, 1986). At the time of collecting the semen sample, each subject should record the duration of abstinence, time of semen collection, and any information regarding spillage. Providing a label on the jar facilitates the recording of this information.

Semen analyses can be conducted in two phases. The initial evaluation of the sample should be conducted when the sample arrives at the laboratory (or

Table 9-1 Endocrine profile for assessing reproductive toxicant effects

	Fluid for measurement		
Hormone	Saliva	Blood	Urine
Luteinizing hormone (LH)		X	X
Follicle-stimulating hormone (FSH)		X	X
Testosterone	X	X	X
Prolactin		X	

Table 9-2 Semen profile for assessing reproductive toxicant effects

Sperm concentration
Sperm viability
Vital stain
Hypoosmotic swelling
Sperm motility
Percent motile
Curvilinear velocity
Straight-line velocity
Linearity
Lateral head amplitude
Beat cross frequency
Sperm size and shape
Morphology
Morphometry
Semen parameters
pH
Volume
Marker chemicals from glands
Toxicant or metabolite concentration

field site), and should consist of recording the temperature, turbidity, color, liquefaction time, volume, osmolality, and pH of the semen. Video recordings for motility assessments, viability estimates, and sperm counts, the preparation of slides, and preservation of seminal plasma should also be conducted at this time. Morphologic and morphometric analyses of sperm on slides and motility and velocity analyses of the sperm recorded on videotapes [either manually or using a computer-assisted sperm analysis (CASA) system] may be conducted at a later time. If CASA is used, several sperm motility variables can be measured (see Table 9-2). These variables provide useful information on both the progression of sperm cells (curvilinear velocity, straight-line velocity, and linearity) and the sperm motility pattern (lateral head amplitude and beat cross frequency).

Sperm viability may be determined by two methods, eosin Y stain exclusion (Eliasson and Treichl, 1971) and hypoosmotic swelling (HOS assay) (Jeyendran et al, 1984). These techniques test respectively for the structural and functional integrity of the cell membrane (Schrader et al, 1986). Sperm concentration and motility characteristics should be measured in a chamber at least 10 μm deep in order for the sperm to move freely in all planes.

Measurements of sperm motility and velocity should be conducted using a microscope stage warmed to 37°C. An attempt to record 100 motile sperm per sample is desirable if one is interested in the distribution of velocity measurements, but 50 motile sperm will suffice if means are to be compared. If the videotapes are being used to calculate the percentage motility, one should avoid "hunting" for motile sperm. All fields examined or searched should be included in the calculations. Therefore, recording a certain number of arbitrary fields is advised. Some researchers have found it useful to record a given number of fields to determine sperm concentration and percentage motility and then extra fields for velocity estimates. If a CASA system is being used for motility and velocity estimates, the number of sperm per field needs to be reduced to minimize cell collisions. When a 10- to 20-μm-deep chamber is used, the sperm concentration should be less than 40 million/ml. Diluents (including seminal plasma), however, alter sperm velocity up to a dilution of about 1:1. The current recommendation is to dilute *all* samples 1 part semen in 1 part isoosmotic buffer. If this dilution does not reduce the sperm concentration below 40 million/ml, then an additional dilution in the same buffer should be performed on those concentrated samples.

Sperm morphology should be estimated on air-dried, stained semen smears. During the past 30 years, several schemes have been presented for the assessment of normal and abnormal sperm morphologies. Variation in sperm size and shape are not distinct, but rather a continuum. This provides a challenge within and especially among laboratories to establish a repeatable system for morphologic classification (MacLeod, 1951; Freund, 1966; Fredricson, 1979; Hanke, 1981). With recent advances of computerized image analyses, several methods of sperm morphometry have been introduced (Schmassmann et al, 1979; Katz et al, 1981; Schrader et al, 1984; Jagoe et al, 1986; DeStefano et al, 1987; Turner et al, 1988; Moruzzi et al, 1988). These morphometric analysis systems provide objective assessments of individual sperm head size and shape. Comparisons of measurements between different analysis systems should be avoided. Sperm morphometry is now routinely used as part of the assessment of reproductive hazards to the male worker (Schrader et al, 1987).

SEMEN PARAMETER AND TOXICANT SITE

Testes

Sperm count, sperm morphology, and sperm head morphometry all provide indices of the integrity of spermatogenesis and spermiogenesis. Thus, the number of sperm in the ejaculate is directly correlated with the number of germ cells per gram of testis (Zukerman et al, 1978), while abnormal morphology is probably a result of abnormal spermiogenesis. Azoospermia is probably the most severe observation as it is often an indication that type A spermatogonia have been lost and recovery is unlikely. The promising new methods, DNA stability and DNA adducts, will hopefully provide information about spermatogenesis at the genetic level.

Some toxicants have been shown to exhibit an effect at the testis–spermatogenesis–spermiogenesis site. Exposure to dibromochloropropane (DBCP) reduced sperm concentration in ejaculates to 46 million cells/ml in exposed workers compared with a median of 79 million cells/ml in unexposed men (Whorton et al, 1979). When the workers were removed from the exposure, those with reduced sperm counts experienced a partial recovery, while men who had been azoospermic remained sterile. Testicular biopsy revealed that the target of DBCP was the spermatogonium. This substantiates the severity of the effect when stem cells are the target of toxicants. There were no indications that DBCP exposure of men was associated with adverse pregnancy outcome (Potashnik and Abeliovich, 1985). Another example of a toxicant targeting the testis was the study of workers exposed to ethylene dibromide (EDB). They had more sperm with tapered heads and fewer sperm per ejaculate than did controls (Ratcliffe et al, 1987).

Genetic damage is difficult to detect in human sperm. Epidemiologic studies of large populations have demonstrated increased frequency of spontaneous abortions in women whose husbands were working as motor vehicle mechanics (McDonald et al, 1989). Such studies indicate a need for methods to detect genetic damage in human sperm.

Excurrent Duct System

Seminal plasma is not essential for fertilization. Thus, the artificial insemination of sperm collected from the epididymis results in conception. On the other hand, seminal plasma contributes importantly to the normal coitus–fertilization scenario. Seminal plasma serves as a vehicle for sperm transport, a buffer from the hostile acidic vaginal environment, and an initial energy source for the sperm. Cervical mucus prevents passage of seminal plasma into the uterus. Some constituents of seminal plasma, however, are carried into the uterus to the site of fertilization by adhering to the sperm membrane.

The viability and motility of spermatozoa in seminal plasma are typically a reflection of seminal plasma quality. Alterations in sperm viability, as measured by stain exclusion or by hypoosmotic swelling, or alterations in sperm motility parameters would suggest an effect on the excurrent duct system.

Biochemical analysis of seminal plasma provides insights into the function of the excurrent duct system. Chemicals that are secreted primarily by each of the glands of this system are typically selected to serve as a marker for each respective gland. For example, the epididymis is represented by glycerylphosphorylcholine (GPC), the seminal vesicles by fructose, and the prostate gland by zinc. Note that this type of analysis provides only gross information on glandular function and little or no information on the other secretory constituents. Measuring semen pH and osmolality provides additional general information on the nature of seminal plasma.

Seminal plasma may be analyzed for the presence of a toxicant or its metabolite. Heavy metals have been detected in seminal plasma using atomic absorption spectrophotometry (Stachel et al, 1989), while halogenated hydrocarbons have been measured in seminal fluid by gas chromatography after extraction[32] or protein-limiting filtration (Zikarge, 1986).

A toxicant or its metabolite may act directly on accessory sex glands to alter the quality or quantity of their secretions. Alternatively, the toxicant may enter the seminal plasma (Mann and Lutwak-Mann, 1982) and, thereby, affect the sperm, affect the body of the female partner after intercourse, or be carried to the site of fertilization on the sperm membrane and affect the ovum or conceptus.

There are few reports of toxicant effects on the excurrent duct system in humans. Ethylene dibromide (EDB) is one example of a toxicant that exerts posttesticular effects. Short-term exposure to the toxicant reduced sperm velocity and semen volume (Schrader et al, 1988). Chronic exposure decreased sperm motility and viability, decreased seminal fructose levels, and increased semen pH (Schrader et al, 1988). An EDB metabolite was present within the semen of some exposed workers (Zikarge, 1986). Other potential toxicants that have been detected in semen include lead, cadmium, hexachlorobenzene, hexachlorocyclohexane, dieldrin, and polychlorinated biphenyls (Stachel et al, 1989).

SEXUAL FUNCTION

Human sexual function refers to the integrated activities of the testes and secondary sex glands, the endocrine control systems, and the CNS-based behavioral and psychological components of reproduction (libido). Erection, ejaculation, and orgasm are three distinct, independent, physiologic and psychodynamic events that normally occur concurrently in men. If detail on function or mechanisms is desired, several reviews and in-depth reports are available (deGroat and Booth, 1980; Thomas, 1983; Krane, 1989).

Assessment of occupational exposure-induced anomalies of sexual function is difficult. The researcher usually must rely on the testimony and recall of the worker regarding his sexual function. This testimony may often be confounded by the bias of the individual to guard his ego or masculine image or to attribute a preexisting libido problem to exposures at work.

Burris and colleagues (1989) recently reported application of a monitor for assessing erection at home. If these results are substantiated, this may provide a convenient, nonclinical means by which to evaluate erectile function of exposed workers.

The assessment of ejaculate volume may provide information on the integrity of the emission phase of ejaculation. This is, of course, complicated by effects on accessory sex glands' secretory capacity. Thus, a semen sample of reduced volume but with a normal ratio of constituents (marker chemicals) supports a diagnosis of an emission phase defect.

PROMISING NEW METHODS

Several methods being evaluated for assessing the effects of toxicants on the male reproductive system are listed in Table 9-3. The sperm penetration assay (SPA), also known as the zona-free hamster egg penetration assay, provides information on sperm function. To perform this assay, sperm are capacitated and then incubated with hamster eggs with the zona pellucida removed. The frequency of sperm penetration is then calculated. This procedure has become quite routine in several laboratories, and with the development of an overnight capacitation system, samples could be sent to an analysis laboratory by overnight delivery (Rogers, 1988). Incubating the sperm–egg complex until the formation of the male pronucleus may be useful in sperm chromosome karyotyping (Martin, 1983).

The DNA stability assay may provide information regarding genetic damage to sperm. To perform the assay, the sperm DNA is stressed thermally or chemically and then stained with acridine orange. Double-stranded DNA is stained green while single-stranded DNA is stained red (very little if any RNA is present in mature sperm) (Evenson et al, 1980). This analysis is performed with a flow cytometer. Animals exposed to known mutagens have an increase in single-stranded DNA (Evenson et al, 1986, 1989a, b) indicating an increase in genetic damage. The fertility rate of bulls is correlated with the percentage of double-stranded DNA (Ballachey et al, 1987). A recent report indicates that the DNA stability assay is highly repeatable between ejaculates from the same man (Evenson et al, 1991). This procedure has been developed in only a few laboratories at this time; however, the sperm can be frozen on dry ice and shipped to the analysis laboratory without affecting the results (Evenson et al, 1991).

Methods to evaluate DNA adducts and DNA probes are on the horizon

Table 9-3 Promising new methods for assessing reproductive toxicant effects

Sperm penetration assay (SPA)
DNA stability assay
Sperm chromosome analysis
Sperm DNA adducts
Home nocturnal penile tumescence monitor

and may be useful in the future. DNA adducts, the complex formed between a toxicant or its metabolites and the sperm nucleotide, may be carried in the sperm chromatin to the site of fertilization. New methods for labeling nucleotides may permit the detection of these adducts in ejaculated sperm. DNA probes are monoclonal antibodies to distinct modifications of the DNA molecule (Hecht, 1987). These antibodies serve as specific probes that may allow the detection of subtle changes in the DNA code.

SUMMARY

Toxicants may attack the male reproductive system at several sites. An arsenal of laboratory assessment procedures is available for detecting the effects of toxicants at many of these sites (Table 9-4). Insults to the neuroendocrine system can be determined by quantifying reproductive hormones in either the blood or the urine. Effects on the reproductive organs can best be detected using semen analysis. While most of the methodologies were originally developed and validated for the clinical assessment of infertility, they have proved to be of great value in the assessment of cohorts of exposed men in population-based studies.

Table 9-4 Assessment of reproductive function

Assessment method	Endocrine system	Testes	Posttesticular	Sexual function
FSH	X			
LH	X			
Testosterone	X			
Prolactin	X			
Sperm count		X		
Sperm morphology and morphometry		X		
Sperm viability (vital stain and HOS)			X	
Sperm motility (% motile and velocity)		?	X	
Semen biochemistry			X	
Semen pH			X	
Sperm penetration assay		X	X	
Nocturnal penile measurements				X
Personal history				X

REFERENCES

Apostoli P, Romeo L, Peroni E, Ferioli A, Ferrari S, Pasini F, Aprili F: Steroid hormone sulphation in lead workers. *Br J Ind Med* 46:204–208, 1989.

Ballachey BE, Hohenboken WD, and Evenson DP: Heterogeneity of sperm nuclear chromatin structure and its relationship to bull fertility. *Biol Reprod* 36:915–925, 1987.

Bardin CW: Pituitary–testicular axis, in Chen SSC, Jaffe RB (eds): *Reproductive Endocrinology.* Philadelphia, WB Saunders, 1986, pp 177–199.

Burris AS, Banks SM, Sherins RJ: Quantitative assessment of nocturnal penile tumescence and rigidity in normal men using a home monitor. *J Androl* 10:492–497, 1989.

Collecting A Semen Sample. Cincinnati, National Institute for Occupational Safety and Health, 1986.

deGroat WC, Booth AM: Physiology of male sexual function. *Ann Intern Med* 92:329–331, 1980.

Désir D, Van Cauter E, L'Hermite M, Refetoff S, Jadot C, Caufriez A, Copinschi G, Robyn C: Effects of "jet lag" on hormonal patterns: III. Demonstration of an intrinsic circadian rhythmicity in the plasma prolactin. *J Clin Endocrinol Metab* 55:849–857, 1982.

DeStefano F, Annest JL, Kresnow MJ, Flock ML, Schrader SM: Automated semen analysis in large epidemiologic studies. *J Androl* 8:24, 1987.

Eliasson R, Treichl L: Supravital staining of human spermatozoa. *Fertil Steril* 22:134–137, 1971.

Evenson DP, Darzynkiewicz Z, Melamed MR: Relation of mammalian sperm chromatin heterogeneity for fertility. *Science* 210:1131–1133, 1980.

Evenson DP, Baer RK, Jost LK, Gesch RW: Toxicity of thiotepa on mouse spermatogenesis as determined by dual-parameter flow cytometry. *Toxicol Appl Pharmacol* 82:151–163, 1986.

Evenson DP, Baer RK, Jost LK. Long-term effects of triethylenemelamine exposure on mouse testis cells and sperm chromatin structure assayed by flow cytometry. *Environ Mol Mut* 14:79–89, 1989a.

Evenson DP, Janca FC, Jost LK, Baer RK, Karabinus DS: Flow cytometric analysis of effects of 1,3-dinitrobenzene. *J Toxicol Environ Health* 28:81–98, 1989b.

Evenson DP, Jost L, Baer R, Turner T, Schrader S: Individuality of DNA denaturation patterns in human sperm as measured by the sperm chromatin structure assay. *Reprod Toxicol* 5:115–125, 1991.

Frantz AG, Kleinberg DL, Noel GL: Studies on prolactin in man. *Rec Prog Horm Res* 28:527–590, 1972.

Fredricson B: Morphologic evaluation of spermatozoa in different laboratories. *Andrologia* 11:57–61, 1979.

Freund M: Standards for the rating of human sperm morphology. *Int J Fertil* 11:97–180, 1966.

Hanke LJ: *Comparison of Laboratories Conducting Sperm Morphology.* NIOSH report TA78-28. US Government Printing Office, 1981.

Hecht NB. Detecting the effects of toxic agents on spermatogenesis using DNA probes. *Environ Health Prospect* 74:31–40, 1987.

Ishizuka B, Quigley ME, Yen SSC: Pituitary hormone release in response to food ingestion: Evidence for neuroendocrine signals from gut to brain. *J Clin Endocrinol Metab* 57:1111–1116, 1983.

Jagoe JR, Washbrook NP, Hudson EA. Morphometry of spermatozoa using semiautomatic image analysis. *J Clin Pathol* 39:1347–1352, 1986.

Jeyendran RS, Van den Ven HH, Perez-Palaez M, Crabo BG, Zaneveld LJD: Development of an assay to assess the functional integrity of the human sperm membrane and its relationship to other semen characteristics. *J Reprod Fertil* 70:219–228, 1984.

Katz DF, Overstreet JW, Pelprey RJ: Integrated assessment of the motility, morphology, and morphometry of human spermatozoa. *INSERM* 103:97–100, 1981.

Krane RJ, Goldstein I, de Tejada IS: Impotence. *N Engl J Med* 321:1648–1659, 1989.

McDonald AD, McDonald JC, Armstrong B, Cherry NM, Nolin AD, Robert D: Fathers' occupation and pregnancy outcome. *Br J Ind Med* 46:329–333, 1989.

MacLeod J: Semen quality in 1000 men of known fertility and in 800 cases of infertile marriage. *Fertil Steril* 2:115–139, 1951.

Mann T, Lutwak-Mann C: Passage of chemicals into human and animal semen: Mechanisms and significance. *CRC Crit Rev Toxicol* 11:1–14, 1982.

Martin RH: A detailed method for obtaining preparations of human sperm chromosomes. *Cytogenet Cell Genet* 35:252–256, 1983.

Moruzzi JF, Wyrobek AJ, Mayall BH, Gledhill BL: Quantification and classification of human sperm morphology by CAIA. *Fertil Steril* 50:142–152, 1988.

Plant TM: Puberty in primates, in Knobil E, Neill JD (eds): *The Physiology of Reproduction.* New York, Raven, 1988, pp 1763–1788.

Potashnik G, Abeliovich D: Chromosomal analysis and health status of children conceived to men during or following dibromochloropropane-induced spermatogenic suppression. *Andrologia* 17:291–296, 1985.

Quigley ME, Ropert JF, Yen SSC: Acute prolactin release triggered by feeding. *J Clin Endocrinol Metab* 52:231–234, 1981.

Raid-Fahmy D, Read GF, Walker RF, Griffiths K: Steroids in saliva for assessing endocrine function. *Endocrinol Rev* 3:367–395, 1982.

Ratcliffe JM, Schrader SM, Steenland K, Clapp DE, Turner T, Hornung RW: Semen quality in papaya workers with long term exposure to ethylene dibromide. *Br J Ind Med* 44:317–326, 1987.

Rogers BJ: Use of SPA in assessing toxic effects on male fertilizing potential. *Reprod Toxicol* 2:233–240, 1988.

Sassin JF, Frantz AG, Weitzman ED, Kapen S: Human prolactin: 24-hour pattern with increased release during sleep. *Science* 177:1205–1207, 1972.

Schmassmann A, Mikuz G, Bartsch G, Rohr H: Quantification of human sperm morphology and motility by means of semi-automatic image analysis systems. *Microscopica Acta* 82:163–178, 1979.

Schrader SM, Turner TW, Hardin BD, Niemeier RW, Burg JR: Morphometric analysis of human spermatozoa. *J Androl* 5:22, 1984.

Schrader SM, Platek SF, Zaneveld LJD, Perez-Pelaez M, Jeyendran RS: Sperm viability: A comparison of analytical methods. *Andrologia* 18:530–538, 1986.

Schrader SM, Ratcliffe JM, Turner TW, Hornung RW: The use of new field methods of semen analysis in the study of occupational hazards to reproduction: The example of ethylene dibromide. *J Occup Med* 29:963–966, 1987.

Schrader SM, Turner TW, Ratcliffe JM: The effects of ethylene dibromide on semen quality: A comparison of short term and chronic exposure. *Reprod Toxicol* 2:191–198, 1988.

Sokol RZ: Endocrine evaluations in the assessment of male reproductive hazards. *Reprod Toxicol* 2:217–222, 1988.

Stachel B, Dougherty RC, Lahl U, Schlosser M, Zeschmar B: Toxic environmental chemicals in human sperm: Analytical method and case studies. *Andrologia* 21:282–291, 1989.

Thomas AJ Jr: Ejaculatory dysfunction. *Fertil Steril* 39:445–454, 1983.

Turner TW, Schrader SM, Simon SD: Sperm head morphometry as measured by three different computer systems. *J Androl* 9:45, 1988.

Whorton D, Milby TH, Krauss RM, Stubbs HA: Testicular function in DBCP exposed pesticide workers. *J Occup Med* 21:161–166, 1979.

Zikarge A: *Cross-Sectional Study of Ethylene Dibromide–induced Alterations of Seminal Plasma Biochemistry as a Function of Post-Testicular Toxicity with Relationships to Some Indices of Semen Analysis and Endocrine Profile.* Dissertation to the University of Texas Health Science Center, Houston, Texas, 1986.

Zukerman Z, Rodriguez-Rigau LJ, Weiss DB, Chowdhury AK, Smith KD, Steinberger E: Quantitative analysis of the seminiferous epithelium in human testicular biopsies, and the relation of spermatogenesis to sperm density. *Fertil Steril* 30:448–455, 1978.

CHAPTER 10

Assessing Reproductive Potential in Women

Anthony R. Scialli

THE NATURE OF THE PROBLEM

Fertility is a term that has multiple meanings. For much of the public, it refers to the ability of a woman to become pregnant, a meaning closest to the Latin derivation, *fero*, "to bear." Clinicians and epidemiologists, however, consider fertility to be characteristic of a couple, using *fecundity* (from *fecundus*, "fruitful") for the individual's ability to conceive. Others make a distinction between fertility as the ability to conceive and fecundity as the ability to bear a live-born child (Leridon, 1977). Under this definition, a woman with habitual abortion would be infecund, not infertile.

A term related to fecundity is *fecundability*, which can be described as the proportion of couples conceiving in a given month of attempting pregnancy. Normal fecundability is 0.20 to 0.25; in other words, about one of every four or five couples will achieve pregnancy in a month of trying to conceive.

The different ideas about terminology demonstrate some of the difficulties in talking about assessing women who may be exposed to reproductive toxicants. For example, if pregnancy is used as an endpoint in evaluating reproductive potential, what is to be done about considering the role of male partners in producing pregnancies? What of pregnancies that are conceived but lost before they are clinically apparent? If fecundability is quantitated as a population measure, can it be applied to an individual? That is, if a woman comes from a population with a fecundability of 0.20, must she have a 20 percent chance of conceiving in a given month? Some of these questions require an epidemiologic approach, discussed in Chap. 11. This chapter will consider primarily approaches that may be used to evaluate reproductive potential in individual women. In most cases, these approaches are independent of pregnancy as an endpoint and independent of couple effects. Methods that do not use pregnancy as an endpoint usually involve assessment of cycle normalcy, because cyclicity is so dependent on the proper functioning of the neuroendocrine axis. It is assumed that if a normal cycle is present, and if normal sperm are delivered to the genital tract at an appropriate time, reproductive function will be normal.

In considering methods of assessment, we should be clear that we will be looking for ways to uncover disruptions in female reproduction that have been induced by toxicants. This means an approach based on monitoring, rather than a diagnostic exercise in an individual woman. The diagnostic approach involves a woman presenting with an endpoint, for example, the inability to make a pregnancy test turn positive. The workup of this complaint proceeds in an orderly fashion and includes elements of history, physical examination, laboratory studies, and other tests with which clinicians are familiar. Monitoring is derived from the Latin verb *monere,* "to warn," and implies having a woman or a group of women under observation for changes that may warn of reproductive dysfunction.

The monitoring approaches entails the need to decide what kinds of changes are acceptable and what kind are not. Biological systems have a great deal of inherent variability, and failure to understand this will lead to alarms being triggered when there is nothing wrong. A relevant clinical example is blood pressure monitoring. If an asymptomatic patient presents with a single reading of 150/100, we know better than to start antihypertensive medication because we understand the variability of blood pressure readings. We ask for several blood pressure readings over the course of days to weeks before we are willing to conclude that the blood pressure is abnormally high.

In evaluating reproductive potential in women, we must ask proposed assessment schemes to be just as permissive of normal variation in reproductive endpoints. At the same time, we would like assessment protocols to be sensitive enough to detect important changes so we can provide meaningful protection for women exposed to xenobiotic agents.

CYCLICITY

The cyclic events underlying normal ovulation and disruption of these events are discussed in Chaps. 2 and 6. A woman with no cycle, that is, with no menstrual bleeding or erratic menstrual bleeding, is likely to have reduced fertility potential. Certainly, if a woman or group of women could associate acyclicity with exposure to a chemical, reproductive toxicity of that agent would be assumed; however, lesser disruptions of cyclicity may be just as important. How can these be assessed, particularly in monitoring schemes appropriate for large groups of exposed women?

Menstrual Records

Because the cycle is marked in most women by obvious genital tract bleeding, the simplest of ideas is to ask women about menstrual cycle length and about other features of the bleeding episodes. Normative data on menstrual cycle characteristics were collected by Treloar and his coworkers at the University of Minnesota (Treloar et al, 1967). They recruited university women beginning in 1934, and by 1938, some 2500 women in this cohort were keeping menstrual calendars. Additional women were recruited over subsequent years, and large numbers of these women remained in the study for decades. The investigators were able to describe normal cycle characteristics and variability over the course of time. It was found, for example, that 90 percent of presumably normal women have cycle lengths between 22 and 35 days, with 2 to 8 days of flow. Some shortening of cycle length occurs normally with age. Cycle variability is a normal feature in most women; a study of more than 30,000 cycles in 2316 women noted variability to be highest in women less than 25 years old. Mean cycle length was 29.1 days with a standard deviation of 7.46, giving a coefficient of variation of 26 percent (Chiazze et al, 1968).

Although excellent normative data are available, the use of menstrual history in monitoring efforts is unreliable. In a study following up on the University of Minnesota cohort, women were interviewed some years after they collected menstrual cycle data. The women correctly recalled their age at menarche and menopause 75 percent of the time but were very inaccurate in their recall of cycle length and variability (Bean et al, 1979). These were women who had been focused during their reproductive lives on recording cycle information, giving us great concerns about the accuracy of information recalled by randomly interviewed women.

Even with accurate, prospectively recorded calendars, there arises a question of how to interpret variations in cyclicity. For example, is it biologically important if there is a change in cycle length from 30 to 27 days? What if there is no change in average cycle length but an increase in cycle variability, from 28–32 days to 26–34 days? Answers to these questions are not available because the correlation between these subtle changes and fertility is unknown. Finally, there are factors other than exposure to a putative toxicant that may influence cycle characteristics. Age, weight, and stress levels are examples of parameters that change with time and that influence cyclicity (Shortridge, 1988). It may be difficult to tell if cycle changes seen after employment of a woman are due to one or more of these factors, rather than to toxicant exposure.

Ovulation Prediction

Menstrual normalcy is an indirect measure of the function of the ovarian follicle and subsequent corpus luteum. The use of methods designed to detect ovulation or to monitor the biochemical events associated with ovulation may be a more precise way to evaluate cyclicity. Strictly speaking, detection of ovulation requires the recovery of an ovum from the genital tract or the achievement of a pregnancy; therefore, the methods used to evaluate ovulation are only predictive or suggestive that ovulation is about to occur or has occurred.

Basal Body Temperature

The dependence of basal body temperature (BBT) on time of the menstrual cycle was known as early as 1904 (Moghissi, 1980), and the relationship between ovulation and a shift in BBT has been described in the scientific literature for nearly half a century (Martin, 1943; Holbaecht, 1945). Although newer instruments offer women more rapid and convenient ways to record their first morning temperatures, the correlation between the pattern of BBT shift and ovarian events is imperfect. The temperature rise reflects a thermogenic response to progesterone, but the magnitude and timing of this response is variable. Attempts have been made to standardize interpretation of BBT charts by using defined criteria. For example, a temperature rise of at least 0.3 °F for 3 days appears to be a useful criterion for identifying the BBT shift (McCarthy and Rockette, 1983). Even with standardized approaches to interpretation, the prediction of ovulation based on BBT characteristics is inexact; there may be several days' discrepancy between suggestion of ovulation by BBT and hormonal or ultrasound evidence of the same event (Hilgers and Bailey, 1980; Bauman, 1981; Leader et al, 1985). In addition, monophasic BBT charts, which are normally interpreted as consistent with anovulation, have been described in women with ovulatory cycles (Moghissi, 1976; Bauman, 1981; Leader et al, 1985).

Because BBTs do not accurately identify the day of ovulation, they cannot be used to measure the length of the follicular or luteal phases of the cycle. Similarly, because the thermogenic response to progesterone is variable, the magnitude of the BBT shift does not indicate the adequacy of luteal production of progesterone. Identification of subtle abnormalities of ovulation would not, therefore, be expected using this method, limiting its effectiveness as a monitoring technique. It is possible, however, that in women with acceptable baseline BBT charts, a more dramatic toxicant-induced disruption in ovarian function would be detected. For example, if the elevated phase of the BBT chart, corresponding to the luteal phase, were shortened to 9 days from a normal of 12 to 14 days, inadequacy of corpus luteum function would be suspected. An adverse effect of this sort would not be readily apparent with a simple menstrual calendar but would be of clinical importance.

Hormone Assays

The levels of steroid and peptide hormones of the ovulatory cycle can be assayed in serum and provide a useful method for predicting ovulation in clinical studies and fertility work (Moghissi, 1980). Measurement of these hormones in serum is not generally suitable for monitoring efforts, however, because the pulsatile release of the hormones requires frequent venipuncture to establish the pattern of hormone secretion. Of more utility in monitoring is the use of hormone measurements in body fluids more easily collected, such as urine and saliva.

Urinary LH measurement has become a simple matter, with kits marketed for use by women at home. Identification of urinary LH by such kits is said to predict ovulation within the next 36 h. One such kit was found to predict ovulation accurately in 87.5 percent of spontaneous cycles (Vermesh et al, 1987). Collection of daily urine specimens around the time of the anticipated LH surge would permit a monitoring program to identify failure of the appearance of LH within the expected time. Urine LH levels presumably reflect the cumulative amount of LH filtered at the glomerulus; therefore, subtle changes in LH pulse amplitude or frequency would not be detectable with urine monitoring.

The use of urine for the measurement of conjugated steroid hormones has gained recent attention with interest in developing a home test for estrone-3-glucuronide. The rise in the level of this hormone appears to give more warning of the approaching fertile period of the cycle than LH does, making the conjugated estrone a more useful adjunct to natural family planning programs (Albertson and Zinaman, 1987). In fact, urinary hormone measurement to evaluate ovulation is not new. In 1949, Georgeanna Seegar Jones, then at Johns Hopkins University, published a classic paper on the evaluation of infertility of nonanatomic causes. She used BBTs, cervical mucus, endometrial biopsy, and levels of urinary pregnanediol, a progesterone metabolite, to determine whether ovulation had occurred (Seegar Jones, 1949).

The monitoring of urine hormone levels has been shown to be practical in relatively large epidemiologic surveys of the female reproductive cycle. The pioneering efforts in this application of urine monitoring were made by Allen Wilcox and his coworkers (1987). These investigators recruited 301 women, 230 of whom had discontinued oral contraceptives in an effort to get pregnant, 40 of whom used intrauterine contraceptive devices, and 31 of whom had had tubal ligations. The women collected their first morning urine each day, storing the samples in their freezers at home. The stored urine samples were picked up every 1 to 2 weeks, and assays were performed for conjugated estrogens (estone and estriol), pregnanediol, LH, and human chorionic gonadotropin (hCG). These measurements permitted profiles of the ovulatory cycles with identification of early pregnancies and early pregnancy losses. Women recruited for this study were believed to be normal, and this monitoring scheme has not been validated in women exposed to putative reproductive toxicants; however, the potential for monitoring such at-risk populations is encouraging.

The evaluation of steroid hormones in saliva has been extensively investigated. In general, it has been believed that salivary measurements reflect unbound and therefore biologically active hormone (Vining and McGinley, 1986). There has not, however, been a good correlation between salivary progesterone and unbound serum progesterone levels (Evans, 1986), although elevations of salivary progesterone levels in the luteal phase of the cycle will still indicate antecedent ovulation. In addition, although salivary progesterone concentrations cannot predict serum progesterone concentrations, normative data have been collected for saliva. Midluteal values have been determined to be 300 to 800 pmol/liter, with good reproducibility from one cycle to the next in the same woman (Lenton et al, 1988). This suggests that salivary progesterone levels might be monitored to detect luteal phase inadequacy. In fact, a study of recreational runners identified a subtle defect in salivary progesterone concentrations in the luteal phase associated with exercise (Ellison and Lager, 1986), and salivary progesterone levels appear to be useful in identifying clinically important luteal phase dysfunction (Riad-Fahmy et al, 1982). Although findings such as these are interesting, the use of salivary progesterone levels as a luteal phase monitor for reproductive toxicants cannot be considered to have been validated.

Electric Resistance

The ionic chemistry of cervical mucus and saliva is cycle-dependent, with concomitant alterations in the electric conductivity of these secretions. Devices have been marketed for the measurement of electric resistance in the vagina and the oral cavity. Resistance has been noted to decrease during the preovulatory portion of the cycle, with a low point reached during the day prior to ovulation. Following ovulation, an increase in resistance occurs. With these commercial devices, ovulation prediction has been shown to be accurate using LH or ultrasound evidence of ovulation as standards (Albrecht et al, 1985; Jacobs et al, 1989). Whether measurement of electric resistance will be a practical or sensitive monitor for reproductive toxicity has not been investigated.

Cervical Mucus

The changes in cervical mucus that characterize the preovulatory portion of the cycle have formed the basis for a method of natural family planning popularized by John and Evelyn Billings (Albertson and Zinaman, 1987). The thin, slippery, stretchy cervical mucus is believed to be an estrogen effect, in contrast to the tacky, cloudy mucus that appears with the production of luteal progesterone. Although the character of the mucus may be a reliable indicator of ovulation in some women, there have been no studies on use of mucus self-evaluation in monitoring programs. The changes in mucus character are also accompanied by changes in chemical composition of the mucus, and there has been limited investigation into measures of mucus components as an indicator of ovulatory status. For example, guaiacol peroxidase levels decrease in cervical mucus 4 to 5 days prior to ovulation, rising thereafter. In a study of 45 cycles in 31 women, one woman with luteal phase deficiency showed marked variability in luteal phase levels of this peroxidase (Tsibris et al, 1986). Whether serial measurements of guaiacol peroxidase, or other mucus components, will show utility for monitoring has yet to be investigated.

Ultrasound

Evaluation of follicle growth and subsequent collapse is often used in clinical research on ovulation. It is assumed that the disappearance of the follicle on sonography is coincident with follicle rupture and ovum extrusion (Queenan et al, 1980). Although ultrasound appears to be a valuable tool in identifying ovulation in individual subjects, it requires serial scanning of women over several days using expensive equipment and expert technicians. It is not likely, therefore, that ultrasound will be useful in large-scale monitoring programs.

SEXUAL FUNCTION

Coitus is an important component of reproductive normalcy, and a reproductive toxicant that altered sexual function might well have important clinical consequences. Evaluation of sexual function can be performed by questionnaire or by examining diaries. In evaluating this information, it is important to know if cycle characteristics alter sexual behavior. There is some evidence, in fact, that self-identified sexual interest and/or activity is highest at midcycle (Matteo and Rissman, 1984; Harvey, 1987; Stanislaw and Rice, 1988). The cyclic variation in sexual interest can be attenuated by oral contraceptives; women on these agents experience less variability from one day to the next in their sexual interest (Warner and Bancroft, 1988). Although a relationship between cyclic hormones and sexual interest is plausible, not all studies agree. In one report, for example, sexual activity was found to be predicted more by whether it was a weekend or weekday than by time of the menstrual cycle (Strauss and Appelt, 1983).

Possible toxicant-induced sexual function in women has been evaluated for only a few agents, including cyproterone acetate (Appelt and Strauss, 1984), ethanol, marijuana, and tobacco (Lex et al, 1988; Harvey and Beckman, 1986). Whether questionnaire surveys on sexual interest or sexual behavior will be useful for monitoring is unknown.

PREGNANCY

Perhaps the ultimate outcome parameter for assessing reproductive potential is the achievement of pregnancy; however, use of pregnancy in monitoring studies is restricted to noncontracepting women. In addition, the use of pregnancy as an outcome requires that a sensitive diagnostic strategy be applied to all members of the population of interest. As discussed above, the work of Wilcox and coworkers represents an important effort in the evaluation of daily urine specimens for evidence of early pregnancy. Use of this technique for monitoring is possible; however, it has been shown that early pregnancy loss is common in normal, presumably nonexposed, women (Wilcox et al, 1988). This makes the use of suitable controls important when evaluating exposed women for an increase in miscarriage incidence.

As attractive as it may be to evaluate daily urine samples for hCG levels, it is more common to see pregnancy evaluation performed by asking men and women about their reproductive histories. It is possible to get an idea about the prevalence of infertility in exposed individuals by asking them if they have had difficulty conceiving. For example, the role of dibromochloropropane in producing male infertility was investigated as a result of reports of infertility by the wives of exposed workers (Levine, 1989). Although dramatic reductions in pregnancy rates may be identified by questioning exposed individuals, there is concern that more subtle reductions in fertility will be missed. Such reductions in fertility might produce an identifiable delay in achieving a pregnancy, rather than absolute infertility. The use of *time-to-pregnancy* as a monitoring technique for occupationally exposed people has shown promise (Baird, 1988). There are, however, two important problems. The first is that accidental pregnancies cannot be used, because there is no reliable way to measure how long it took a couple to conceive. Because unplanned pregnancies may occur more commonly in couples of high fertility, exclusion of these individuals may make the studied population unrepresentative of the population that is intended to be monitored. The second problem is the use only of pregnant couples in time-to-pregnancy studies. If a couple is sterile, there will be no pregnancy and, therefore, no time-to-pregnancy data. It is hoped that if a toxicant produces sterility in some exposed individuals, the effect will be identified by a less sensitive segment of the population that will have a detectable delay in conception.

MENOPAUSE

If a toxicant causes menopause to occur at an early age, it would be correct to infer reproductive toxicity. For example, onset of menopause in 30-year-old women has important adverse effects on the reproductive potential of those women. It is also important, however, to know if an agent advances the age at menopause by even a few years. If, for example, an exposure advanced the average age of menopause from 51 to 45, it would be desirable for a monitoring program to register a warning. This is not because most 45-year-old women are concerned about reproductive potential; instead, it is because the earlier age at menopause implies ovarian toxicity and this ovarian toxicity may be clinically important in young women attempting to conceive.

One strategy for seeking an association between an exposure and age at menopause is to ask postmenopausal women to give their age at menopause and to give an estimate of their exposure to the agent of interest. This relies on accurate recall of both age information and exposure information. The problems inherent in relying on memory can be addressed by sampling a population of women in the perimenopausal age group and determining how many of them are menopausal at that time. This was performed, for example, in an investigation of the reproductive effects of smoking in a study using women aged 44 to 53 years (Jick and Porter, 1977). More smokers than nonsmokers were identified as postmenopausal in each age group at the time of questioning. It is interesting to note that adverse effects of smoking on fertility in younger women were subsequently identified using a time-to-pregnancy approach (Baird and Wilcox, 1989), creating a consistent picture of human ovarian toxicity for this agent.

THE CHALLENGE FOR THE FUTURE

Monitoring programs face problems on two fronts. First, cost is a limiting factor, because regardless how laudable the goals of the program, someone has to pay for it. Second, compliance of the monitored population with the monitoring protocol will be difficult if the protocol is too complex, inconvenient, or distasteful. For example, storing urine in a home freezer may be unacceptable to some women, even if the exercise is said to be for their protection.

It is likely that successful monitoring schemes will incorporate a battery of tests in order to provide an overall cycle profile. One such battery has already been proposed, consisting of salivary and vaginal electric resistance, dipstick urine testing for LH, and luteal phase salivary progesterone measurement (Hughes, 1988). This group of tests would be applied during the first few cycles after employment of a woman and reapplied periodically, particularly if the woman reported menstrual disturbances or infertility.

The real test of a monitoring system is whether it can warn of a reproductive toxicant in time to intervene and prevent reproductive damage. Validation studies will be needed for the evaluation of how effectively competing test schemes can provide warning. In the meantime, it is desirable to investigate the relationship between the tests described above and tests of fertility, such as time-to-pregnancy, in women with exposures that may put them at risk of reproductive dysfunction. This has already been accomplished to an extent with regard to nurses handling cytotoxic drugs (Shortridge, 1988); however, many other candidate reproductive toxicants remain to be investigated.

REFERENCES

Albertson BD, Zinaman MJ: The prediction of ovulation and monitoring of the fertile period. *Adv Contracept* 3:263–290, 1987.

Albrecht BH, Fernando RS, Regas J, Betz G: A new method for predicting and confirming ovulation. *Fertil Steril* 44:200–205, 1985.

Appelt H, Strauss B: Effects of antiandrogen treatment on the sexuality of women with hyperandrogenism. *Psychother Psychosom* 42:177–181, 1984.

Baird DD: Using time to pregnancy data to study occupational exposure: Methodology. *Reprod Toxicol* 2:205–207, 1988.

Baird DD, Wilcox AJ: Cigarette smoking associated with delayed conception. *Am J Epidemiol* 129:1072–1078, 1989.

Bauman JE: Basal body temperature: Unreliable method of ovulation detection. *Fertil Steril* 36:729–733, 1981.

Bean JA, Leeper JD, Wallace RB, Sherman BM, Jagger H: Variations in reporting of menstrual histories. *Am J Epidemiol* 109:181–185, 1979.

Chiazze L Jr, Brayer FT, Macisco JJ Jr, Parker MP, Duffy BJ: The length and variability of the human menstrual cycle. *JAMA* 203:89–92, 1968.

Ellison PT, Lager C: Moderate recreational running is associated with lowered salivary progesterone profile in women. *Am J Obstet Gynecol* 154:1000–1003, 1986.

Evans JJ: Progesterone in saliva does not parallel unbound progesterone in plasma. *Clin Chem* 32:542–544, 1986.

Harvey SM: Female sexual behavior: Fluctuations during the menstrual cycle. *J Psychosom Res* 31:101–110, 1987.

Harvey SM, Beckman LJ: Alcohol consumption, female sexual behavior and contraceptive use. *J Stud Alcohol* 47:327–332, 1986.

Hilgers TW, Bailey AJ: Natural family planning: II. Basal body temperature and estimated time of ovulation. *Obstet Gynecol* 55:333–339, 1980.

Holbaecht J: Ovarian function and body temperature. *Lancet* 2:688–689, 1945.

Hughes CL Jr: Monitoring of ovulation in the assessment of reproductive hazards in the workplace. *Reprod Toxicol* 2:163–169, 1988.

Jacobs MH, Blaso L, Sondheimer SJ: Ovulation prediction by monitoring salivary and vaginal electrical resistance with the peak ovulation predictor. *Obstet Gynecol* 73:817–822, 1989.

Jick H, Porter J: Relation between smoking and age of natural menopause. Report from the Boston Collaborative Drug Surveillance Program, Boston University Medical Center. *Lancet* 1:1354–1355, 1977.

Leader A, Wiseman D, Taylor PJ: The prediction of ovulation: A comparison of the basal body temperature graph, cervical mucus score, and real-time pelvic ultrasonography. *Fertil Steril* 43:385–388, 1985.

Lenton EA, Gelsthorp CH, Harper R: Measurement of progesterone in saliva: Assessment of the normal fertile range using spontaneous conception cycles. *Clin Endocrinol* 28:637–646, 1988.

Levine RJ: Detecting toxicity to the human reproductive system using epidemiological methods, in Working PK (ed): *Toxicology of the Male and Female Reproductive Systems.* New York, Hemisphere Publishing, 1989, pp 323–337.

Leridon H: *Human Fertility: The Basic Components.* Chicago, University of Chicago Press, 1977.

Lex BW, Palmieri SL, Mello NK, Mendelson JH: Alcohol use, marihuana smoking, and sexual activity in women. *Alcohol* 5:21–25, 1988.

McCarthy JJ, Rockette HE: A comparison of methods to interpret the basal body temperature graph. *Fertil Steril* 39:640–646, 1983.

Martin PL: Detection of ovulation by basal temperature curve with correlating endometrial studies. *Am J Obstet Gynecol* 46:53–62, 1943.

Matteo S, Rissman EF: Increased sexual activity during the midcycle portion of the human menstrual cycle. *Horm Behav* 18:249–255, 1984.

Moghissi KS: Accuracy of basal body temperature for ovulation detection. *Fertil Steril* 27:1415–1421, 1976.

Moghissi KS: Prediction and detection of ovulation. *Fertil Steril* 34:89–98, 1980.

Queenan JT, O'Brien GD, Bains LM, Collins PO, Simpson J, Collins WP, Campbell S: Ultrasound scanning of ovaries to detect ovulation in women. *Fertil Steril* 34:99–105, 1980.

Riad-Fahmy D, Read GF, Walker RF, Griffiths K: Steroids in saliva for assessing endocrine function. *Endocrine Rev* 3:367–395, 1982.

Seegar Jones GE: Some newer aspects of the management of infertility. *JAMA* 141:1123–1129, 1949.

Shortridge LA: Assessment of menstrual variability in working populations. *Reprod Toxicol* 2:171–176, 1988.

Stanislaw H, Rice FJ: Correlation between sexual desire and menstrual cycle characteristics. *Arch Sex Behav* 17:499–508, 1988.

Strauss B, Appelt H: Psychological concomitants of the menstrual cycle: A prospective longitudinal approach. *J Psychosom Obstet Gynecol* 2:215, 1983.

Treloar AE, Boynton RE, Behn BG, Brown BW: Variation of the human menstrual cycle through reproductive life. *Int J Fertil* 12:77–125, 1967.

Tsibris JCM, Virgin SD, Khan-Dawood FS, Langenberg PW, Thomason JL, Spellacy WN: Cervicovaginal peroxidases: Markers of the fertile period. *Obstet Gynecol* 67:316–320, 1986.

Vermesh M, Kletzky OA, Davajan V, Israel R: Monitoring techniques to predict and detect ovulation. *Fertil Steril* 47:259–264, 1987.

Vining RF, McGinley RA: Hormones in saliva. *Crit Rev Clin Lab Sci* 23:95–146, 1986.

Warner P, Bancroft J: Mood, sexuality, oral contraceptives and the menstrual cycle. *J Psychosom Res* 32:417–427, 1988.

Wilcox AJ, Baird DD, Weinberg CR, Armstrong EG, Musey PI, Wehmann RE, Canfield RE: The use of biochemical assays in epidemiologic studies of reproduction. *Environ Health Perspect* 75:29–35, 1987.

Wilcox AJ, Weinberg CR, O'Connor JF, Baird DD, Schlatterer JP, Canfield RE, Armstrong EG, Nisula BC: Incidence of early pregnancy loss. *N Engl J Med* 31:189–194, 1988.

CHAPTER 11

Toxic Exposures and Reproduction: A View of Epidemiology and Surveillance

Grace K. Lemasters and Sherry G. Selevan

The first widespread recognition of the potential for toxicants to cause reproductive and developmental effects came in the 1960s. This was due to recognition of the association of thalidomide with rather unusual birth defects in children exposed during a narrow time span in early pregnancy. Since then, and especially in the last decade or so, increasing concern has been expressed about effects of environmental and occupational exposures on reproduction.

The concern intensified for women because of their increasing presence in the workplace: Almost 70 percent of women are employed outside the home during their predominant childbearing years (ages 20 to 34). In 1970, only 40.8 percent of married women were employed outside the home compared with 56.5 percent in 1988 (Bureau of the Census, 1990).

For men, the possibility of effects of workplace exposures on male reproduction was generally recognized in the late 1970s, when male infertility was associated with exposure to a pesticide (Whorton et al, 1977). The effects on semen characteristics were so dramatic that they emphasized the potential for effects directly on male reproduction, and the possibility of male-mediated developmental effects.

We are now well aware that a couple's reproductive success depends on a delicate physiochemical balance within and between the paternal, maternal, and fetal system. Any disruption of this balance may result in a broad range of effects including infertility, adverse pregnancy outcome, childhood cancer, or heritable alterations affecting future generations. This chapter focuses on effects of exposure on the male and female reproductive systems (semen parameters, subfertility, menstrual abnormalities). In addition, it includes a general description of epidemiologic and surveillance methods for examination of occupational and environmental toxicants.

REPRODUCTIVE OUTCOMES

Fertility

Population statistics show that approximately 16 percent of married women between the ages of 15 and 44 years have fecundity impairments not associated with surgery (National Center for Health Statistics, 1982). This rate is considerably higher in blacks (23 percent) than in whites (15 percent). The definition applied affects the prevalence of infertility observed (Marchbanks et al, 1989): In a recent study, the age-adjusted prevalence of infertility, defined as no conception after 2 years of trying, was 12.5 percent (crude prevalence was 16.1 percent). However, if the definition required consultation with a physician, the rate was 9.6 percent, and if it required a physician to diagnose a specific problem, the rate was 6.1 percent (Marchbanks et al, 1989). Potential risk factors for infertility, and potential confounders, include age, race, marital status, socioeconomic status, gravidity, adverse pregnancy outcome, birth control methods, interval since last pregnancy, body mass, certain infectious or chronic diseases, cigarette smoking and alcohol use, irradiation, and other home, environmental, or occupational exposures to reproductive toxicants such as lead, heat, glycol ethers, or carbon disulfide (Baird et al, 1986).

Infertility or subfertility may be thought of as a non-event: A couple is unable to have children within a specific time frame. Therefore, the epidemiologic measurement of reduced fertility is typically indirect and is accomplished by evaluating "reproductive success," that is, comparing birth rates or time intervals between births or pregnancies. These evaluations estimate the couple's joint ability to procreate.

An approach suggested by Wong and coworkers (1979) and further discussed by others (Levine et al, 1980, 1981, 1983; Levine, 1983; Starr and Levine, 1983; Starr et al, 1986) compares the birth rates in an exposed group to those expected based on some standard, unexposed population. This approach is limited because birth probabilities may be affected by many factors (e.g., calendar time period, age, parity, marital status) that may not be available, especially for the standard population, typically national birth data. Other important factors, not considered in the analysis, include contraceptive history, functional infertility, and other risk factors and potential confounders. This method, known as the *standardized birth ratio* (SBR; also called the *standardized fertility ratio*), computes expected births based on the person–years of observation within categories of calendar time period, age, race, marital status, parity, and contraceptive use (if available).

Analysis of the time between recognized pregnancies or live births is another indirect measure of fertility (Baird et al, 1986; Dobbins et al, 1978; Weinberg and Gladden, 1986). The length of time to pregnancy, estimated by determining the number of noncontracepting cycles required for conception after complete termination of birth control, is compared for exposed and unexposed employees. A significantly increased time to pregnancy is an indirect measure of interference with the reproductive process.

The approach for analyzing infertility depends on the study design. If only general information on individuals is available and a similar referent population is not available, the SBR analysis is used. If considerable details are known on potential confounders and a referent group is available, the time-to-pregnancy or more detailed SBR (Starr et al, 1986) approach may be preferred.

Male Reproductive Studies

Throughout the process of spermatogenesis, any one of the developing cell types may be susceptible to toxic exposures. Ejaculated semen, the product of the male reproductive process, provides a direct measure of male reproductive impairment and, possibly, an indirect measure of genetic damage to the next generation. More details on these assays are in Chap. 4.

Direct effects on sperm include azoospermia (absence of sperm), oligospermia (fewer than 20 million sperm per milliliter of semen), reduced viability of semen, reduced motility, altered velocity, altered morphology, changes in DNA content, and increases in numbers of degenerative cells. The relationships among these parameters are not well understood. At this time, we can only speculate which endpoint(s) is most sensitive and best reflects changes in "reproductive success." Alterations in sperm morphology may be associated with decreases in fertility, fetal loss, and heritable genetic abnormalities (Wyrobek et al, 1984). Sperm motility and morphology are considered two of the most sensitive indicators of fertility potential (Jouannet et al, 1988; Grunert et al, 1989). Automated semen analysis techniques have been developed to enhance consistency of evaluation (Schrader et al, 1987; Moruzzi et al, 1988; Boyers et al, 1989).

In prospective cohort studies (described in the section on epidemiologic study designs), samples are collected at different times in the study. In studies of men with acute exposure, such as after an accidental spill, an informative approach obtains several semen samples timed to characterize damage at various stages of sperm development. To determine the sampling time after this acute, short-lived exposure, the semen sample should be produced on the schedule shown in Table 11-1.

Two studies of semen quality compared statistical variation of mean sperm concentration, sperm count (millions per ejaculate), semen volume, sperm velocity, percentage motile sperm, and percentage with normal morphology (Schrader et al, 1987, 1988b). Both suggested that sperm count lacks precision

Table 11-1 Time delay needed to assess sensitive developmental stages of sperm

Weeks after exposure	Stage of sperm development
1–3	Spermatozoa
3–8	Spermatids and spermatocytes
10–13	Spermatogonia

for detecting population differences, though it is the most frequently measured characteristic in occupational field studies. The average coefficient of variation of sperm count for a given individual is high, 44 percent. The mean sperm count for all samples was 47.43 million/ml with a between-man standard deviation of 29.47 and a within-man deviation of 22.9. Sperm velocity, morphology, and viability, on the other hand, have low coefficients of variation, allowing better precision for detecting trends and population differences.

Elevated scrotal temperature may cause lowered sperm count because of limited thermoregulatory capacity of the scrotum. Changes in semen parameters have been observed in workers exposed to high temperatures. In a recent study of 131 outdoor workers in Texas, a significant decrease in sperm concentration, total sperm count per ejaculate, and concentration of motile sperm was observed during the summer months compared with winter months (Levine et al, 1990). After adjustment for other factors, the mean percentage reductions in sperm concentration, count, and motile sperm values for summer compared with winter months were 32, 28, and 24 percent, respectively. Of particular interest was the finding that men with the lowest sperm count, and thus those most likely to be subfertile, were at greatest risk for the seasonal effect.

Examples of Semen Studies

The first published reports of reproductive effects of an industrial chemical occurred in the late 1800s. Unusually high rates of infertility, spontaneous abortion, stillbirth, neonatal death, macrocephaly, and convulsions in offspring were recorded in European lead-working communities (Rom, 1980). Lancranjan and her coworkers (1975) found that moderately increased absorption of lead in male workers resulted in asthenospermia, hypospermia, and teratospermia. More recent studies show that lead battery workers are experiencing elevations in lead levels in blood, urine, and semen; these levels are associated with a decrease in sperm count that is not associated with changes in levels of gonadotropins, prolactin, or testosterone (Assennato et al, 1987). These findings suggest that effects are not mediated via the hypothalamic–pituitary axis.

It was not, however, until the late 1970s, after workers' exposure to a nematocide, 1,2-dibromo-3-chloropropane (DBCP), showed a striking effect on reproduction, that male reproductive effects generated wider concern. In addition to sterility, workers' exposure was associated with an increased incidence of spontaneous abortions in their wives (Whorton et al, 1977; Kharrazi et al, 1980). Effects of exposure on sperm count were most pronounced in workers with longer employment (Whorton et al, 1977). Histologic examinations from testicular biopsies showed a decrease in spermatogenic cells and in cellularity within the seminiferous tubules (Whorton, 1982).

Other human studies support effects of ethylene dibromide (EDB) on male fertility. A longitudinal study of timber fumigators with short-term exposure to EDB demonstrated that sperm velocity was decreased when pre- and postexposure samples were compared (Ratcliffe et al, 1984). A cross-sectional study examined 46 men employed in the papaya fumigation industry with an average employment duration of 5 years and an 88-ppb exposure to EDB [8-h time-weighted average (TWA)]. Results included a significant decrease in sperm count, viability, and motility, and morphologic abnormalities (Ratcliffe et al, 1987; Schrader et al, 1988a). However, no differences were observed in chromosomal aberrations or sister chromatic exchanges in these workers (Steenland et al, 1986).

Because of their ubiquitous usage, solvents are of increasing concern. Glycol ethers [e.g., 2-methoxyethanol (2ME) and 2-ethoxyethanol (2EE) and their acetates] have been associated with male reproductive effects in animal experiments. Studies of rats, mice, rabbits, and dogs have associated 2EE with testicular atrophy, degeneration of seminiferous tubules, severe oligospermia, abnormal sperm morphology, and reduced sperm motility. Two human studies suggest that 2EE decreased sperm count at a metal castings company and in a group of painters (Ratcliffe et al, 1989; Welch et al, 1988). No other semen parameters were significantly associated with exposure.

Concerns about reproductive toxicity of perchloroethylene have been raised. This is another solvent widely used in industry, particularly in dry cleaning establishments (Van der Gulden and Zielhuis, 1989). In a recent study comparing 34 dry cleaners with 48 laundry workers, no significant differences in sperm concentrations or overall percentage of abnormal forms were observed (Eskenazi et al, in press). However, more subtle, but significant, dose-related changes in sperm shape and lateral head displacement were found. Another study of male reproductive effects of mixed solvent exposure found a decrease in FSH levels with toluene exposure (Morck et al, 1988).

In a study that attempted to determine indirectly the effects of Agent Orange, a lower mean sperm concentration among the Vietnam veterans (64.8 million/ml) was observed than among non-Vietnam veterans (79.8 million/ml) (DeStefano et al, 1989). The Vietnam veterans were twice as likely to have sperm concentrations less than 20 million/ml, 15.9 versus 8.1 percent, respectively (OR = 2.7; 95 percent CI = 1.3, 5.7). The two groups were not significantly different for mean percentage of motile and morphologically normal cells. These differences were not sufficient to decrease the fecundity of the veterans but could suggest some irreversible damage (e.g., to stem or Sertoli cells).

Female Reproductive Studies

Studies of reproductive effects of toxic exposure in female worker populations are unique because potentially two are at risk—the woman and, if she is pregnant, her developing offspring. Exposure to either may result in a wide range of adverse responses. Female exposure to a toxicant may cause infertility, menstrual disorders, illness during pregnancy, chromosomal aberrations, breast milk alteration, early onset of menopause, and suppressed libido. Adverse fetal outcomes include preterm delivery, fetal loss, perinatal death, lowered birth weight, altered sex ratio, congenital malformations, childhood malignancies, infant or childhood illnesses, chromosomal aberrations, and developmental disabilities.

Reproductive effects may result from developmental exposures. For example, environmental exposures may result in oocyte toxicity, in which a loss of primordial oocytes irreversibly affects the woman's fertility. The exposures of importance may occur during the prenatal period and beyond. Oocyte depletion is difficult to examine directly in women because of the invasiveness of the tests required; however, it can be studied indirectly through evaluation of the age at menopause (Everson et al, 1986).

This discussion will be limited to effects on the woman. This section deals with the potential effects of toxic exposures on reproductive integrity, especially menstrual cycle variability.

Menstrual Cycle Variability

Although women constitute about one-half of the work force, the impact of exposures on menses has rarely been examined. Toxic exposures may alter the menstrual patterns by inhibition or damage to the follicles, effects on the CNS, (which can, in turn, affect the endocrine system), damage to the hormone-secreting organs, or disruption of the delicate hormonal balance that regulates ovulation and the menstrual cycle.

In humans, between 3 and 4 million follicles are present at birth. By puberty fewer than 400,000 are present. During each ovarian cycle a number of follicles start to mature, but most fail. After menopause, few, if any, follicles are present in the ovary. This sensitive and complex process of ovulation represents several potential targets for damage by industrial chemicals. Premature reproductive senescence may occur if xenobiotics block oogenesis in the fetus, or destroy oocytes at any age prior to menopause (Mattison, 1983). Damage to the ovulatory process may be expressed in disorders of menstruation that, in turn, may be a surrogate for other health effects (e.g., decrease in the individual's fertility potential) (Wilcox et al, 1988).

Menstrual abnormalities can be examined by (1) cycle length or rhythm, (2) characteristics of bleeding patterns, and (3) the presence of pain. Studies or surveillance programs concerned with the effects of exposure on menstruation must account for the myriad of risk factors associated with these conditions. These factors include age, weight, anorexia nervosa, amenorrhea, vigorous exercise, not bearing children, lactation, pathology of the genital tract, systemic illness (e.g., renal disease and thyroid disease), use of oral contraceptives or intrauterine devices, stress, smoking, and alcohol use.

Examples of Menstrual History Studies

Relatively few studies have attempted to assess alterations in menstrual cycles related to occupational exposures. Workers exposed to formaldehyde in occupations such as biology, ink making, and embalming report menstrual disorders more frequently than unexposed workers (47.5 versus 9.2 percent) (Griesemer et al, 1982). Studies of women exposed to solvents such as benzene, toluene, and xylene have shown menstrual disturbances primarily associated with abnormal bleeding (Beskrovnaja, 1979; Michon, 1965; Butarewicz et al, 1969; Syrovadko et al, 1973). Styrene was reported to be associated with irregular cycles and hypermenorrhea in one study (Zlobina et al, 1975), but these findings were not confirmed in another study (Lemasters et al, 1985). Women working in dry cleaning businesses and exposed to perchloroethylene solvents were compared with laundry workers. Dry cleaners experienced a significant excess of menstrual disorders including cycle length alterations, menorrhagia, dysmenorrhea, and premenstrual syndrome (Zielhuis et al, 1989). In petroleum workers (exposed to hydrocarbons), 50 percent of women over 40 reported some type of menstrual disorder compared with 18.7 percent of the aged-matched controls. Eighty-three percent of workers exposed to trinitrotoluene in explosive manufacturing experienced changes in their menstrual cycle with symptoms diminishing to 65 and 20 percent after 1 and 2 years postexposure, respectively (Gesher, 1967). The manufacture or handling of certain drugs may place the female worker at risk; examples include estrogen (Harrington et al, 1978) and cytotoxic drugs (Shortridge, 1988).

Exposure to physical hazards has been linked to abnormal menses. Women experiencing vibration in occupations such as textile manufacturing are three times more likely to develop hyperpolymenorrhea and severe dysmenorrhea (50.7 versus 16.4 percent) (Lotis et al, 1962). Studies on effects of shift work on the menses have conflicting findings, with two reporting irregular menses (Tasto et al, 1978; Uehata and Sasakawa, 1982) and another reporting no effect on cycle length, duration, and flow (Kuchinski, 1989). Airline stewardesses experience a variety of menstrual disorders possibly due to several factors (e.g., vibration, disruption of circadian rhythms, altitude changes, and solar radiation) (Iglesias et al, 1980; Prill and Maar, 1971); menstrual function, however, reverts to preflight status with increasing jet experience (Cameron, 1969).

Surveillance of Menstrual History

Many challenges occur with any type of surveillance program of menstrual disorders in working populations. When menstrual history information is collected using questionnaire data, some studies have shown poor reliability of reports (Matsumoto et al, 1962; Bean et al, 1979). In a more recent survey of nurses, however, some menstrual variables were found to have high reliability (e.g., age at menarche and menopause, history of severe irregularity, uterine fibroids, ovarian cysts, endometriosis, pelvic inflammatory disease, and use of oral contraceptives and intrauterine devices) (Shortridge, 1988). Reports of dysmenorrhea, hypermenorrhea, clotting, irregularity, and spotting have fair to poor reliability. Methods to improve data collection include restricting reports to recent events (e.g., the previous 3 months) or the use of specific diaries or daily logs, but obtaining compliance with the maintenance of these logs may be low (Kuchinski, 1989). The use of hormonal measures on (at least) a subpopulation could be used to estimate validity.

EPIDEMIOLOGIC STUDY DESIGNS

The following is a brief introduction to the types of epidemiologic studies that could be used to evaluate reproductive effects of toxic exposures. This description is followed by some general design issues to aid in the evaluation of epidemiologic literature. Typical epidemiologic study designs include ecological, cross-sectional, case–control, and cohort.

1. Ecological studies typically assign exposure to groups by residential history and then compare rates of the outcome in each group. The lack of individual data leads to less information on exposure–outcome relationships. These data are useful for generating hypotheses. These studies require fewer resources and time than other studies. Environmental studies are more likely than occupational studies to use the ecological study design and may be particularly useful with registry data such as the Centers for Disease Control's Birth Defects Monitoring Program.
2. Cross-sectional studies measure exposure and outcome at the same point in time. If the exposure relevant to the outcome is in the recent past, the exposure–response relationship has a greater likelihood of being representative, especially if the exposure levels are relatively consistent over time (Lemasters and Selevan, 1984). Chronic exposure may be associated either with acute damage attributed to certain steady-state levels of the toxicant, potentially occurring with or without bioaccumulation of the toxicant, or

with the cumulative effects of repeated cellular insult resulting in the accumulation of injury (Lemasters and Selevan, 1984; Lemasters and Zenick, 1985; Selevan, 1989). If the health effects studied affect the likelihood of remaining in the work force, these data may not be representative. For example, a study of menstrual dysfunction could result in biased results if women with severe dysmenorrhea are more likely to leave the workplace.

3. Case–referent studies assign individuals or events to a group based upon the presence (case) or absence (referent or control) of the health outcome of interest. This approach could use in-house medical records or insurance claims. Exposure histories are compared in the two groups. These studies are frequently used to examine rarer events (such as childhood cancers and birth defects) in a large population, by selecting all qualifying cases and some sample of the noncases. Since only the specific outcomes selected are studied, these studies may miss effects of exposure that a cohort study, which can examine multiple outcomes, might identify. In addition, exposure typically is assessed retrospectively, which carries with it problems of accurately determining exposures.
4. Cohort studies define groups by exposure and evaluate subsequent health characteristics of study members. These groups may be either defined in the past (retrospective) and followed to the present time, or defined in the present time and followed into the future (prospective). With the second scenario, investigators may have an opportunity to obtain exposure measurements at the etiologic time for the outcome of interest. Such studies generally require more resources, both in time and money. Studies of industrial exposures and reproductive outcomes frequently make use of occupational cohorts.

General Design Considerations

Many occupational and environmental studies examine similar exposures. However, differences in patterns of exposure and characteristics of study groups affect the designs typically used. The major characteristics that may vary in the two types of settings include size of the study groups (and consequently the power of the study to detect various outcomes) and the magnitude and frequency of exposure.

Occupational populations are typically smaller than groups of people potentially exposed to environmental toxicants. These size limitations affect outcomes studied. Rare events are generally harder to study in occupational groups. The power, or ability of a study to detect a true effect depends on the size of the study group, the frequency of the outcome in the general population, and the level of excess risk to be identified. In case–referent studies, study sizes are dependent upon the frequency of exposure within the source population. The power of the study to detect meaningful differences affects the confidence one has in the results of a study with negative findings.

ESTABLISHMENT OF SURVEILLANCE PROGRAMS

This chapter reviewed the biological and epidemiologic issues in evaluating reproductive hazards in the workplace. Table 11-2 reviews the types of surveillance programs and the advantages and disadvantages of each. Careful definition of goals and uses of surveillance is necessary to ensure effective use of resources.

Surveillance of reproductive workplace hazards is dependent on the workers' willingness to provide this information on a routine basis. One key to good participation in a surveillance program is good communication. Management and workers need to be well informed about the purpose and objectives of the program. Measures used to inform the work force include articles in the plant newsletter, small group meetings, development of brochures to be given out during wellness programs, videos about the program, and inclusion of information about the programs during orientation. The worker often looks to his or her immediate supervisor for information. With that under-

Table 11-2 Types of reproductive evaluation programs

Study type	Advantages	Disadvantages
Registry (of pregnancies, infertility, etc.)	Provides a service and promotes health. Crude exposure assessments.	Will not answer scientific questions of relationship of exposure to adverse outcome. May have errors or incomplete registration of events.
Collection of retrospective reproductive histories	Requires a one-time evaluation. Can address questions of exposures.	Possible misclassification of exposure. Possible over or under reporting problems.
Update of recognized reproductive information	Recall error reduced. Facilitates exposure linking. May be an early indicator of problems.	Requires periodic contact. Participation over time may decrease.
Monitoring of reproduction using biological measurements (e.g., hormone levels)	Objective measure can be related directly to exposure. No reliance on recall. Eliminates potential for reporting bias.	Requires a high degree of compliance.

standing, it is therefore essential to include these individuals in the planning stages. "Ownership" in a surveillance program must be achieved at all levels.

Part of a good communication package includes explaining the potential benefits. The primary benefit to the employer is that he or she will have a data base to answer "impact" questions about possible effects when there have been changes in the process, product, or chemical usage. It also provides a tangible way of demonstrating to employees that the company is concerned about their well-being. The surveillance program can go a long way in ensuring that the work environment is safe or to direct the company toward needed changes. The employee also benefits by receiving systematically collected information that replaces paternalistic "trust me" attitudes.

Another key element in reproductive surveillance programs is confidentiality. We are asking individuals to share very confidential information including sensitive data on infertility, induced abortions, alcohol, and drug use. Workers may not provide this information if they feel that company personnel have access to their records. One way to ensure confidentiality to employees is the development of a unique identifier. The worker may be involved in the development of his or her identifier. Limited access is essential to protect the confidentiality of these data. All reproductive data must be secured in locked files or secured computer data bases.

Ideally, the nurse, physician, epidemiologist, toxicologist, industrial hygienist, and safety officer will serve as a team in the recognition and prevention of reproductive hazards. The mission of this health team includes informing both management and labor on the nature of the workplace hazards. Most convincing is objective information obtained from a review of the literature on a particular hazard and/or health data collected through a company's surveillance program. Certainly there is no reason to repeat the DBCP tragedy of not preventing a potential hazard that should have been suspected years earlier.

REFERENCES

Assennato G, Paci C, Baser ME, Molinini R, Candela RG, Altamura BM, Giorgino R: Sperm count suppression without endocrine dysfunction in lead exposed men. *Arch Environ Health* 42:124–127, 1987.

Baird DD, Wilcox AJ, Weinberg CR: Use of time to pregnancy to study environmental exposure. *Am J Epidemiol* 124:470–480, 1986.

Bean JA, Leeper JD, Wallace RB, Sherman BM, Jagger H: Variations in the reporting of menstrual histories. *Am J Epidemiol* 109:181–185, 1979.

Beskrovnaja NJ: Gynecological morbidity in women workers in the rubber industry. *Gig Tr Prof Zabol* 8:36–38, 1979.

Boyers SP, Davis RO, Katz DF: Automated semen analysis, in *Current Problems in Obstetrics, Gynecology and Fertility* (vol 12). Chicago, Year Book Medical, 1989, pp 173–200.

Bureau of the Census, US Department of Commerce: *Statistical Abstract of the United States* (110th ed). 1990, p 384.

Butarewicz L, Gosk S, Gluszczopma M: Examination of the health of female workers in the leather industry, especially from the gynecological viewpoint (abstract). *Med Pr* 20:137–140, 1969.

Cameron RG: Effect of flying on the menstrual function of air hostesses. *Clin Aviat Aerospace Med* 40:1020–1023, 1969.

DeStefano F, Annest JL, Kresnow M, Schrader SM, Katz D: Semen characteristics of Vietnam veterans. *Reprod Toxicol* 3:165–173, 1989.

Dobbins JG, Eifler CW, Buffler PA: The use of parity survivorship analysis in the study of reproductive outcomes. *Am J Epidemiol* 108:245, 1978.

Eskenazi B, Wyrobek AJ, Fenster L, Katz D, Sadler M, Lee J, Hudes M, Rempel DM: I. A study of the effects of perchloroethylene exposure on semen quality in dry cleaners. *Am J Ind Med* 20:593–600, 1991.

Everson RB, Sandler DP, Wilcox AJ, Schreinemachers D et al: Effect of passive exposure to smoking on age at natural menopause. *Br Med J* 293:792, 1986.

Gesher G: Changes in the menstrual cycle of women occupationally exposed to trinitrotoluene (translation). Parva Nacionalna Koferenciza Na Aspirantite. Presented at Sofia, Medicina I Fizkultura, April 1967, pp 159–262.

Griesemer RA, Ulsamer AG, Arcos JC, Beall JR, Blair AE, Collins TFX, deSerres FJ, Everson RB, Gamble JF, Gaylor DW, Groth DH, Kang HK, Keenlyside RA, Lloyd JW, Nettesheim P, Saffiotti U, Weisburger EK: Report of the federal panel on formaldehyde. *Environ Health Perspect* 43:139–168, 1982.

Grunert JH, deGeyter C, Bordt J, Schneider HPG, Nieschlag E: Does computerized image analysis of sperm movement enhance the predictive value of semen analysis for in vitro fertilization results? *Int J Androl* 12:329–338, 1989.

Harrington JM, Stein GF, Rivera RO, deMorales AV: The occupational hazards of formulating oral contraceptives—A survey of plant employees. *Arch Environ Health* 33:12–15, 1978.

Iglesias R, Terres A, Chavarria A: Disorders of the menstrual cycle in airline stewardesses. *Aviat Space Environ Med* 51:518–520, 1980.

Jouannet P, Ducot B, Fenleux D, Spira A: Male factors and the likelihood of pregnancy in infertile couples: I. Study of sperm characteristics. *Int J Androl* 11:379–394, 1988.

Kharrazi M, Potashnik G, Goldsmith JR: Reproductive effects of dibromochloropropane. *Isr J Med Sci* 16:403–406, 1980.

Kuchinski BB: *The Effect of Shift Work on the Menstrual Characteristics of Nurses*. Dissertation. Baltimore, Johns Hopkins University, 1989.

Lancranjan I, Popescu HI, Gavanescu O, Klepsch I, Serbanescu M: Reproductive ability of workmen occupationally exposed to lead. *Arch Environ Health* 30:396–401, 1975.

Lemasters GK, Selevan SG: Use of exposure data in occupational reproductive studies. *Scand J Work Environ Health* 10:1–6, 1984.

Lemasters GK, Zenick H: Assessing reproductive effects of workers exposed to hazardous wastes. Presented at the Seventh Annual Rocky Mountain Conference for Occupational and Environmental Health, October, 1985.

Lemasters GK, Hagen A, Samuels SJ: Reproductive outcomes in women exposed to solvents in 36 reinforced plastics companies: I. Menstrual dysfunction. *J Occup Med* 27:490–494, 1985.

Levine RJ: Methods for detecting occupational causes of male infertility: Reproductive history versus semen analysis. *Scand J Work Environ Health* 9:371–376, 1983.

Levine RJ, Symons MJ, Balogh SA, Arndt DM, Kaswandik NT, Gentile JW: A method for monitoring the fertility of workers: I. Method and pilot studies. *J Occup Med* 22:781–791, 1980.

Levine RJ, Symons MJ, Balogh SA, Milby TH, Whorton MD: A method for monitoring the fertility of workers: II. Validation of the method among workers exposed to dibromochloropropane. *J Occup Med* 23:183–188, 1981.

Levine RJ, Blunden PB, DalCorso DR, Starr TB, Ross CE: Superiority of reproductive histories to sperm counts in detecting infertility at a dibromochloropropane manufacturing plant. *J Occup Med* 25:591–597, 1983.

Levine RJ, Matthew RM, Chenault CB, Brown MH, Hurt ME, Bentley KS, Mohr KL, Working PK: Differences in the quality of semen in indoor workers during summer and winter. *N Engl J Med* 323:12–16, 1990.

Lotis VM, Solovieva IP, Tartakovskya LY: The effect of general vibrations on the female sexual apparatus (abstract). *Vopr Okhran Materinstva Detstva* 10:62–66, 1962.

Marchbanks PA, Peterson, HB, Rubin GL, Wingo PA, Cancer Steroid Hormone Study Group: Research on infertility: Definition makes a difference. *Am J Epidemiol* 130:259–267, 1989.

Matsumoto S, Nogami Y, Ohkuri S: Statistical studies on menstruation: A criticism on the definition of normal menstruation. *Gunma J Med Sci* 11:294–318, 1962.

Mattison DR: *Reproductive Toxicology*. New York, Alan R Liss, 1983.

Michon S: Connection between aromatic hydrocarbons and menstrual disorders analyzed (translation). *Pol Tyg Lek* 20:1648–1649, 1965.

Morck HI, Winkel P, Gyntelberg F: Health effects of toluene exposure. *Dan Med Bull* 35:196–200, 1988.

Moruzzi JF, Wyrobek AJ, Moyall BH, Gledhill BL: Quantification and classification of human sperm morphology by computer-assisted image. *Analysis Fertil Steril* 50:142–152, 1988.

National Center for Health Statistics, US Department of Health and Human Services: *Reproductive Impairments among Married Couples: United States.* OHHS no 83-190, ser 23, no 11. Hyattsville, MD, 1982.

Prill HJ, Maar H: Changes in the menstrual cycle caused by altitude. *Med Klin* 66:986–989, 1971.

Ratcliffe JM, Schrader SM, Meinhardt TJ, Steenland K, Clapp DE, Turner T: *Semen Evaluation of Timber Fumigators Exposed to Ethylene Dibromide.* Report no TA-83-244. Cincinnati, OH, NIOSH, 1984.

Ratcliffe JM, Schrader SM, Steenland K, Clapp DE, Turner T, Hornung RW: Semen quality in papaya workers with long term exposure to ethylene dibromide. *Br J Ind Med* 44:317–326, 1987.

Ratcliffe JM, Schrader SM, Clapp DE, Halperin WE, Turner TW, Hornung RW: Semen quality in workers exposed to 2-ethoxyethanol. *Br J Ind Med* 46:399–406, 1989.

Rom WN: Effects of lead on reproduction, in Infante PF, Legator MS (eds): *Proceedings of a Workshop on Methodology for Assessing Reproductive Hazards in the Workplace.* Government Printing Office, 1980, pp 33–42.

Schrader SM, Ratcliffe JM, Turner TW, Hornung RW: The use of new field methods of semen analysis in the study of occupational hazards to reproduction: The example of ethylene dibromide. *J Occup Med* 29:963–966, 1987.

Schrader SM, Turner TW, Ratcliffe JM: The effects of ethylene dibromide on semen quality: A comparison of short-term and chronic exposure. *Reprod Toxicol* 2:191–198, 1988a.

Schrader SM, Turner TW, Breitensteien MJ, Simon SD: Longitudinal study of semen quality of unexposed workers: I. Study overview. *Reprod Toxicol* 2:183–190, 1988b.

Selevan SG: Methods for environmental quantitative risk assessment. Presented at the Symposium on Environmental Epidemiology, University of Pittsburgh, April 1989.

Shortridge LA: Assessment of menstrual variability in working populations. *Reprod Toxicol* 2:171–176, 1988.

Starr TB, Levine RJ: Assessing effects of occupational exposure on fertility with indirect standardization. *Am J Epidemiol* 118:897–904, 1983.

Starr TB, Dalcorso RD, Levine RJ: Fertility of workers: A comparison of logistic regression and indirect standardization. *Am J Epidemiol* 123:490–498, 1986.

Steenland K, Carrano A, Ratcliffe J, Clapp D, Ashworth L, Meinhardt TJ: A cytogenetic study of papaya workers exposed to ethylene dibromide. *Mutat Res* 170:151–160, 1986.

Syrovadko ON, Skormin VF, Pron'kova EN: Effect of working conditions on the health and some specific functions in female workers exposed to white spirit [translation]. *Gig Tr Prof Zabol* 16:5–8, 1973.

Tasto KL, Colligan MJ, Skjei EW, Polly SJ: *Health Consequences of Shift Work*. DHEW (NIOSH) publication no 78-154. Government Printing Office, 1978.

Uehata T, Sasakawa N: The fatigue and maternity disturbances of night work on women. *J Hum Ergol* (suppl) 2:465–474, 1982.

Van der Gulden JWJ, Zielhuis GA: Reproductive hazards related to perchloroethylene: A review. *J Int Arch Occup Environ Health* 61:235–242, 1989.

Weinberg CR, Gladen BC: The beta-geometric distribution applied to comparative fecundability studies. *Biometrics* 42:547–560, 1986.

Welch LS, Schrader SM, Turner TW, Cullen MR: Effects of exposure to ethylene glycol ethers on shipyard painters: II. Male reproduction. *Am J Ind Med* 14:509–526, 1988.

Whorton D: Male occupational reproductive hazards. *West J Med* 137:521–524, 1982.

Whorton D, Krauss RM, Marshall S, Milby TH: Infertility in male pesticide workers. *Lancet* 2:1259–1261, 1977.

Wilcox AJ, Weinberg CR, O'Connor JF, Baird DD, Schlatterer JP, Canfield RE, Armstrong EG, Nisula BC: Incidence of early loss of pregnancy. *N Engl J Med* 319:189–194, 1988.

Wong O, Utidjian HMD, Karten VS: Retrospective evaluation of reproductive performance of workers exposed to ethylene dibromide (EDB). *J Occup Med* 21:98–102, 1979.

Wyrobek AJ, Watchmaker G, Gordon L: An evaluation of sperm test as indicators of germ-cell damage in men exposed to chemical or physical agents. *Teratogenesis Carcinog Mutagen* 4:83–107, 1984.

Zielhuis GA, Gijsen R, van der Gulden JWJ: Menstrual disorders among dry cleaning workers. *Scand J Work Environ Health* 15:238, 1989.

Zlobina NS, Izyumora AS, Ragulie NY: The effect of low styrene concentrations on the specific functions of the female organism. *Gig Tr Prof Zabol* 18:21–25, 1975.

INDEX

Page numbers followed by t and f indicate tables and illustrations, respectively.